VOLUME EIGHTY TWO

ADVANCES IN QUANTUM CHEMISTRY

Quantum Boundaries of Life

VOLUME EIGHTY TWO

Advances in QUANTUM CHEMISTRY

Quantum Boundaries of Life

Edited by

ROMAN R. POZNAŃSKI
Faculty of Informatics and Computing
UniSZA, Terengganu, Malaysia

ERKKI J. BRÄNDAS
Department of Quantum Chemistry
Uppsala University, Uppsala, Sweden

Academic Press is an imprint of Elsevier
50 Hampshire Street, 5th Floor, Cambridge, MA 02139, United States
525 B Street, Suite 1650, San Diego, CA 92101, United States
The Boulevard, Langford Lane, Kidlington, Oxford OX5 1GB, United Kingdom
125 London Wall, London, EC2Y 5AS, United Kingdom

First edition 2020

ISBN: 978-0-12-822639-1
ISSN: 0065-3276

For information on all Academic Press publications
visit our website at https://www.elsevier.com/books-and-journals

Publisher: Zoe Kruze
Acquisitions Editor: Sam Mahfoudh
Editorial Project Manager: Shellie Bryant
Production Project Manager: Denny Mansingh
Cover Designer: Alan Studholme

Typeset by SPi Global, India

Contents

Contributors

Erkki J. Brändas
Department of Chemistry, Uppsala University, Uppsala, Sweden

Danko D. Georgiev
Institute for Advanced Study, Varna, Bulgaria

Chaim Gilon
Institute of Chemistry, Hebrew University Jerusalem, Jerusalem, Israel

James F. Glazebrook
Department of Mathematics and Computer Science, Eastern Illinois University, Charleston, IL, United States

Janusz E. Jacak
Department of Quantum Technology, Wrocław University of Science and Technology, Wrocław, Poland

Witold A. Jacak
Department of Quantum Technology, Wrocław University of Science and Technology, Wrocław, Poland

Liaofu Luo
Faculty of Physical Science and Technology, Inner Mongolia University, Hohhot, China

Jun Lv
Center for Physics Experiment, College of Science, Inner Mongolia University of Technology, Hohhot, China

Gerard Marx
MX Biotech Ltd., Jerusalem, Israel

Akihiro Nishiyama
Graduate School of System Informatics, Kobe University, Nada-ku, Kobe, Japan

Alfredo Pereira Jr.
Goldsmiths, University of London, London, United Kingdom; São Paulo State University (UNESP), Botucatu, São Paulo, Brazil

Roman R. Poznański
Faculty of Informatics and Computing, Universiti Sultan Zainal Abidin, Terengganu, Malaysia

Christopher J. Rourk
Jackson Walker LLP, Dallas, TX, United States

Shigenori Tanaka
Graduate School of System Informatics, Kobe University, Nada-ku, Kobe, Japan

Jack A. Tuszynski
Department of Oncology, Cross Cancer Institute; Department of Physics, University of Alberta, Edmonton, AB, Canada

Preface

The boundaries between organic and live matter are sites of the most significant interactions and transformations in science from biology through chemistry and physics. In this volume, we are happy to present our readers with a unique thematic volume 82 of the *Advances in Quantum Chemistry* devoted to the theme *Quantum Boundaries of Life*. We present a transdisciplinary exploration of the quantum boundaries relating to a molecular basis of life and consciousness, where the integration process begins at the molecular level, grounded on the research agenda, concepts, and shared values of quantum chemistry. There is a boundary to the integration process, but it provides the reality of a deeper foundation that is the quintessential mechanism of life where reality isomorphically aligns with consciousness.

The evolutionary biologist Ernst Mayr contended three principal classes of scientific reduction in biology, i.e., the ontological, the epistemological, and the methodological. Molecular catalysis is central in molecular evolution, like all other teleological phenomena, advanced by Darwin's theory of evolution. Hence, it follows that molecular reductionism provides the ground level for the most straightforward kind of life, based on the evolution of the prokaryotic to the more complex eukaryotic cell, where the wire-like flow of charges, protons, ions, and other molecular constituents of the microenvironment, including elements of the cytoskeleton, extends to the cellular membrane itself representing the quantum machinery of life and consciousness. However, as all quantum physicists know, the reading of "*Pioneering Quantum Mechanics*" contains, among all versions, the von Neumann-Wigner interpretation "consciousness causes collapse of the wave function." Irrespective of the copiousness of these views, the reduction argument spirals back to fundamental biological concepts within the life sciences.

The present volume editors represent the fields of biological applications of quantum chemistry in the broadest sense. First, it is important to recognize the subtle difference between quantum physics and quantum chemistry. The former is strictly reductionistic, using quantum mechanics and field theory to hierarchically formulate the fundamental subsystems of nature, whereas the latter attempts to use the "quantum platform" to build more complex systems while setting fundamental goals for the discipline. Second, to take the step from quantum chemistry to quantum biology, one must permeate a crucial barrier, namely how to account for functionality. This functional interaction is nonlocal and across scale as opposed to the classical concept of levels, which is a continuous notion. So, microscale, mesoscale, and macroscale assume level continuity—a kind of biological nonlocality. Third, we have two types of demarcations: (1) the quantum boundaries of life, e.g., when quantum chemistry becomes quantum biology and (2) the often-discussed quantum-classical interface. The quantum-classical boundary has, for a long time, been the concern of theoretical physics and will only indirectly be connected with the quantum boundary of life. Physicists suggest that the quantum-classical transition should be linked with John Wheeler's quantum foam or *spacetime fluctuations*. At the same time, chemically oriented scientists believe that the interface hides in the process of decoherence, and hence the quantum boundary of life must be of thermoquantal origin, derived from the entropy-temperature duality.

How do we go about this boundary, this demarcation sector or frontier zone? What will we find? Life should, first of all, always be inside such a boundary—or as is suggested in the final contribution that it is more relevant to view Life as a quantum phenomenon intrinsic to Nature—such as all biological organisms that have acquired an intrinsic function. Teleological notions in biology adumbrate that living organisms have *intrinsic* purposes that begin with simple physical interactions between entities of what we denote as nonliving matter evolving into more complex correlative communicative exchanges of anticipation and information. What are the physical attributes of such intrinsicality? One should remember that the mechanisms of teleological causation are "hidden"; we fail to perceive reality as a conjugate link between matter and experience. As mathematicians, biologists, and chemists, including ourselves, scientists seek to discover the isomorphic connection between matter, life, and consciousness.

Several fundamental stumbling blocks need to be unraveled. The most striking one is how to handle the threat of decoherence, destroying molecular wave interferences due to incoherent scattering between atomic, ionic, and molecular constituents, ubiquitous aqueous solutions, etc. in the wet and hot environment of a human brain. The thermal noise, about 0.025 eV at 310 K, might wash away subtle quantum effects making the latter seemingly obsolete. In other words, quantum decoherence becomes an unavoidable obstacle in the organization of energy to maintain order and overcoming entropy production in living systems. Another difficulty relates to the notion of energy-time scales, conjugated through the uncertainty principle, with the consequence that integrated information, as a synergy of emerging information, becomes impossible. This is so due to the irreducible character of the degenerate physical representation that derives from the law of self-reference suggesting a holistic view, i.e., an informational holarchy, which to some extent reminds of Arthur Koestler's intemperate critique of the classical citadel of orthodoxy culminating in his contentious notion of holons. Accordingly, our present conception is founded on an interrelation-informational structure, where the whole is nonsynergistically affected by nonintegrated information. This relationalism provides the key to understanding how nonintegrated information holistically subsumes concrescence while conferring negentropic information as a transformation process of quantum nature.

As a consequence of the above, temperature dependencies must be addressed in a serious quantum-theoretical formulation of life processes that should concern systems evolving far from equilibrium, with the latter dissipating energy and entropy to the environment. Our warm brain is an example of this, suggesting that neural processes are thermal and, therefore, claimed to be nonquantum. However, there is a disparity between bound quantum states and quantum effects associated with the continuum and rigorous extensions to open system quantum dynamics. The chapters will deal with various situations and circumstances related to quantum theory in the broadest sense, such as tunneling and resonance formation, including density matrices and associated generalizations.

Assuming that animate and inanimate systems are subject to the same physical principles, one might wonder what sets them apart at the quantum boundary? To answer this question, one cannot avoid appealing to a teleological notion of function, which, according to Jacques Monod, constitutes "a profound epistemological contradiction" effectively exhibiting a central problem in biology. We are left with the question of etiology, and it is here where the quantum boundaries of life are espoused. The formulation must also include the macroscopic scale to encompass the teleological functions from quantum chemistry to quantum biology. However, as ventilated above, quantum effects, asserted to be essential for life processes, cannot survive as eigenfunctions to the Schrödinger equation, i.e., will not be coherent in the thermal medium, since its wave properties cannot resist decoherence by thermal perturbations. The density matrix provides a more general representation of the state of a quantum system, reflecting upon its nonlocality, impending localization, like classical particles in a biological medium, such as the microstructure of neurons.

Although modern quantum chemistry goals are to accumulate chemical data from, e.g., the physical constants and atomic numbers employing the

Schrödinger equation and its Liouville generalizations, its base is quantum physics, for instance, treating momentum-space and energy-time as fundamental conjugate variables-operators. Imbricating chemical physics adds fundamental technological capabilities facing the atomic and molecular levels that will restrict the concept of a quantum boundary of life toward quantum delocalization and long-range correlative information. In what follows, the invited authors have examined various problems related to their expertise in the stride to uncover the gap between the function of the cells, such as neurons, in a living organism—the easy problem—and the conscious experience, how it is like to be—the hard problem. Cognitivism mistakes consciousness for mindless neural network computations. As is the case with naive realism, i.e., with its primary focus on visual perception and a total lack of a self-referential basis, it does not work when dealing with the hard problem of consciousness.

The first chapter complements this Preface, answering the question already posed by Erwin Schrödinger, "What is Life?" The conclusion is that life is a quantum phenomenon to be further explicated in what follows. The next contribution presents an interesting confronting view that pioneering quantum mechanics is incapable of formulating emotive mental states (feelings). While admitting that quantum mechanical concepts may have different meanings for each discipline, the authors review several unsolvable problems for the consideration of transcendental mental states rationalizing cognitive abilities such as memory. In response, they develop a tripartite neural mechanism with molecular underpinnings, fusing psychology with biochemistry. An appealing conclusion is a suggestion that the encoding of emotive states should be expressed by a more complex bit, labeled the *cuinfo*. The authors' view is that the quantum scale is too refined to describe the physiology of neural nets, and hence is poorly suited to describe the emotive nervous system, thus becoming an irrelevant motif in the ocean of physical correlations dominated by physiological processes and phase transitions from which mental states emerge.

In the following contribution, the author discusses a novel hypothetical mechanism of neuron communication via quantum electron transport in the basal ganglia and brainstem, based on the idea that iron storage proteins, electronically similar to quantum dots, can be understood only in terms of quantum mechanics. To back the hypothesis, the author argues that the literature for quantum dots supports electronic communication between ferritin molecules separated by order of 50–100 nm length scales. This is an order higher than previous studies have indicated for electron transport based on spectroscopy. Probing conductive AFC (atomic force microscopy) tests display variational conduction related to the presence of, e.g., ferritin. However, the data interpretation is still complicated and requires further study to understand if such long-range tunneling can be supported in the nematode *Caenorhabditis elegans*.

Chapter 4 continues with the deliberation on the saltatory conductivity mechanisms extending between Ranvier nodes in the electrolyte component of myelinated segments of axonal cytosol (output parts of nerve cells) over considerable regions and the topological concept of memory functioning. The authors advocate a new wave-type model, based on the wave plasmon-polariton. Although the quantum character is quite moderate, the collective plasmon modes have synchronic dipole oscillations of myelinated sectors that go back to Bohm and Pines' theoretical work, an important forerunner of the microscopic BCS (Bardeen–Cooper–Schrieffer) mechanism of superconductivity.

A more esoteric framework is employed next. The authors provide an overview of quantum brain dynamics within a quantum electrodynamics setting. The nonequilibrium quantum brain dynamics suggest a quantitative framework incorporating Schrödinger-like equations for electron dipole fields, the Klein-Gordon equation for coherent electric fields, and the Kadanoff-Baym equations for quantum fluctuations. Some numerical simulations are presented to promote nonequilibrium dynamics and to provide macroscopic order, but the conclusion is that further extensions and modifications are necessary.

In Chapter 6, the focus is shifted to the very complex and demanding problem of protein folding. The authors make a careful analysis of the various dynamical variables at work. In particular, torsion motion depends on the physiological temperature and the assumption of time scale separation, where molecular stretching and bending are treated as fast variables. Nevertheless, the authors stress that the dynamical mechanism of protein folding is still unclear, noting that the torsion potential has several minima and shows a peculiar non-Arrhenius behavior indicating possible weaknesses and inabilities for adiabatic theory to fully decide whether the folding obeys a classical or quantum law.

The classical-quantum interplay is discussed further in what follows. In addition to mainly focusing on the neuronal brain system, the author has recurrently emphasized the importance of the extracellular composition of neural tissues. The aim is to examine negentropic mechanisms and possible transient quantum processes. An interesting by-product is the revival of the old classical Grotthuss effect that has enjoyed a somewhat unexpected quantum interpretation. The boundary between the classical-quantum is ambiguous regarding its higher-up experiential nature of consciousness. The present physiological, mental, unconscious, and conscious aspects of human brain activities satisfy the metaphysical belief known as triple aspect monism.

In Chapter 8, the authors consider the protein foldamer hypothesis within a natural selection scenario. In particular, they concentrate on protein α-helices. While stating that full quantum simulations are in principle within the capabilities of modern quantum chemistry applications, they decided on a different route, i.e., studying the generalized Davydov-Hamiltonian with its robust soliton solutions and its collective attributes in delivering free energy at desired active sites. The article is a very careful demonstration of the richness and flexibility of what can be established in the soliton factory. Although the simulations could not effectively constitute the quantum-to-classical transition, the approach is auspicious, particularly in shedding new light on the abiotic origin of life.

In the final chapter, the authors communicate a new materialistic framework based on recent advances in quantum chemistry and neurosciences. The presentation pursues two paths: (i) Schrödinger-Liouville dissipative dynamics and (ii) Bohmian complexified action. The former embodies long-range quantum thermal correlations suggesting a specific universal grammar for (proto)communication, while the latter provides an alternative vernacular for holistic nonintegrated information. It is important to note that the present transition to the self-organizing phase is distinguished from the topological structure forming phase transitions in condensed matter, for instance compared to the deeper understanding of the properties of superconductivity and superfluid helium. In the present formulation, the thermodynamic instability affects the thermodynamic region, determined by de Broglie's wavelength, rather than emerging as self-organized criticality whence spatial and temporal scale invariance sets in. As a result, the thermoquantal information is represented by a transition matrix of quantum states, fine-tuned to the temperature, and the associated timescales and code-protected from decoherence.

Panexperiential materialism advances a quantum holistic viewpoint that denies emergence, instead advocating a holarchical irreducible thermodynamic domain founded on the conjugate relation between temperature and entropy as a basis for thermo-qubital communication. Furthermore, the mind and the brain are isomorphically linked through the Fourier-Laplace transform providing a foundational self-referential constitutional law for biological evolution.

Finally, as editors of this collection of contributions, we offer the sincere reflections of the invited authors on some of the most fantastic puzzles of our time. Although we are satisfied with the positive responses and contributions to this enterprise, we have not succeeded in providing a realistic gender distribution. For this, we apologize. Yet we hope that "Quantum Boundaries of Life" will transmit the same respect and veneration as the contributors, and we experienced during the preparation of this volume. In particular, we have discovered that the topic of *quantum boundaries of life in the brain* is an exciting area of advancing modern quantum neurobiology that should draw attention to and be confronted by multidimensional and interdisciplinary efforts.

Roman R. Poznański
Erkki J. Brändas

CHAPTER ONE

Is life quantum Darwinian?

Erkki J. Brändas[a,*] and Roman R. Poznański[b]
[a]Department of Chemistry, Uppsala University, Uppsala, Sweden
[b]Faculty of Informatics and Computing, Universiti Sultan Zainal Abidin, Terengganu, Malaysia
[*]Corresponding author: e-mail address: erkki.brandas@kemi.uu.se

Contents

Abstract

This is a short introduction as a complement to the Preface, for those not familiar with the current view of pre-existing quantum biology as a re-evolving subject and its influences concerning the theme "quantum boundaries of life." We suggest that the information-based quantum Darwinism encapsulates macroscopic quantum potentialities via self-referential amplification. It is shown to play a pivotal role, giving support to the signature of life and consciousness at the quantum-classical transition zone where long-range correlative information is cultivated by energy-entropy dissipation in organisms at multiple levels of hierarchical and functional organization.

The full understanding of a cell or organism, at the microscopic level, is still a distant dream, whether one wants to penetrate the inner workings of the cell or proliferate evolution with a direct quantum approach. Nevertheless, there have been successful molecular interpretations that help, promote, and support the understanding and the pragmatic findings of the various phenomena in biology. For example, the celebrated Watson-Crick helical model of DNA conveys the detailed code for the transfer of "genetic information" between cells and providing their functions. Meanwhile, from biology to physics and chemistry mainstream orthodox science has no leads regarding the origin of the occurring transfer mechanism of *information* between the molecular constituents. There is furthermore no idea of the teleological evolutionary *historical trait* of a cell/organism, such as the emergence, existence, aging, and death, as well as many other characteristic features of "living matter." Even if quantum theory is the default in explaining the atomic and molecular structure, it is usually believed that this microscopic order juxtaposed with classical science is hardly ready to explain

Advances in Quantum Chemistry, Volume 82
ISSN 0065-3276
https://doi.org/10.1016/bs.aiq.2020.09.001

phenomena in the living world. Hence, an organism seems to be simply an "impossible" thing. The latter's *distinctive* features, such as *flexibility, responsiveness, adaptation,* and *teleonomic/memory/history* traits incorporating self-organization, growth, development, aging, and death, are beyond the scope of conventional science.

In the last century, classical science has undergone a radical paradigm shift to cover a wide range of natural phenomena, including the dynamics of atoms, molecules, chemical reactions, coherent states of electrons and atoms, the solid-state, condensed matter, and many other phenomena, opening a tremendous range of new exciting quantum electronic and optical technologies. In biochemistry, the molecular models based on balls and springs have been replaced by atomic and molecular wave functions of immense complexity, building on the fundamental understanding and original text[1] by the Nobel Laureate Linus Pauling. A new scientific subdiscipline, quantum chemistry that falls between the historically established areas of mathematics, physics, chemistry, and biology, was fathered, established, and consolidated by Per-Olov Löwdin, starting already with his groundbreaking thesis in 1948, on a quantum mechanical solution of the failure of the classical Cauchy problem for the elastic constants of ionic crystals. For a recent historical essay on the development of quantum chemistry, we refer to.[2] This development also touched upon biology, where quantum effects entered explicitly in the discussions of the quantum mechanical tunneling of protons in the hydrogen bonds between the base pairs in the Watson-Crick model of DNA, suggesting a mechanism for the production of pairs of *tautomeric* bases.[3]

Quantum chemistry, which originally dealt with the simplest, non-biological objects, has significantly expanded into the landscape of modern biology. Although the original quantum biological calculations have been found too crude and simplistic to provide reliable mechanisms, yet computational and theoretical advances have today fostered novel applications in cell and evolutionary biology, and in particular, two directions have survived time: delocalized electrons, i.e., the π-electrons in the so-called conjugated systems, and mobile/quantum *protons* in the hydrogen bonds—both being of fundamental importance in the dynamics of both DNA and RNA. The first one has been critical in understanding aromaticity, i.e., a broad category of aromatic compounds or molecular derivatives of benzene, as well as metal aromaticity from organic compounds extended to metals, with both being fundamental in biology and biochemistry. The second one renders a long history from proton transfer and high excess conductivities in water and aqueous solutions to *pH*-dependencies in the cytosol to the genetic message in the proton-electron pair code in the cell. These observations are closely related as the highly mobile π-electrons, a critical part of the nucleosides, are very much in control of the barriers for the tunneling of protons.

The quantum substrate, embracing anomalous, nonclassical, proton dynamics,[4] plays a fundamental role *in an* organism. As established by Pauling and exploited in his celebrated structure of proteins, the quantum effects underpinning the hydrogen bonds were further extended to study the tunneling arrangements in the double helix by P.-O. Löwdin.[3,5] Löwdin also coined the term "quantum biology" as well as founding the present book series. Since then, quantum effects and their role in living matter have been extensively discussed, particularly since hydrogen bonds are *common* to almost any organic matter. Due to their quantum properties, the electron-proton formation of hydrogen bonding provides both stability, preserving the genetic code, and transiency, giving rise to mutations. Such a structure exhibits a *degenerate* energy spectrum with many states of the same energy, suggesting quantum transitions and interesting thermal correlative dynamics.

Whether a cell, under thermal attack, might find itself in a *quantum superposition* state is a crucial issue under intensive discussions. For example, consider all protons in the hydrogen bonds capable of tunneling between base pairs, creating a *highly degenerate* quantum state of the double helix's polynucleotide chains. The high symmetry and degeneracy become utterly sensitive to thermal perturbations. For instance, the environment imposes significant *constraints* that easily *breaks the symmetry* of a molecular aggregate's organizational order. Symmetry violations are often used in quantum mechanics in connection with, e.g., the famous Jahn-Teller effect, see,[6] the proceedings of 16th Jahn-Teller Meeting dedicated to Frank Ham, yielding a geometric distortion of a non-linear molecular system reducing its symmetry and energy. Not only are the degenerate situation vulnerable to decoherence by the thermal environment, but stability and degeneracy do not seem to go hand in hand. This is a significant problem that a molecular cell dynamics of life processes has to overcome.

General quantum/tunneling effects are well-known for the *lightest* of atoms, H, the most abundant element widely available in any organic substrate. However, other *light* atoms also display interesting quantum dynamical properties. For example, the physiologically active molecules XH_3 (X=N, P, As) display the so-called umbrella *inversion* moving the X-molecule through the H_3 plane, tunneling back and forth through the

X atom barrier with the frequencies from 30 GHz for the lightest atom, more slowly for X = P, and up to one in two years for the heavy atom X = As. In other words, for AsH_3, the inversion occurs quite rarely, so one can select one of the two components of the entangled state. In biological systems, racemization and enantiomer separation are crucial conceptions as exemplified by the thalidomide medical disaster.

Parity violation in chiral and achiral molecules are under intensive theoretical and experimental scrutiny. Recent resolutions covering time scales from yoctoseconds to seconds and longer, including the analysis of the very complex tunneling-rotation-vibration spectra, have been carried out in the laboratory of Martin Quack.[7] Another indication of the importance of quantum effects in life processes is the experiments, where one replaces the proton with its heavier sibling, the deuteron, for instance, in *Euglena Gracilis*[8] displaying a very distinctive structure and function compared to the regular protium microorganism. Interestingly, the D atom has chemical properties almost identical to the H atom. Still, their nuclei have quite different physical properties, such as mass and quantum statistics – the deuteron is a boson instead of the fermionic proton, see, e.g., their disparate molecular conductance properties in H_2O/ D_2O mixtures as a function of deuterium mole fractions.[9]

So, there are good reasons to believe that organic matter provides a legitimate *quantum substrate* that explains its unique properties, such as *flexibility, responsiveness,* and the traits of *memory/history*. Neglecting its quantum nature does not seem to be a feasible alternative to understand those features that are critical for the life forms appearing in our environment and including ourselves. From this perspective, modern quantum biology is not only more realistic, but also a necessary consequence of our subjective observations. Macroscopic organic matter as a quantum substrate should be considered in the same sense or at least along the same macroscopic timeline as superconductivity and superfluidity exhibiting Yang's celebrated concept of Off-Diagonal Long-Range Order, ODLRO.[10] Although macro-quantities, e.g., those of tin (Sn) or liquid helium ^{4}He are typical quantum substrates, exhibiting marvelous coherence features at very low temperatures, a high priority issue concerns the transduction of long-range phase correlations to room or human body temperatures far above any, so far detected, Bose condensates.

As emphasized, a quantum approach yields options of *direct communication* between molecular aggregates, between organism's genome/DNA, between cells, and between the cell and its environment, mostly missing in the vernacular of traditional biology. This allows for an organism to monitor and record the environment's dynamics for its further *reuse* to build *consistent* responses to incoming challenges, McClintock.[11] Even if her brilliant discovery of mobile genetic elements did not explicitly use quantum theory, the gradual understanding from isolating DNA and ultimately the discovery of DNA-RNA generated a necessary vernacular for making and transmitting novel discoveries within a semiclassical framework. Although the genome/DNA appears to be *energetically* stable with respect to its environment, separating an organism's ambiance hierarchically, there is a flow of negentropy generated by entropy production under steady-state conditions. The DNA/genome and the individual cell is an open system allowing for real-time updates and self-organization—all in agreement with the second law.

The issue of DNA origins and its openness has been a topic for disagreements and still constitutes a watershed between evolutionary theorists, such as Lynn Margulis and neoDarwinists, viz. Richard Dawkins and others. Like McClintock, Margulis emphasized the importance of a genome as a dynamic organ of the cell monitoring the environment and responding to its challenges by proper *restructuring* itself. Her endosymbiosis theory,[12] is considered an outstanding achievement even by her opponents, yet viewing life as a giant network of social connections, *symbiogenesis*, needs to be set on a more fundamental platform which demands a deeper quantum physics-chemistry life principle.

The topic of this book, *Quantum Boundaries of Life*, has a universal aim to show that one needs quantum mechanics, not only as the default mechanism mentioned above but also as a necessary platform for the quantum deportment in the hot and wet conditions that prevails in a human brain. This attitude leads to *quantum mechanistic* approaches that open new exciting horizons for understanding life dynamics far beyond the scope of present quantum- and non-quantum biology. Another goal is to better understand the *quantum boundary of life*. In general terms, to what degree quantum effects and related quantum states impact life? Although a quantum state is not *directly* observable, it determines the nature of related quantum effects and their dynamics. Quantum mechanics describes quantum effects in terms of *quantum transitions* between the initial and final quantum states that include the transition operator connecting the initial and final quantum states. For instance, the famous Fermi's Golden Rule describes the quintessential quantum characteristics, appropriating the transition rate from an energy eigenstate to a group of energy states in the continuum resulting from a weak perturbation. Not only does Fermi's formula predict the preferential spread of available energy over multiple states, but it also provides a negative imaginary part, $i\epsilon$, of the energy as a consequence of the Cauchy principle value of the associated reaction operator. This limiting procedure, rendering the state's lifetime proportional to the inverse of ϵ, can be interpreted as the first step to perform a rigorous analytic continuation onto the so-called unphysical Riemann sheet. This extension of pioneering quantum mechanics provides a *key feature* in determining the specifics characterizing the quantum boundary of life.

Quantum dualism, or its conjugate relationships, depicts quantum states in the brain as vectors governed by quantum operators, such as generators of time evolution and their resolvents. Classical physical variables correspond to quantum observables, together with their transformation laws, and conjugate entities entail consequences for predictions in quantum measurements. Such Fourier transform duals occur in the Hilbert space connecting the spectral domain with time and similarly between the momentum operator and space. Quantum information also respects the wave-particle dualism, not to be confused with the dualistic view that mental phenomena are non-physical or that the mind and body are distinct and separable, suggesting that mind-matter be correlated with preconscious experientialities on the premise of quantum indeterminism. As a result, elementary particles are endowed with the propensity to make choices and entanglement of quantum states across a spectrum of neuronal microstructure generating unitary dynamics in combination as a mark of a physical theory of consciousness.[13] This becomes a problem, stemming from the idea that entangled quantum states in brains, as storing sites for long-term memories as well as the notion that function is a causally ineffective epiphenomenon limited to the deterministic laws of classical physics.

The spectrum of teleofunctionalism suggests that physicalism should not be restricted to force-based action only, but should be complementary in understanding the nature of physical feelings at the quantum boundary of life, insofar as raw feelings originate from potentialities in the quantum realm. This view is commensurate with the conjugate Fourier-Laplace relations embodying in addition entropy and temperature.[14] The "potentialities in the quantum realm" is one of the *key features* of the *quantum boundary of life*. Classical physics, like traditional bound state quantum mechanics, exhibits interesting energy degeneracies due to the existence of physical conservation laws and related fundamental symmetries; however, in an open system with a thermal environment, a classical system, due to its unlikelihood for accidental degeneracies, does not allow for such potentialities.[15]

Can potentialities enrich our explanatory power of how the mind originates in the brain? Classical studies cannot invoke this possibility, i.e., that the nature of consciousness has a teleological origin rather than being phenomenological. The latter opinion brings about the consequence of explaining away consciousness. For instance, the view of the phenomenological process as anticipated before the evolution of cognition[16] contends that thirst, hunger, and pain, involving primordial feelings, are the likely precursors to consciousness. However, from a quantum perspective, raw feelings are not based on emotional motifs nor perception, but instead on potentialities[14]—thereby rejecting perception and cognition as the underlying mechanism of physical feelings at the quantum boundary.

Since the theme of this volume concerns the question of whether quantum features do play a significant role in the essential processes of life, it is necessary to briefly return to two issues that have been amply belabored in the book. One is the Quantum Darwinistic aspects, advanced by Wojciech Zurek,[17] related to the selection of states of a quantum system, and second, how does an etiological perspective of the so-called quantum potentialities[14] fit in?

At first, one notes that environment-induced decoherence is an often-suggested solution to the infamous measurement problem.[17] The ambient proliferation of selected quantum states through einselection, where so-called pointer states persist, is habitually advanced as an explanation of why localized states survive. Although this is an interesting and promising line of development, there have also been some words of dissent, see, e.g.,[18] The main issue comprises the worry whether an observer-independent physical partitioning is commensurate with the subjective choice of the system versus environment and the restriction to unitary-only dynamics during the whole decoherence process. Boltzmann's attempt to derive irreversible thermodynamics from reversible laws has been brought up as an analogy for asserting that the Quantum Darwinism program has a touch of circularity.[18] The first problem can be avoided by viewing the partial tracing as an associated projection operator, which, in principle, can be extended to an exact sub-dynamical formulation.

The second is trickier since it demands a quantum mechanical extension to open systems and requires more than decoherence, einselection and envariance. It relies on a fusion of thermal and quantum fluctuations, thermo-qubits, which permits storage and communication of information. Quantum Darwinism is therefore impossible in an entirely equilibrated environment whose subsystems cannot store phase information.[19] Decoherence is the process of the *loss of phase coherence* caused by the interaction between the system and the environment.[20] Although the decoherence program successfully demonstrates that quantum mechanics is indeed consistent, one might further consider the arguments above and compare the state of the matter in relation to the theoretical basis underpinning Panexperiential Materialism,[14] namely the incorporation of conjugate Fourier duals between energy-momentum and time-space, as well as entropy-temperature and phase and number of particles, saving fundamental spatio-temporal correlations without any restrictions such as coarse grainings, molecular chaos, etc.

The present work, for details, see, e.g.,[21] is based on a rigorous extension to dissipative dynamics, which does not obey unitary-only propagation, neither depends subjectively on actual system-environment partitionings, nor the assumption of Boltzmann's assumption of molecular chaos. The temperature-temporal imprints are objectively instigated through precise boundary conditions providing natural thermodynamic constraints. The result yields, through the steady state condition $dS = 0$, a simple relation between entropy production and negentropic gain, where dS is the entropy exchange in the open system. Instead of considering a trace-projected reduced density matrix as a tool for maintaining relevant information, the entropy-temperature conjugate relationship is a theoretical consequence derived from the transition density matrix.

$$\rho_{\mathrm{tr}} = \sum_{k=1}^{n-1} |\psi_k\rangle\langle\psi_{k+1}|$$

where the phase-locked states $\{\psi_k\}$; $k = 1, 2, \ldots n$ define the Off-Diagonal Long-Range Correlative Information, ODLCI, cf.,[10] and n the characteristic dimension for the functional assignments of the cellular organization, yielding the syntax for molecular and higher-order communication.

In photosynthesis, the "*quantum Goldilock effect*" implies natural selection chooses just the right temperature and time scales for quantum transport to be maximized (see, e.g.,[22]). This principle has recently been proposed, within a connectionist framework, to apply also to consciousness.[23] These effects are reminiscent of current compelling applications of free-energy minimization for homeostasis as a defining measure of consciousness.[24] However, in the brain, there is no consensus on whether brains exhibit quantum coherence, entanglement, or not. Can long-range correlations overcome decoherence in the chaotic milieu of the material brain? In this thematic volume, it is suggested that a different principle is operational in brains and living systems at the quantum level. While dissipation, as employed above, is used to enable the transport of energy in photosynthetic complexes, we have taken decoherence as a fundamental quantum-thermal mechanism of long-range correlative information, i.e., as a teleological process of quantum Darwinism that operates as an information-based action through energy-entropy dissipation.

Is quantum Darwinism influenced by the selective pressures brought upon by the environment? From the viewpoint of univocal decoherence, the result is quantum-classical transduction. In our present formulation, however, the fine-tuned Goldilock effect is orchestrated by thermodynamic constraints, such as adequate temperature-time boundary conditions, generating negentropic gains, and the syntax for teleonomic processes, governed by an evolved program, rather than declining to aimless randomness, chance, and complete disorder. It should be stressed that this strategy undermines neural Darwinism, advocating a Bayesian dynamics with a semantic flavor, conceptually applying selection at the level of neuronal groups, see, e.g.[25]

Organisms are dissipative systems, with their material organization crucially dependent on energetic and entropic exchanges with the environment. The intrinsic information within the organism continuously modifies the material substrate that has generated it. Quantum potentialities exchange energy and entropy to act on the material substrate under steady state conditions according to the second law of thermodynamics. In a wider context, information-based quantum Darwinism encapsulates, via self-referential amplification, macroscopic quantum potentialities to play a pivotal role in the transduction from condensed matter to animate matter, differentiating life in an organism from the non-living mechanistic world. So, life as information transformation, via quantum Darwinism, increases the entropy of informational pathways. The physical interactions through quantum correlations, thermo-qubits and a more complex syntax for communication extend quantum Darwinism through entropy production, surprisal and free energy reorganization. The transformation of information creates information pathways, balancing the relative disorder of the whole, increasing the layers of specialization, exploiting evolution as a teleological process and a driver of more complexity to meet the requirement of the whole organism's survival adaptability in its environment. In this version of abiogenesis one obtains an evolutionary mechanism of intrinsic information via negentropic entanglement supporting the extended view that life is quantum Darwinian.

Acknowledgment

The impetus for this work was conceived from discussions with Dr. Val Bykovsky.

References

1. Pauling, L. *The Nature of the Chemical Bond and the Structure of Molecules and Crystals*; Cornell University Press: Ithaca, New York, 1945.
2. Brändas, E. J. Per-Olov Löwdin–Father of Quantum Chemistry. *Mol. Phys.* **2017**, *115* (17–18), 1995–2024.
3. Löwdin, P.-O. Proton Tunneling in DNA and its Biological Implications. *Rev. Mod. Phys.* **1963**, *35* (3), 724–732.
4. Hertz, H. G. Diffusion and Conductance in Ionic Liquids. A Linear Response Treatment. *Z. Phys. Chem. Suppl.-H* **1982**, *1*, 1–151.
5. Löwdin, P.-O. Correlation Problem in Many-Electron Quantum Mechanics. I. Review of Different Approaches and Discussions of Some Current Ideas. *Adv. Chem. Phys.* **1959**, *2*, 207–322.
6. Ceulemans, A.; Chibotaru, L.; Kryachko, E. Manifestations of Vibronic Coupling in Chemistry and Physics. *Adv. Quant. Chem.* **2003**, *44*, 1–673.
7. Quack, M.; Seyfang, G.; Wichmann, G. Fundamental and Approximate Symmetries, Parity Violation and Tunneling in Chiral and Achiral Molecules. *Adv. Quant. Chem* **2020**, *81*, 51–104.
8. Mandeville, S. E.; Crespi, H. L.; Katz, J. J. Fully Deuterated Euglena Gracilis. *Science* **1964**, *146* (3645), 769–770.
9. Chatzimitriou-Dreismann, C. A.; Brändas, E. J. Proton Delocalization and Thermally Activated Quantum Correlations in Water: Complex Scaling and New Experimental Results. *Ber. Bunsenges. Phys. Chem.* **1991**, *95*, 263–272.
10. Yang, C. N. Concept of off-Diagonal Long-Range Order and the Quantum Phases of Liquid Helium and of Superconductors. *Rev. Mod. Phys.* **1962**, *34*, 694–704.
11. McClintock, B.; Nobel Lecture, Dec 1983. The Significance of Responses of the Genome to Challenge. Science 226. *No* **1984**, *4*, 792–801.
12. Margulis, L.; Sagan, D. *Acquiring Genomes: A Theory of the Origin of Species*; Basic Books: New York, 2002.
13. Georgiev, D. D. Quantum Information Theoretic Approach to the Mind-Brain Problem. *Prog. Biophys. Mol. Biol.* **2020**, S0079-6107(20)30078-X. https://doi.org/10.1016/j.pbiomolbio.2020.08.002.
14. Poznanski, R. R.; Brändas, E. J. Panexperiential Materialism: A Physical Exploration of Qualitativeness in the Brain. *Adv. Quant. Chem.* **2020**, *82*, 301–367.
15. McIntosh, H. V. Symmetry and Degeneracy. *Group Theory and its Applications* **1971**, *2*, 75–144.
16. Denton, D. A. *The Primordial Emotions: The Dawning of Consciousness*; Oxford University Press: Oxford, 2006.
17. Zurek, W. H. Quantum Darwinism. *Nat. Phys.* **2009**, *5*, 181–188.
18. Kastner, R. E. Classical Selection and Darwinism. *Physics Today* **2015**, *68*, 8–9.
19. Zurek, W. H. Classical Selection and Quantum Darwinism. *Phys. Today* **2015**, *68*, 9–10.
20. Zurek, W. H. Quantum Theory of the Classical: Quantum Jumps, Born's Rule and Objective Classical Reality Via Quantum Darwinism. *Philos. Trans. A* **2018**, *376*, 20180107.
21. Brändas, E. J. Abiogenesis and the Second Law of Thermodynamics. In *Advances in Quantum Systems in Chemistry, Physics and Biology*; Mammino, L., Ceresoli, D., Maruani, J., Brändas, E. J., Eds.; Progress in Theoretical Chemistry and Physics, Vol. 32; Springer: Switzerland, 2020; pp. 393–436.

22. Lloyd, S. Quantum Coherence in Biological Systems. *J. Phys.: Conf. Ser.* **2011**, *302*, 012037.
23. Guevara, R.; Mateos, D. M.; Velazquez, J. L. P. Consciousness as an Emergent Phenomenon: A Tale of Different Levels of Description. *Entropy* **2020**, *22*, 921.
24. Solms, M.; Friston, K. How and why Consciousness Arises: Some Considerations from Physics and Physiology. *Journal of Consciousness Studies* **2018**, *25*, 202–238.
25. Sporns, O.; Tononi, G., Eds. *Selectionism and the Brain*; Academic Press: New York, 1994.

CHAPTER TWO

Quantum considerations of neural memory

Gerard Marx[a] and Chaim Gilon[b,*]
[a]MX Biotech Ltd., Jerusalem, Israel
[b]Institute of Chemistry, Hebrew University Jerusalem, Jerusalem, Israel
*Corresponding author: e-mail address: chaimgilon@gmail.com

Contents

Abstract

To explain how a mental state can be achieved and remembered by cells employing inanimate components, some have called on quantum mechanics (QM, including quantum chemistry, quantum physics and quantum field theory) to provide an explanatory rationale.

In that context, we consider how the terms "uncertainty," "entanglement" and "memory" are defined by physicists and neuroscientists, as they have different meanings for each discipline.

Emotions are core existential states for neural nets. But QM is incapable of formulating such. Following the opinion of Medawar[37]: *"Chemistry is the only physical science which offers a pathway to understanding animate biology."*

We review a tripartite mechanism with neurotransmitters (NTs), as providing an explanatory process which clarifies the molecular underpinnings of neural memory and emotions.

1. The challenge of memory

All admit to the material basis of the brain, which like all tissues, is composed of atoms and molecules, organized as groupings of cells. But the unique challenge to cognitive scientists is to conceive of a process that

Advances in Quantum Chemistry, Volume 82
ISSN 0065-3276
https://doi.org/10.1016/bs.aiq.2020.04.003

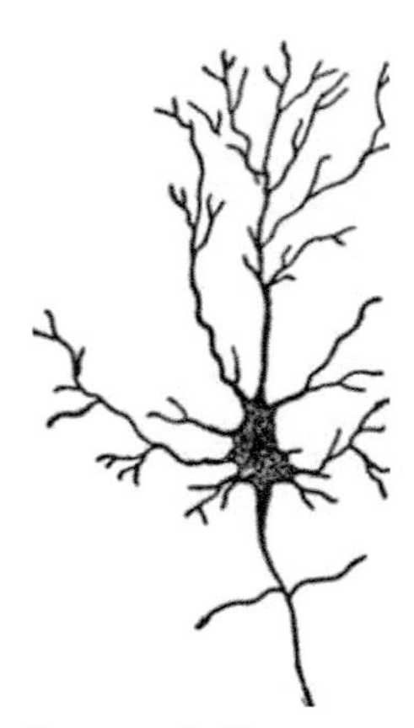

Fig. 1 A drawing of a neuron, as if suspended in empty space (i.e., naked). The surrounding web of aqueous glycans and proteins is ignored.

explains the emergence of psychic states, like memory and emotions, from the interactions of atoms and molecular comprising neurons in the brain.

Physicists and computer scientists have attempted to address the enigmas of brain consciousness and memory.[1–6] In search of a mechanism that could rationalize the unique ability of neural nets to generate psychic states which are remembered, many adopted the electrodynamic theories proposed by Cajal and later generations of neuroscientists.[7,8] Mainly, the neural ensemble was considered as a circuit of neurons in synaptic, electrodynamic contact with one another, like an electronic circuit at the heart of all our devices (chips, phones, etc.). Neurons were represented as "naked," as if they were floating in space (Fig. 1).

The late Marvin Minsky referred to the brain as a "meat machine".[9] Indeed, it is "meat" like other biological organs (liver, kidney, etc.), powered by metabolism, activities circumscribed by the strictures of physiology and rules of chemistry. But the brain is certainly not a "machine." It is not manufactured nor does it convert linear force into circular movement, or transform magnetic force into electric current or light. Rather, it seems that the brain can generate a new dimension termed "mentality."

2. Mental states and quantum mechanics (QM)

To explain how a mental state can be achieved and remembered by cells employing inanimate components, some have called on quantum mechanics

(QM, including quantum chemistry, quantum physics and quantum field theory) to provide an explanatory rationale.[1,10–15] Some refer to "quantum cognition." These approaches are beset by linguistic ambiguities as well as conceptual enigmas relating to mental states. For example, inanimate matter, as well as the organic matter of the neural net, can all be considered as "quantum substrates." But what distinguishes the neural net that permits it to generate mentality?

The terms "uncertainty" and "entanglement" are employed by both the QM physicist and the cognitive scientist, though they mean quite different things for each.

The QM physicist defines "uncertainty" with a formula originated by Heisenberg, to describe the energy and location of the electron.[11,12]

$$\text{Uncertainty} = \Delta \times \Delta p \sim h$$

where x = position; p = momentum calculated as (distance/time) × mass; h = Planck's constant. Time, as well as space and momentum, are inherently uncertain, though linked by Planck's constant.

For the cognitive scientist, "uncertainty" relates to the inability to describe mental states such as emotions analytically. There is no formula, calculus or algorithm for such, only verbal metaphors, poems, songs. But without a code for emotive states, one is at a loss to analytically describe a neural experience.

"Entanglement" for the QM physicist, relates to the perfect temporal coupling of physical properties such as position, momentum, spin or polarization of particles, even when separated by a large distance (spooky interactions).[13–15]

For the cognitive scientist, "entanglement" relates to the intrinsic coupling of psychic states (i.e., emotions such as fear, love, anxiety, etc.) with physiologic reactions (i.e., pulse, temperature, dilation, etc.). Unhappily, the cocurrent use of these same terms does not clarify their purpose for either discipline.

Similarly the term "memory." For computers, it must be physical[16] as related to the disposition of spins, holes or dopants in a matrix, in a manner that translates into a "demotive" (lacking emotions) binary code of information. Humans are masters of such, witness the manufacture of computers discs and memory chips. It might have been less muddling had the term "storage" been applied to such devices, rather than the term "memory."

For the cognitive scientist, neural "memory relates to the recall of past experience saturated with emotive overtones. Notwithstanding great efforts in psychoanalysis,[17] all are still at a loss to analytically describe mental states. One cannot employ the metrics of physics or the Information Theory of computers to formulate emotions ,[18–21] citing only a few). The fatal flaw of quantum computing is that it cannot clarify the processes underlying neural memory.

To address these challenges, one needs to conceive of a process that describes how psychic states, like memory and emotions emerge from the molecular interactions of atoms, molecules and cells in the brain. Like medical clinicians diagnosing the workings of malfunctioning biologic organs, we adopt biochemically scaled processes to evaluate the processes of the brain that generate mental states, emotions and memory.

3. Facts

Certain facts of neurobiology must be acknowledged. For example, neurons are not "naked" as represented by Cajal and following generations of neuroscientists.[22,23] Neurons (including glial and astrocyte cells)[24–26] are surrounded by glycosaminoglycans (GAGs), which comprise an aqueous extracellular matrix (nECM) through which both electric and chemical signals pass.[27–32] This nECM is not a "soup," but forms a filigree of polysaccharides which exerts structural as well as chemodynamic effects. It could be considered as a "quantum substrate," but so could every other biologic tissue or inorganic material. Such an approach is too all encompassing and not explanatory.

Moreover, neural and glial cell signaling is not only synaptic, but also occurs by ephaptic pathways through the nECM.[33–36] Notwithstanding, most modern neurobiologists follow Hebb and attempt to model learning and memory with changes of synapse number and function.

Scientists attempting to describe mental states face near insurmountable difficulties. There are no objective tools for measuring the subjective aspects of mentality. The computer and QM models crumble under the weight of physiologic considerations. Neither the binary-coded "Information Theory" nor QM apply to mental states, as they cannot formulate the physiologic basis of emotions or how they are instigated, encoded or recalled.

Following the opinion of Medawar[37]:

> "Chemistry is the only physical science which offers a pathway to understanding animate biology."

We suggest the following mechanism as the molecular underpinnings of neural memory and emotions.

4. Tripartite mechanism of neural memory

The *tripartite* mechanism[38–42] involves the following:

- *Neurons*—Morphologically spindly, arborized, synaptically and non-synaptically connected signaling cells (including glial cells), all suspended in nECM, some with synaptic contacts between them.
- *nECM*—Neural extracellular matrix, a hydrogel lattice, comprised of anionic polymers of glycosaminoglycans (GAGs) and proteins enswathing all neurons. It has been visualized by histologic stains, antibody staining and electron microscopy (SEM, TEM). It serves as the neuron's "memory material."
- *Dopants* (metals and neurotransmitters (NTs) are released from neural vesicles into the nECM during signaling, which form metal-centered complexes, used by the neurons to encode cognitive information. Dopants in the brain have been detected and analyzed by HPLC-MS, X-ray fluorescence, neutron activation and immunofluorescence, among other techniques.

Like all other materials, the nECM can be considered as a "quantum substrate." But this does not lead to a clarifying insight. Rather, we adopt a biochemical description. The tripartite mechanism establishes a "physical connection" between neurochemical components and mentality. A chemographic notation describes metal-centered complexes of *cognitive units of information (cuinfo)* (Fig. 2), which are formed by the interactions of dopants with the nECM, whereby neurons encode, memory. The neuro-transmitters (NTs)[44] serve dual functions of "entangling" physiologic with psychic effects: Effectively the NTs are the molecular elicitors and encoders of emotive states.

The *cuinfo* is a quantal unit of cog-info. It is more complex than the "bit" in that it has emotive "flavoring." The "bit" can be expressed in many media (electronic, magnetic, light, sound, holes). But the *cuinfo* can only be expressed around neural tissue.

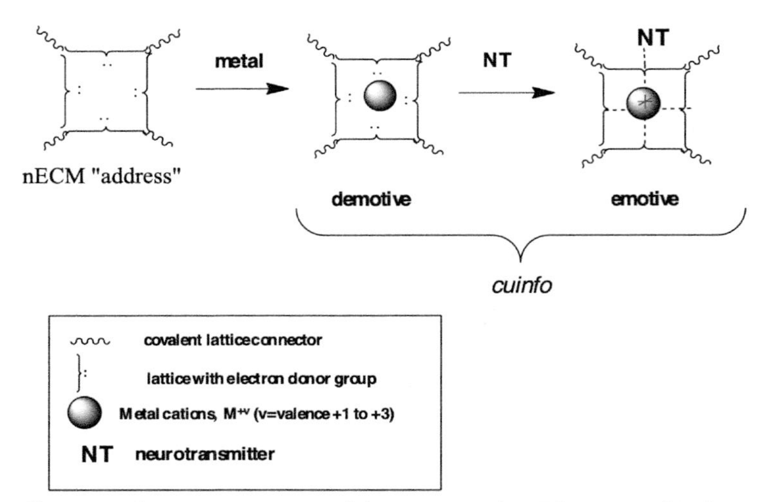

Fig. 2 Chemographic representations of the reaction of a nECM anionic binding site for a metal cation, an "address." The binding of an electron-rich neurotransmitter (NT) to the metal-centered cognitive unit of information (*cuinfo*) confers emotive context.

This chemo-electric mechanism (reviewed and detailed in refs. 38–41) is consonant with experimental observations relating to the composition of the brain and the morphology of the highly branched neurons, as it relates to their interactions with their surrounding nECM.[42] Also, it involves diffusible materials available to neurons, notably NTs[43] and trace metal cations.

5. Conclusion

The quest to comprehend mental processes like memory, drives many to forward hypotheses based on one or another thesis. Some forwarded a transcendental approach (reviewed by Searle[44]) suggesting that memory and mind may "exist independent of the physical body".[45] Some philosophers debunked "spirits," "souls" and "ghost".[46] Others offered quantum mechanical (QM) rationales,[47–49] though these cannot describe emotive states. Similarly, efforts aimed at ascribing mental states to "information processing" suffer the similar lack.[50]

Following the opinion of Medawar (cited above), we have embarked on a quest to identify the biochemical underpinnings of memory, as a clue to all mental processes.

Mental states inextricably link (entangle) physiologic sensations (feelings) with mental states (emotions), all simultaneously instigated by NTs. One can

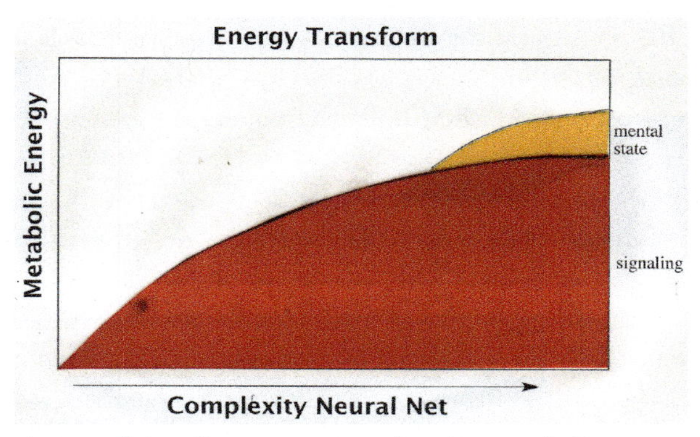

Fig. 3 The evolution of signaling processes, whereby metabolic energy of ever more complex neural nets is phase-changed to achieve a new mental dimension.

assume that the neural net signaling system evolved from primitive biologic signaling. This implies that the increasingly complex interactions of matter around the evolving neural nets instigated a novel aspect of biochemical energy, a mental state experienced as memory. The evolution of matter around neurons led to the emergence of a new dimension of biologic energy, a mental state manifest as "feelings," recalled as emotive memory (Fig. 3). Thus, the tripartite mechanism presents a fusion of psychology with chemistry.

"Meaning" is inconsequential in a binary or QM system. Only neural creatures can establish the "meaning" of any event or stimuli, invariably tied to an emotive state. According to the laws of QM, information may never be destroyed. But the cognitive information of the neural net is destroyed when the creature dies. It can be only vaguely resuscitated in the memory of the survivors, the offspring or the acquaintances of the deceased.

Undoubtedly, QM serves as a basis for describing the physical nature of biologic tissues as well as innate material (see Linus Pauling). But its scale is too refined to describe the physiology of neural nets resulting in mental states. It is poorly suited for describing the nonlinear continuous time dynamics of the emotive nervous system. Though QM required a fundamental modification of all ideas regarding the absolute character of physical phenomena,[51] it is "demotive," incapable of formulating an emotive state. One is left to fall on neuro-chemical descriptions to allude to the mental states instigated by neural nets.

QM descriptions are challenged by enigmas that touch on core issues:

- Undularity or corpuscularity (wave/mass), analogous to Descarte's Mind/Body conundrum.

- For the QM consideration of the life or death state of Schrodinger's cat, the (external) observer determines the cat's final state. The QM rationale is that the observation itself collapses the QM wavefunction. But for consciousness, the (internal) observer (the self) is the only determinant of subjective experience, being continuous until death.
- "Density matrices" and "Liouville formulations" of phase space do not furnish a transcendental algebra of mental states.

Rather, the emotive dimensions of mentality are based on physiologic reactions accompanied by psychic states. Chemicals (neurotransmitters (NTs) and drugs) are the only molecular entities capable of eliciting such responses from animate beings.

We should be courageous in confronting the enigmas of QM and admit that the metrics of physics are not adequate to describe mental states; that "consciousness" occupies a novel role in the laws of Nature. A new paradigm is needed to account for the transformations of metabolic energies into new phases (Fig. 3), which account for mental dimensions manifest as subjective experiences entangled with physiologic reactions.

"Criticality" refers to the appearance of erratic fluctuations in a dynamical system, and a transition through a "critical point" is called a phase transition.[52] The emergence of mental states from the metabolism of neural nets can be considered as such a phase change. However, it is unique in that it does not correspond to a phase change of matter (i.e., ice/water/steam) or the transformation of energy from one form to another (i.e., electrical into light into heat) and cannot be detected instrumentally. Instead, it relates to the emergence of mental states from the metabolism of ever more complex neural nets. It is manifest in the subjective experience of the neural creature.

The tripartite mechanism provides a hypothetical description of a possible biochemical process for such emergence. It invokes physiologically relevant materials available to the neuron. It presents the neurotransmitters (NTs) as the molecular elicitors and encoders of emotive states, necessarily recalled for survival. Without memory, there are no emotions and vice versa without emotive qualities, memory fades. Emotions and memory are inherently entangled properties of active mentality.

Acknowledgments

G.M. is a founder of MX Biotech Ltd., involved in developing biotechnologies for wound healing and new memory materials.

C.G. is emeritus professor of HU, but is active in developing and patenting peptide-based tools for surgery and pharmacology.

Notwithstanding, the ideas forwarded here are scientifically genuine and presented in good faith, without commercial clouding of the concepts expressed herein.

Conflict of interest

This work received no funding.

References

1. Lotka, A. J. *Elements of Mathematical Biology*. Dover: New York, 1956.
2. Lange, M. *The Philosophy of Physics: Locality, Fields, Energy, and Mass*. Blackwell Publishing: London, 2002.
3. Churchland, P. M. *Matter and Consciousness*, 3rd ed.; MIT Press: Cambridge MA, 2013.
4. Muller, R. A. *Now: The Physics of Time*. Norton Company: NY, 2016.
5. Becker, A. *What Is Real? The Unfinished Quest for the Meaning of Quantum Physics*. John Murray: London, 2018.
6. Erdi, P.; Esposito, A.; Marinaro, M.; Scarpetta, S., Eds.; *Computational Neuroscience: Cortical Dynamics*; Springer: Berlin, 2004.
7. Hebb, D. O. *The Organization of Behavior*. Wiley: New York, 1949.
8. Kandel, E. R.; Dudai, Y.; Mayford, M. R. The Molecular and Systems Biology of Memory. *Cell* **2014**, *157*, 163–186.
9. Levy, S. *Marvin Minsky's Marvelous Meat Machine*. Wired, 2016; January 2016.
10. Bergmann, E. D.; Pullman, A. *Molecular and Quantum Pharmacology: Proceedings of the 7th Jerusalem Symposium on Quantum Chemistry and Biochemistry Held in Jerusalem, March 31st-April 4th, 1974*. Springer: Berlin, 1975.
11. Lloyd, S. Quantum Computing. Computation from Geometry. *Science* **2001**, *292*, 1669.
12. Rosenblum, B.; Kuttner, F. *Quantum Enigma: Physics Encounters Consciousness*. Oxford University Press: NY, 2011.
13. Pereria, A. The Quantum Mind-Classical Brain Problem. *NeuroQuantology* **2003**, *1*, 94–118.
14. Zu, C.; Wang, W. B.; He, L.; Zhang, W. G.; Dai, C. Y.; Wang, F.; Duan, L. M. Experimental Realization of Universal Geometric Quantum Gates with Solid-State Spins. *Nature* **2014**, *514*, 72–75. https://doi.org/10.1038/nature13729.
15. Carleo, G.; Troyer, M. Solving the Quantum Many-Body Problem with Artificial, Neural Networks. *Science* **2017**, *355*, 602–605. https://doi.org/10.1126/science.aag2302.
16. a Landauer, R. *Information Is Physical*. Physics Today, 1991;23–29 May; b Landauer, R. The Physical Nature of Information. *Phys. Lett. A* **1996**, *217*, 188–193.
17. Solms, M. Issue: Unlocking the Unconscious: Exploring the Undiscovered Self. *Ann. N. Y. Acad. Sci.* **2017**, *1406*, 90–97.
18. Turing, A. M. Computing Machinery and Intelligence. *Mind* **1950**, *49*, 433–460.
19. Sejnowski, T. J.; Koch, C.; Churchland, P. S. Computational Neuroscience. *Science* **1988**, *241*, 1299–1306.
20. Pickering, A. *The Cybernetic Brain*. University of Chicago Press: IL, 2010.
21. Guidolin, D.; Albertin, G.; Guescini, M.; Fuxe, K.; Agnati, L. F. Central Nervous System and Computation. *Q. Rev. Biol.* **2011**, *86*, 265–285.
22. Cajal, R. Y. *Cajal's Histology of the Nervous System of Man and Vertebrates*. Oxford University Press, 1911;1995. ISBN: 9780195074017.
23. Churchland, P. S. *Neurophilosophy: Toward a Unified Theory of Mind*. MIT Press: Cambridge, MA, 1989.
24. Agnati, L. F.; Fuxe, K. Extracellular-Vesicle Type of Volume Transmission and Tunnelling-Nanotube Type of Wiring Transmission Add a New Dimension to Brain Neuro-Glial Networks. *Philos. Trans. R. Soc. Lond. B Biol. Sci.* **2014**, *369*(1652)Pii: 20130505.

25. Giaume, C.; Oliet, S. Introduction to the Special Issue: Dynamic and Metabolic Interactions between Astrocytes and Neurons. *Neuroscience* **2016**, *323*, 1–2. https://doi.org/10.1016/j.neuroscience.2016.02.062.
26. Ashhad, S. Stores, Channels, Glue, and Trees: Active Glial and Active Dendritic. Physiology. *Mol. Neurobiol.* **2019**, *56*, 2278–2299. https://doi.org/10.1007/s12035-018-1223-5.
27. Amit, D. *Hebb Vs Biochemistry: The Fundamentalist Viewpoint.* In *Selected Papers by Daniel Amit (1938–2007)*, 2013 http://www.thefreelibrary.com/Selected+papers+of+Daniel+Amit+(1938-2007).-a0332371120.
28. Arshavsky, Y. I. The Seven Sins of the Hebbian Synapse: Can the Hypothesis of Synaptic Plasticity Explain Long-Term Memory Consolidation? *Prog. Neurobiol.* **2006**, *80*, 99–113.
29. Barros, C. S.; Franco, S. J.; Muller, U. Extracellular Matrix: Functions in the Nervous System. *Cold Spring Harb. Perspect. Biol.* **2011**, *3*, 1–25, a005108.
30. Bukalo, O.; Schachner, M.; Dityatev, A. Modification of Extracellular Matrix by Enzymatic Removal of Chondroitin Sulfate and by Lack of Tenascin-R Differentially Affects Several Forms of Synaptic Plasticity in the Hippocampus. *Neuroscience* **2001**, *104*, 359–369.
31. Padideh, K. Z.; Nicholson, C. Brain Extracellular Space:Geometry, Matrix and Physiological Importance. *Basic Clin. Neurosci.* **2013**, *4*, 282–285.
32. Wlodarczyk, J.; Mukhina, I.; Kaczmarek, L.; Dityatev, A. Extracellular Matrix Molecules, their Receptors, and Secreted Proteases in Synaptic Plasticity. *Dev. Neurobiol.* **2011**, *71*, 1040–1053. https://doi.org/10.1002/dneu.20958.
33. Vizi, E. S.; Fekete, A.; Karoly, R.; Mike, A. Non-synaptic Receptors and Transporters Involved in Brain Functions and Targets of Drug Treatment. *Br. J. Pharmacol.* **2010**, *160*, 785–809.
34. Vizi, E. S. Role of High-Affinity Receptors and Membrane Transporters in Nonsynaptic Communication and Drug Action in the Central Nervous System. *Pharmacol. Rev.* **2013**, *52*, 63–89.
35. Anastassiou, C. A.; Perin, R.; Markram, H.; Koch, C. Ephaptic Coupling of Cortical Neurons. *Nat. Neurosci.* **2011**, *14*, 217–223.
36. Jefferys, G. R. Nonsynaptic Modulation of Neuronal Activity in the Brain: Electric Currents and Extracellular Ions. *Physiol. Rev.* **1995**, *75*, 689–723.
37. Medawar, P. *The Art of the Soluble.* Methnon Co: UK, 1967.
38. Marx, G.; Gilon, C. The Molecular Basis of Memory. *ACS Chem. Neurosci.* **2012**, *3*, 633–642.
39. Marx, G.; Gilon, C. The Molecular Basis of Memory. MBM Pt 2:The Chemistry of the *tripartite* Mechanism.ACS Chem. *Neuroscience* **2013**, *4*, 983–993. https://doi.org/10.1021/cn300237r.
40. Marx, G.; Gilon, C. The Molecular Basis of Memory. Part 3: Tagging with "Emotive" Neurotransmitters. *Front. Aging Neurosci.* **2014**, *6*, 58, eCollection 2014. https://doi.org/10.3389/fnagi.2014.00058.
41. Marx, G.; Gilon, C. The Tripartite Mechanism as the Basis for a Biochemical Memory Engram. *J. Integr. Neurosci.* **2019**, *18*, 181–185. https://doi.org/10.31083/j.jin.2019.02.6101.
42. Marx, G.; Gilon, C. Interpreting Neural Morphology. *Acta Sci. Neurol.* **2020**, *3*, 1–4.
43. Reith, M. E. Ed.; *Neurotransmitter Transporters: Structure, Function, and Regulation/Edition 2*; Springer-Verlag: New York, 2002.
44. Searle, John. R. Minds, Brains, and Programs. *Behav. Brain Sci.* **1980**, *3*(3), 417–457.
45. Grof, S. *Holotropic Mind.* Harper: San Francisco, 1992.
46. Ryle, G. *The Concept of Mind.* Penguin Books: NY, 1949.
47. Penrose, R. *The Emperor's New Mind.* Oxford University Press: New York, 1989.

48. Vannini, A. Quantum Models of Consciousness. *Quantum Biosystems* **2008**, *2*, 165–184.
49. Poznanski, R. R.; Cacha, L.; Latif, A. Z. A.; Salleh, S. H.; Ali, J.; Yupapin, P.; Tuszynski, J. A.; Tengku, M. A. Theorizing how the Brain Encodes Consciousness Based on Negentropic Entanglement. *J. Integr. Neurosci.* **2019**, *1*, 1–10.
50. Tononi, G. An Information Integration Theory of Consciousness. *BMC Neurosci.* **2004**, *5*, 42–64.
51. Wheeler, J. A. Recent Thinking about the Nature of the Physical World: It from Bit. *Ann. N. Y. Acad. Sci.* **1992**, *655*, 349–364.
52. Cocchi, L.; Gotto, L. L.; Zadesky, A.; Breakspear, M. *Criticality in the Brain: A Synthesis of Neurobiology, Models and Cognition.* (EMBO/EMBL Symposia). www.embo-embl-symposia.org/; 2020.

CHAPTER THREE

Functional neural electron transport

Christopher J. Rourk*
Jackson Walker LLP, Dallas, TX, United States
*Corresponding author: e-mail address: crourk@jw.com

Contents

Advances in Quantum Chemistry, Volume 82
ISSN 0065-3276
https://doi.org/10.1016/bs.aiq.2020.08.001

Abstract

Ferritin and neuromelanin appear to form quasi-ordered array structures in the substantia nigra pars compacta and certain other neural structures containing large catecholaminergic neurons. In addition, ferritin has observed properties of quantum mechanical electron transport similar to those of quantum dots, and the quasi-ordered arrays of ferritin and neuromelanin are similar to quantum dot and pi-conjugated polymer structures used in solar photovoltaic cells to generate excitons at room temperature. Based on this information, a hypothesis of functional neural electron transport in these neurons is developed that describes a possible mechanism that would use that behavior to assist with voluntary action selection. Conductive atomic force microscopy test results that are consistent with the hypothesis are also discussed, and a voluntary action selection mechanism that could result from the use of that energy transfer mechanism in the associated neural structures is proposed.

Abbreviations

LC *locus coeruleus*
MPTP 1-methyl-4-phenyl-1,2,3,6-tetrahydropyridine
QD quantum dot
SNc *substantia nigra pars compacta*

1. Introduction

One of the most recent discoveries in electrical energy conduction is the phenomenon of non-classical electron transport in QD solids, as opposed to classical electron movement in conductors, or even coherent electron transport from tunneling in individual molecular structures, such as proteins and bacteria.[1,2] This discovery traces its roots to as early as 1901, when Robert Francis Earhart observed a conduction regime between closely spaced electrodes that did not remain constant, as predicted by Paschen's Law, but which instead decreased linearly with the electrode separation distance.[3] This was the first observation of the ability of what was previously considered to be "particulate" matter, such as electrons, to exhibit wave-like properties. While the phenomenon of matter waves was first explained by Louis de Broglie in 1927, and the extension of that concept to matter outside of the nuclear realm was made by Max Born,[4] it was not until 1951 that electron tunneling in a solid state was discovered by Cornelius Gorter.[5] Electron tunneling is a form of non-classical, quantum mechanical electron transport.

QDs were discovered in 1981 by Ekimov and Onushchenko, who studied color formation in semiconductor doped glass and observed that the absorption frequency of light in such doped glass was lower than expected.

This effect was subsequently found to be caused by the quantum confinement of electron-hole pairs, also called "excitons," in small semiconductor crystals.[6] Quantum confinement occurs when a free electron is trapped in one or more dimensions and is unable to easily escape by classical electron conduction, which allows the electron to exhibit a more wave-like behavior called coherence. In particular, a QD can be said to exist when the Bohr radius of the electron of an exciton is greater than the size of the particle that contains the exciton.

QDs can be spherical and are usually 50 nm in diameter or smaller, although as discussed, the size requirement for a specific QD is related to the Bohr radius of an exciton associated with the QD.[7] The discovery of resonant tunneling in QDs through quantum states arising from lateral confinement was made in 1988.[8] While QDs have been extensively studied since their discovery, there is still much that is unknown about them.

References to quantum effects in biology were first made shortly after the development of quantum mechanics,[9] and the field of quantum biology was established to some extent by the 1960s.[10] Following an initial emphasis on quantum chemical applications to biomolecules like DNA around 1960–70, enthusiasm was subdued because analytical methods were not accurate and reliable enough. New understanding as well as computational advances have opened the field again. The most notable mechanisms of quantum biology are ones that can explain a function that was previously difficult to explain. For example, the mechanism behind photosynthetic transfer of energy from light harvesting molecules such as chlorophyll to the reaction center where the photon energy is converted into chemical energy and stored was known to be highly efficient, but classical theory could not explain such high efficiencies. While there is still some skepticism in the scientific community regarding quantum biology, with the usual objection being that biological organisms are too "warm and wet" for quantum effects to have any functional effect, there is growing acceptance and study of quantum biological mechanisms

One aspect of quantum biology that has not been explored is the potential for a quantum biological function to be performed that is related to the principles of QDs. It has been proposed that quantum confinement may be the principle behind the quantum mechanical behavior of excitons in photosynthesis.[11] In addition, the ability of natural sunlight to induce coherent exciton dynamics has been demonstrated using a Hierarchical Equations of Motion model.[12] However, no specific biological mechanism has yet been identified that uses quantum dot physical principles of operation, other than the one discussed herein.

Neuromelanin and ferritin are found in high concentrations in certain groups of catecholaminergic neurons, such as those of the SNc and the LC. Extensive evidence exists that shows that neuromelanin and ferritin have physical characteristics of QDs, and in the SNc and LC, they also form a quasi-ordered array that could support the formation of one or more electron transport mechanisms that are capable of transferring electrons between neurons. Unlike the short-range tunneling mechanisms within individual molecules that do not result in the formation of excitons,[1,2] these arrays utilize a different physical mechanism to result in electron transport across the extent of the array, even at distances as great as 40 μm.[13] Electron transport would be further facilitated by ferritin in the intercellular fluid between those neurons and in glial cells, in combination with the generation of internal cell voltages and possibly pressures. If present, electron transport could be associated with a function performed by the neuron groups. One possible function would be to cause electrons to transfer to a neuron having an axon that presents the lowest impedance path to ground, where that impedance is a function of the extracellular field of downstream neurons. This hypothesized action would effectively form a gate circuit that continuously senses the impedance of each of the available axon paths to ground and conducts energy to the neuron that is best situated to activate, to assist with formation of the action potential for that neuron, under certain circumstances. The neurological function of this gate circuit would enable multiple parallel processes to be performed by the neural network of the brain and to allow for selection of the "best" of those processes under those circumstances, such as when action potential generation in those neurons is not capable of being driven only by dendritic innervation, and could also correlate to the experience of conscious selection of an action under those conditions. Other possible timing-related functions might also or alternatively be associated with the hypothesized electron transport that could also correlate to the experience of consciousness.

This chapter reviews the technical literature as it relates to the hypothesis set forth above. Each element of the hypothesis has strong support in the technical literature, and it is shown how these diverse pieces fit together. Clinical and laboratory evidence is also discussed that appears to corroborate this theory of function, as well as the results of conductive atomic force microscopy (c-AFM) tests that were performed that also demonstrate the possible presence of electron transport in SNc tissue. The relationship of the hypothesized mechanism to the global workspace theory is also discussed, and an explanation of the number of potential states that could be associated with different activated neurons is presented that can be modeled as a state machine.

2. Physics of ferritin and neuromelanin

This section reviews the current literature as it relates to the hypothesis presented in this chapter. First, the physical properties of ferritin that relate to QD-like behavior are discussed, followed by a discussion of the physical properties of neuromelanin. The concentration of neuromelanin and ferritin in SNc and LC neurons is then discussed, followed by a discussion of the generation of electric and pressure fields in neurons that can facilitate spatial and temporal synchronization of exciton formation. The effect of extracellular electric fields on impedance to ground seen at the soma is discussed, because it pertains to the effective impedance to ground see at the soma by an electron. The electrons are not conducted to ground, though, but instead cause the generation of Fe^{2+} ions and the release of Ca^{2+} following generation of Fe^{2+}, and these topics are reviewed. Additional topics are also covered that support the hypothesis, including the function of iron in SNc and LC neurons, clinical and laboratory studies of damage to SNc and LC neurons, the relationship between dopamine neurons and movement, Bereitschaftspotential and reaction time, the evolutionary history of the basal ganglia, quantum biology and neurology. These topics are interconnected and the discussion of each topic overlaps in many places, but a clearer explanation of the relevance of these topics to the hypothesis and each other is presented later in the chapter. While an effort has been made to identify the most pertinent references for each topic, textbooks of material have been written on each of these individual topics, and the following is at best a cursory overview.

2.1 Ferritin QD properties

The first and perhaps most important aspect of the neural electron transport hypothesis is the QD nature and properties of ferritin. A number of studies in the published technical literature clearly show that ferritin has the physical properties of a QD. Ferritin is a spherical protein complex having a diameter of approximately 12 nm with an inner core of ferrihydrite that is approximately 8 nm, and is found in both plants and animals.[14] In humans, ferritin in the SNc has 24 subunits that include heavy chain subunits or H-chain, and light-chain subunits or L-chain.[15] The ratio of H-chain to L-chain subunits is tissue-specific, but the H-chain subunit is a ferroxidase, whereas the L-chain subunit conducts electrons between the inner iron storage area of the ferritin complex and the external environment. These subunits perform a number of steps in the process of storing intracellular iron in the form of reactive Fe^{2+} ions as inert Fe^{3+} ions, as shown in Fig. 1.

The process by which Fe^{2+} is catalyzed and stored to the ferritin core, and the reverse process by which it is released (not shown) require energy to perform, and probably would not occur in isolated ferritin cores. In those cores, if a free electron is absorbed by the Fe^{3+} ion and it is converted to Fe^{2+}, the associated molecular structure would not be stable, which could result in the subsequent release of the electron at a lower energy, but this process would depend on the specific molecular dynamics of the ferritin core.

It was recognized as early as 1992 that ferritin has measurable macroscopic quantum mechanical effects that are representative of QDs, based on measurements conducted on ferritin isolated from horse spleen.[16] Those measurements were made at low temperatures and did not include the natural environment of

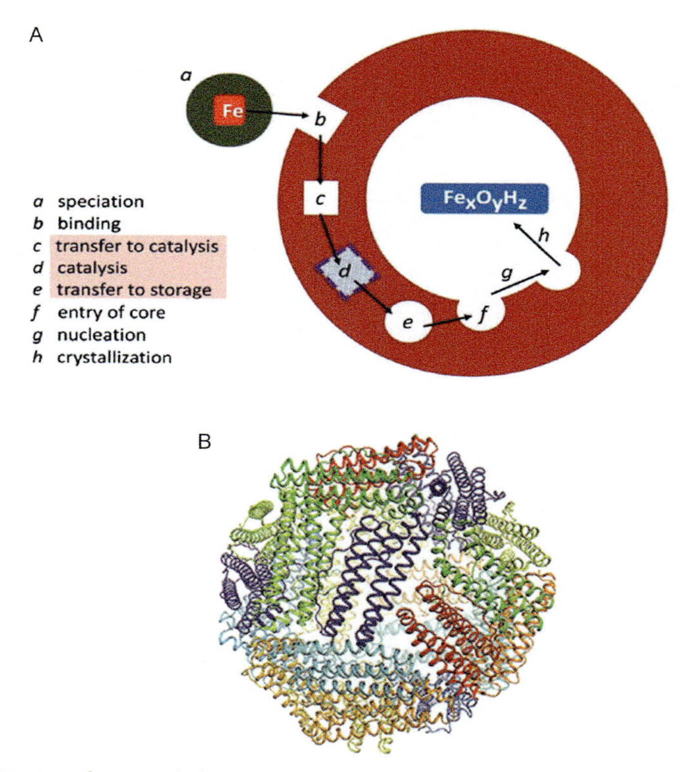

Fig. 1 "Ferritin form and functions." *From Hagen, W.; Hagedoorn, P.; Honarmand Ebrahimi, K. The workings of ferritin: a crossroad of opinions.* Metallomics ***2017,*** 9 *(6), 595–605.*

the ferritin, which can include biomolecules such as ligands that have been found to have the ability to space QDs to achieve optimal electron transport characteristics, as well as to facilitate quantum mechanical behavior. As such, while this early measurement is important, it may have failed to identify electron coherence functions that are performed at higher temperatures in the spleen or other tissues by ferritin.

Additional observations that confirmed the ability of ferritin to generate coherent electron effects were subsequently made. For example, it has been observed that an increase in the magnetic relaxation rate in ferritin occurs at low temperature as the field approaches zero, which is not consistent with classical behavior.[17] Subsequent measurements using the nitrogen vacancy center in diamond as a magnetic field sensor confirmed these low-temperature results.[18] It has also been shown that coherent electron tunneling, sequential electron tunneling, and electron hopping in ferritin occurs at room temperature, and that the type of transport is a function of the iron content inside of the ferritin.[19] Electron transport in QD structures using electron flux, coherent tunneling, sequential tunneling, and hopping (quantum mechanical electron transport mechanisms) has been generated under laboratory conditions using ferritin with differing iron content levels.[19–22]

Ferritin is a magnetic nanoparticle[23] and includes iron in a form that is antiferromagnetic at room temperature.[24] Antiferromagnetism has been shown to extend coherence lifetimes in QDs under certain conditions.[25–29] Ferritin has both indirect and direct electron band gaps, meaning that it can generate excitons either due to an electric field or phonon activity and in the absence of photons (indirect), or as a function of photon energy (direct). Measured band gaps for ferritin range from approximately 2.1 to 3.07 eV and vary as a function of the number of iron atoms stored and the presence of different anionic elements or compounds.[30,31] This prior work has thus established that the physical/electrical properties of naturally occurring ferritin can be used to generate quantum mechanical effects similar to those of man-made QDs; it has also established that ferritin can generate excitons due to an applied electric field, chemical excitation, or due to other mechanisms.

For example, individual ferritin cores were examined using conductive atomic force microscopy, and were found to exhibit a smooth, non-linear current-voltage response.[32] Subsequently, ferritin cores disposed in a quasi-ordered planar array were shown to exhibit the same non-linear current-voltage response but with regions of negative differential resistance.[33] These regions included current deviations of approximately 10 nA, over a period of approximately 200 ms, which corresponds to approximately

$(6.2 \times 10^{18}$ electrons/s$) \times (0.2$ s$) \times (10^{-8}) = 1.24 \times 10^{10}$ electrons. The large number of electrons involved with the negative differential resistance suggests that the measured current variation was due to electron transport between the ferritin core that was being tested and the array of quasi-ordered adjacent ferritin cores, as a result of exciton formation in those adjacent ferritin cores and the subsequent electron transport.

Ferritin was subsequently used to fabricate a multilayer structure on a silicon substrate that was shown to be capable of supporting electron transport over distances as great as 40 μm (equivalent to 3300+ layers) in the direction of two-dimensional order, but which was incapable of conducting electrons over more than four layers in the direction of order.[13] In particular, the Bera device (from[13]) included six layers of ferritin that were sequentially deposited on an SiO_2 substrate between two interdigitated electrodes, as well as on the electrodes themselves. The distribution of the ferritin cores within each layer was disordered (Z direction), such that no order was present from the point of the c-AFM probe over the grounded electrode, but the layer structure provided two-dimensional order (X-Y direction), which was also the direction of the electrical field applied between the electrodes, as shown in Fig. 2.

Current measurements were made (1) using c-AFM probes between the top of each layer as it was deposited and an underlying electrode, as well as (2) between each electrode. The c-AFM current measurements increased slightly with each increasing layer from one to four layers, but then dropped to zero for five or more layers. In contrast, the current measurements

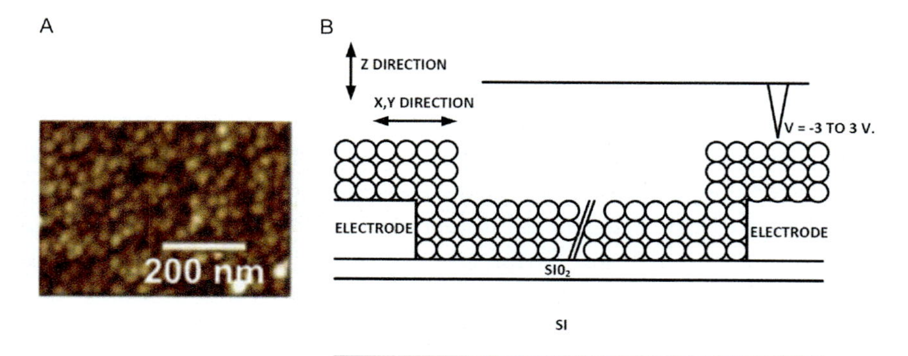

Fig. 2 Integrated circuit with array of ferritin cores; subtitle part (A) Disorder in Z direction; subtitle part (B) Arrangement of c-AFM probe and electrodes. *Panel (A) from Bera, S.; Kolay, J.; Pallabi Pramanik, P.; Bhattacharyyab A.; Mukhopadhyay, R. Long-Range Solid-State Electron Transport Through Ferritin Multilayers.* J. Mater. Chem. C ***2019,*** *7, 9038–9048.*

between the electrodes were non-linear as a function of voltage, and three orders of magnitude greater than the expected value based on impedance characteristics of individual ferritin cores. As discussed, the distance between electrodes was 40 μm, which is equivalent to greater than 3300 layers of 12 nm ferritin cores. There does not appear to be a classical explanation for this directional, non-linear electrical behavior, but as discussed below, this behavior is consistent with that of disordered arrays of quantum dots.

The multilayer structure disclosed by Bera was more dense than the distribution of ferritin in the SNc, but was a disordered array in-plane (one dimension), which was also the direction of the applied field for the c-AFM measurements, and was an ordered array parallel to the planes (in the other two dimensions), which was the direction of the applied field between the interdigitated electrodes. As will be discussed further below, the work of Bera demonstrates that disordered arrays of ferritin are capable of electron transport over distances that are of the same scale as the separation between SNc dopamine neuron soma.

2.2 Neuromelanin QD properties

The presence of high concentrations of neuromelanin gives the SNc the dark hue that resulted in its name, and is highly unusual. Most of the melanin that is present in humans is found in the skin, where it is used to protect skin cells from the damaging effects of ionizing solar radiation. A clear correlation can be seen between the melanin content of skin in regions where there is a high exposure to solar radiation, such as the equatorial zones, versus the low melanin content of skin in regions where there is a low exposure, such as arctic zones, which indicates an evolutionary basis for the function of melanin in skin. As such, it can be inferred that the presence of neuromelanin in human neural tissue also has an evolutionary basis, and that it serves an important function that is associated with the ability of humans to evolve and survive.

Although the technical literature does not appear to have addressed whether neuromelanin is a QD, melanin in sheet form has been studied for its semiconducting properties and for possible use as an organic semiconductor. Neuromelanin, a combination of eumelanin and pheomelanin, is a polymer complex that is approximately 30 nm in diameter and is found in high concentrations in certain catecholaminergic neurons, including dopamine neurons of the SNc and norepinephrine neurons of the LC.[34–37] An example of the distribution of neuromelanin in an unstained normal (non-diseased) dopamine neuron is shown in Fig. 3.

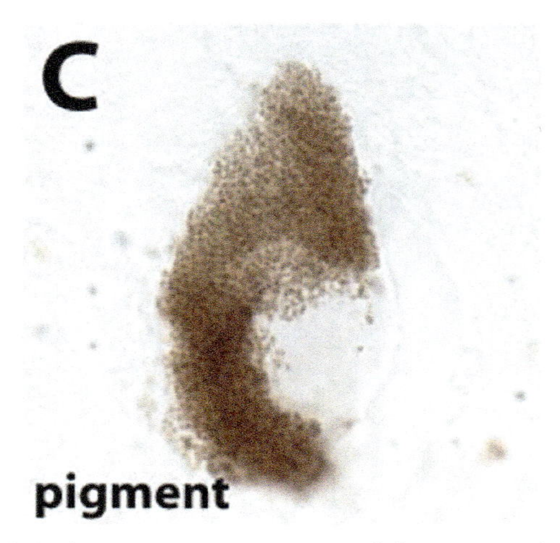

Fig. 3 Neuromelanin in dopamine neuron. *From Halliday, G.; Ophof, A.; et al. A-Synuclein Redistributes To Neuromelanin Lipid in The Substantia Nigra Early in Parkinson's Disease.* Brain ***2005,*** 128 *(11), 2654–2664.*

Neuromelanin has been extensively studied, but no consensus has been reached on its function or even its properties, which is unusual for a material that might have an important biological function that is related to the evolutionary success of humans as a species. Some observers have suggested that it is detritus that accumulates with age,[38] but as will be discussed below, neuromelanin per se does not appear to be present to the same extent in species with an earlier evolutionary provenance than *Homo sapiens*, regardless of their age, despite the fact that the basal ganglia structure traces its roots back 500 million years.[39,40] It has also been suggested that neuromelanin might function to collect heavy metals and other material that might otherwise damage the neuron,[41] but that would seem to have little evolutionary value prior to modern times, and that hypothesis also fails to account for the fact that the vast majority of neurons lack substantial concentrations of neuromelanin but are not adversely affected by such materials.

In regards to electrical properties of neuromelanin, no reported measurements of its electrical characteristics appears to exist, but some studies have estimated the band gap of melanin to range from 2.5 to 3.4 eV,[42,43] and at least one study concluded that the electrical behavior of melanin can be explained as an electronic-ionic hybrid conductor.[44] These observations are consistent with the pi-conjugated structure of melanin, because pi-conjugated polymers can have conductive or semiconductive properties as a function of their specific material properties.[38,45] Melanin has been shown to have a highest occupied molecular orbit (HOMO) with an energy of −5.54 eV and a lowest unoccupied molecular orbit (LUMO) with an energy of −2.54 eV, which are properties that would support QD electrical behavior in nanoscale structures.[46] Neuromelanin has also been shown to generate excitons due to chemiexcitation,[47] which may allow neuromelanin to generate excitons that are used to create an electron transport mechanism as a function of cellular chemistry and not electrical field variations.

Ferritin has been demonstrated to be present in proximity to the neuromelanin of the human SNc in large quantities. In Fig. 4 (from[48]), immunogold markers (small black dots) were used to identify the location of ferritin within SNc dopamine neurons and adjacent to NMOs, but the quantity of immunogold markers was insufficient to indicate the concentration of ferritin.

An indication of a high concentration of ferritin in the vicinity of NMOs in the SNc is provided in Fig. 5 (from[49]). In these images, SNc NMOs were identified using transmission electron spectroscopy, and examination of the same structures using electron spectroscopic imaging to identify the presence of iron (in red) shows that ferritin appears to form a layer around the exterior of NMOs.

As will be discussed below, this data is consistent with c-AFM measurements from SNc tissue, and indicates that ferritin forms a layer around NMOs at a high density.

Neuromelanin is formed by the reaction of iron with excess cytosolic catecholamines not accumulated in synaptic vesicles,[50] such that the presence of Fe^{2+} in the intracellular environment could contribute to the formation and accumulation of neuromelanin. As such, the presence of ferritin in proximity

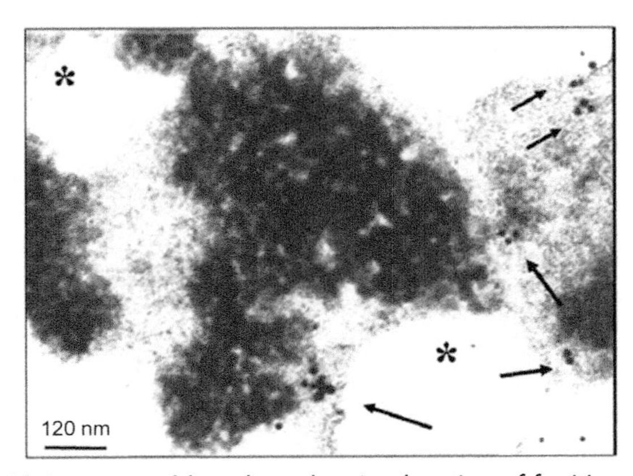

Fig. 4 NMOs with immunogold markers showing location of ferritin.

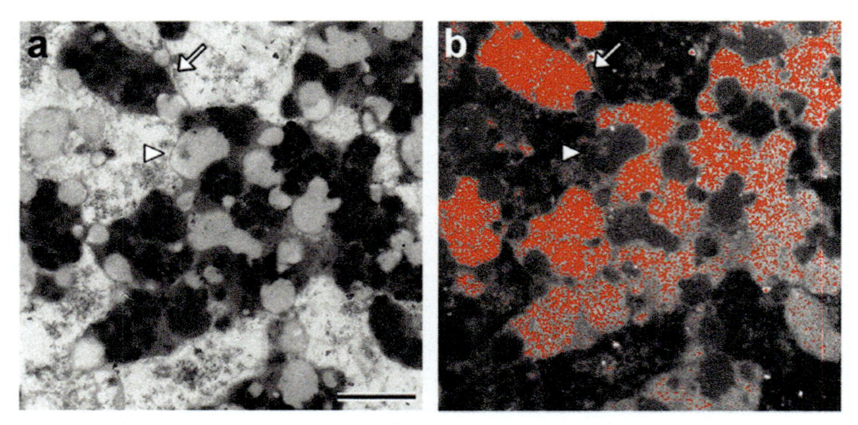

Fig. 5 Left: "SNc NMOs, transmission electron spectroscopy," Right: "SNc NMOs, electron spectroscopic imaging."

to neuromelanin may be related to iron homeostasis in the SNc associated with the dynamic generation of Fe^{2+} ions that are used for action potential generation. The magnetic properties of neuromelanin have also been investigated and documented, and may have a functional influence on the electrical properties of neuromelanin.[51]

2.3 Concentration of ferritin and neuromelanin in catecholaminergic neurons

In order for ferritin and neuromelanin to support electron transport, they would need to be present in a sufficient concentration and also to have a suitable spacing and configuration, and electron transport would also be a function of the associated coherence time and distance. In this regard, the specific configuration of ferritin and neuromelanin is important, and the coherent electron transport behavior could be modeled using an iterative solution of the Kronig-Penney and Schrodinger-Poisson equations.[52,53] However, that model would require substantial computational resources or simplifying assumptions that could have a determinative effect on the results, in addition to requiring details on ferritin and neuromelanin distribution and physical properties that would be difficult to determine. A more efficient way to both determine the distribution and concentration of ferritin and neuromelanin and to verify whether electron transport is supported by that structure is by specialized testing, such as c-AFM testing discussed in greater detail below.

Neuromelanin is found in organelles in catecholaminergic neurons, and it is found in the SNc and LC in greater proportion than in any other areas of the brain.[19] One study reported that neuromelanin organelles (NMOs) make up 50% of the image area of dorsal SNc neurons, where the density of SNc neurons is lower, and 25% of the image area of ventral SNc neurons, where the density is greater.[54,55]

Tyrosine hydroxylase (TH) staining can be used to indicate areas with high levels of iron in neural tissue, because TH is part of the iron homeostasis mechanism. TH staining of SNc tissue shows that the intercellular space between dopamine neurons in the SNc has high concentrations of ferritin or other iron-bearing compounds, as shown in Fig. 6.

These observations suggest that neuromelanin content is lower in dopamine neurons in areas where they have greater density, where a lower density of neuromelanin and ferritin would be required to support electron transport, and greater in areas where these neurons have lower density and would require a higher density of neuromelanin and ferritin to support electron transport. Based on an average estimated number of 1000 NMOs per neuron, 100 neuromelanin molecules per NMO and a cell body diameter of 25 μm, the average distance between neuromelanin molecules within the SNc and LC cell bodies is 50 nm, although it is noted that neuromelanin complexes aggregate in NMOs and are not evenly distributed throughout the cell body. In addition, as shown in the images above, the actual distance between the soma of these neurons is often less than 10 μm.

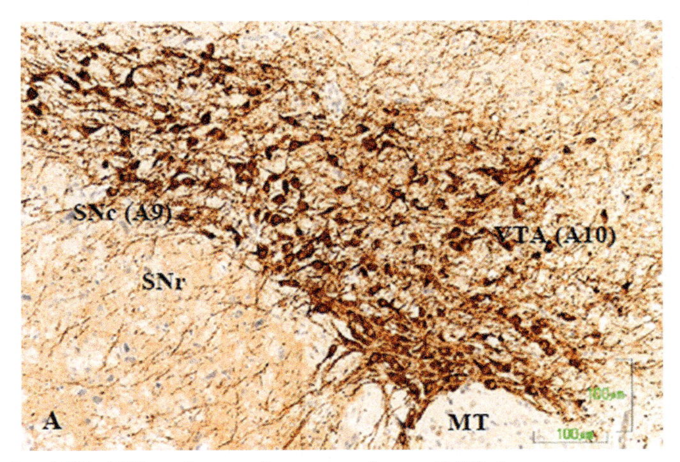

Fig. 6 Stained SNc (A9) and ventral tegmental area (A10) dopamine neurons showing cell body spacing and an indication of intercellular iron accumulations. *Excerpted from Santana, Y.; Montejo, A.L.; Martín, J.; LLorca, G.; Bueno, G.; Blázquez, J.L. Understanding the Mechanism of Antidepressant-Related Sexual Dysfunction: Inhibition of Tyrosine Hydroxylase in Dopaminergic Neurons after Treatment with Paroxetine but Not with Agomelatine in Male Rats.* J. Clin. Med. ***2019,*** 8 *(2),133.*

Based on the TH staining as seen above, it can be inferred that ferritin is disposed between the soma of dopamine neurons in the SNc. As shown above, ferritin has also been observed in large quantities in SNc cell bodies, adjacent to NMOs, using immunogold markers and electron spectroscopic imaging, and with an estimated density greater than neuromelanin.[48] Serum ferritin has a low normal concentration in men ($350\ \text{ng mL}^{-1}$) and women ($150\ \text{ng mL}^{-1}$), which equates to a concentration of approximately $5 \times 10^{11}/2.2 \times 10^{11}$ molecules per mL, or a separation distance of approximately 2 μm between molecules, but the concentration of ferritin is higher in the intercellular fluid of the SNc and LC. For example, one study estimated that the concentration of ferritin in the SNc in healthy subjects is approximately 3 ferritin cores per $0.003\ \mu\text{m}^3$, which works out to a spacing of approximately 100 nm between ferritin cores, on average.[56] This is similar to the spacing of ferritin as studied for application in a qubit-structured QD array,[57] and would support electron transport between these molecules. In addition, the concentration of ferritin in astrocytes, microglia and oligodendrocytes appears to be more than two, three and six times the concentration of ferritin in SNc neurons, respectively.[58] While the specific configuration of ferritin in those cells is not known, it is possible that the ferritin in those cells provides a quasi-ordered spacing that can support electron transport. Further, as shown above, the actual spacing and concentration of both dopamine neurons in the SNc and ferritin in the intercellular space between those neurons is more favorable, and the assumptions made herein regarding the average distribution of neuromelanin and ferritin are conservative assumptions.

2.4 Electric and pressure fields in SNc and LC neurons

The formation of coherent electron bands or other electron transport mechanisms between neuromelanin and/or ferritin in different neurons may be caused by time-varying electric fields and possibly strain fields, which are known to be present in the cell bodies of at least some neurons. For example, a biophysical model for a mechanical action wave that accompanies the electrical component of the action potential has been reported and has been used to model predictions for a giant squid axon, garfish olfactory neurons, crab motor neurons, and rat hippocampal neurons.[59] That mechanical action could be associated with exciton generation.

SNc dopamine neurons exhibit different electrical behaviors as a function of the extent of axonal branching. Neuron models that have fewer branches exhibit autonomous firing of action potentials that invade the entire axonal arbor, but synaptic stimulation was required to create an action potential for neuron models with a large number of branches in a computer simulation of the function of these neurons.[60,61] This difference in behavior of SNc dopamine neurons as a function of axonal branches shows that there are two classes of dopamine neurons in the SNc. The first class is associated with higher frequency autonomous neuronal activity, such as the pacemaker functionality that has been observed in the SNc.[62] This high frequency firing activity involves Na^+ action potential generation. The second class is associated with lower-frequency non-autonomous neuronal activity, such as reflexive or voluntary action,[63,64] which is associated with Ca^{2+} action potential generation. The number of branches in the first class of SNc neurons has been observed to be less than or equal to a transition point (nine branches), whereas the number of branches in the second class is greater than that transition point. Similar pacemaker activity and less extensive axonal branching has also been observed in the norepinephrine neurons of the LC. However, it is noted that the functions and structures of the LC neurons are varied and substantially different from the dopamine neurons of the SNc.[65,66]

The relative number of very large dopamine neurons (having a volume of 36,000 μm^3) is greater with increasing age. As shown in,[67] the percentage of small and medium dopamine neurons is greater in young and middle age controls than in old controls, but the percentage of large and very large dopamine neurons is greater in older controls (the total number of neurons is typically 250,000 or more, and decreases with age). The relative percentage of the size of these neurons was categorized as follows:

Size (μm^3)	**Young**	**Middle age**	**Old**
9000	38	39	28
18,000	43	39	38
27,000	13	17	24
36,000	6	5	10

This data indicates that a relative loss of smaller dopamine neurons and an increase in larger dopamine neurons occurs as people age, which may be due to some of the smaller and medium sized dopamine neurons increasing in size, or due to losses of the larger dopamine neurons being relatively less than the loss of the smaller dopamine neurons.

This change in size distribution also indicates that some dopamine neurons increase in importance with age, which may result from repeated use and the generation of neuromelanin from Fe^{2+} ion generation associated with voluntary action. Iron homeostasis in these neurons may provide protection from damage, and may also result in the formation of additional neuromelanin for improved homeostasis in neurons that are frequently used. This increase could improve the ability of those neurons to participate in the electron transport function, which would be related to the axonic impedance of the neuron, assuming that soma size increases are associated with an increase in the number of axonic synapses and an increase in the number of axonic branches.

2.5 Effect of efferent dendritic extracellular field on electrotonic axon impedance

A key aspect of the functional importance of electron transport in SNc and LC neurons is coherent electron localization, which needs to be associated with a neural function in order for an evolutionary driver to be implicated in the differences between *Homo sapiens* and other animals. For a balanced system of coherent electrons in neurons to experience localization, there needs to be a corresponding change in the system that favors localization in one neuron and drives the generation of action potentials. Dendritic stimulation would not have that effect, because the dendritic potentials cause the membrane potential at the soma to be less negative, and would also create a counter-EMF that needs to be overcome by electron coherence potentials at the soma to impact localization as a function of impedance. In contrast, a lower potential at the axonic synapses would not impact coherent electron generation or create a counter-EMF that would need to be overcome, and would instead function only as a negative impedance. In this regard, the efferent dendritic potentials associated with axon synapses would have the greatest impact on localization, and an increase in the number of axonic synapses would be consistent with greater control over voluntary actions that utilize that neuron.

The non-propagated electrical behavior of neurons can be modeled using electrotonic modeling.[68,69] Electrotonic conduction is different from propagated axonic electric impulses that result from the generation of an action potential in a neuron, which are driven by chemical processes associated with the alternating release of sodium and potassium channels in the axon membrane. Instead, electrotonic currents can be conducted in either direction along the length of the dendrites and the axon.

To cause electron localization from coherent electron bands in one neuron of the neurons that comprise a disordered array of neuromelanin and ferritin formed from the distribution of neuron cell bodies in the SNc and LC, it is necessary for the electrons to have a lower impedance path to ground in one neuron than in any other neuron. Otherwise, the electrons will either localize at the initial position of each electron within the coherent electron bands,[70] or fail to localize and remain in a coherent state, assuming that the mechanism responsible for exciton generation is not removed. For coherent electron bands extending through neuromelanin and ferritin in an array of neuron cell bodies in the SNc or LC, the neuron with the lowest impedance from the cell body to the common reference voltage of the intercellular fluid will present such a lowest impedance path.[68,71,72] An example of the distribution of extracellular voltages is shown below, where it can be seen that the voltage at some dendrites is substantially greater than the common reference voltage in Fig. 7, where red (cell body) is ~100 μV, yellow is ~10 μV and blue is less than 1 μV.

The extracellular field associated with the dendritic voltages at the synapse of each axon branch is seen as a negative resistance from the soma, and functions to lower the instantaneous impedance seen from the soma for that axon.[68] This is referred to as "antidromic" electrotonic behavior, and has

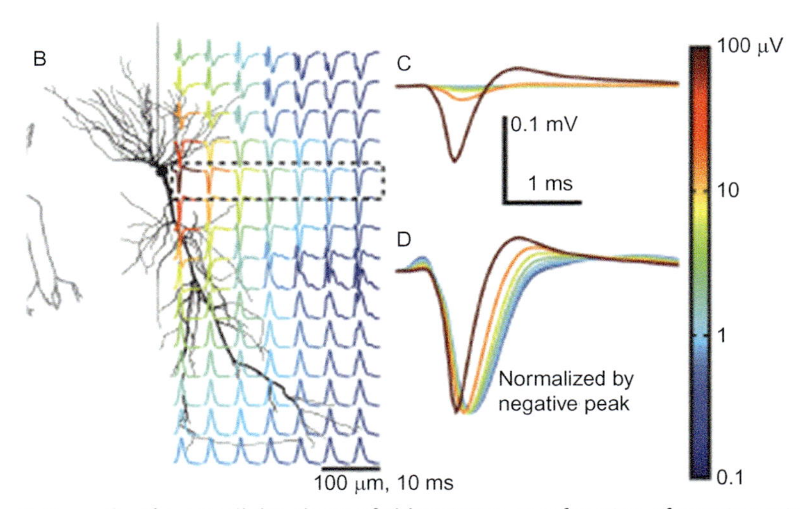

Fig. 7 Example of extracellular electric field variation as a function of axonic position. *From Schomburg, E.; Anastassiou, C.; Buzsáki, G.; Koch, C. The Spiking Component of Oscillatory Extracellular Potentials in the Rat Hippocampus.* J. Neurosci. ***2012,*** 32 *(34), 11798–11811.*

been observed in the LC as well as in other neural structures.[69,73] For the purposes of this analysis, the effect of dendritic impedance has been neglected, because the diameter of axons in SNc neurons appears to be substantially greater than the diameter of dendrites and would have much lower impedance.[74] However, elevated dendritic synapse voltages could potentially reduce the electrotonic impedance to ground seen from the soma relative to dormant dendrites or dendrites that are receiving a lower level of activation, depending on the soma voltage, and could thus contribute the effective impedance seen at the soma in the same manner as the antidromic axonic potentials, independent of the ionic contributions to cell body potential.

In the SNc, extensive axonal arborization is associated with the second class of dopamine neurons and involves hundreds of thousands of synaptic connections to the striatum and other regions,[60,61] including the neocortex, the colliculus and the hypothalamus.[75] An example of the extensive arborization is shown below in Fig. 8, which is an excerpt of a camera lucida drawing of the axonic branches of a large SNc neuron.

These efferent connections provide a large number of antidromic negative impedances when the associated efferent neural structures are activated but have not yet reached an action potential and been discharged, or are otherwise dissipated. The anatomy of the LC is different from the SNc in this regard. The LC includes four different classes of neurons, one of which has a

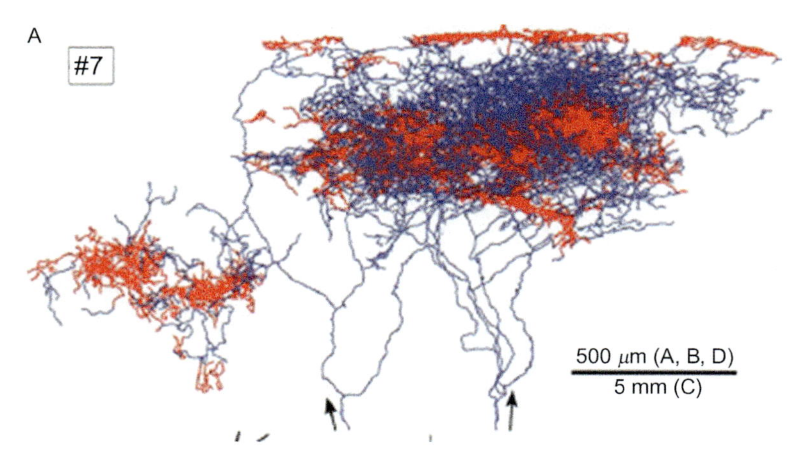

Fig. 8 Example of extensive axonic arbor in SNc dopamine neurons. *From Matsuda, W.; Furuta, T.; Nakamura, K.C.; Hioki, H.; Fujiyama, F.; Arai, R.; Kaneko, T. Single Nigrostriatal Dopaminergic Neurons Form Widely Spread and Highly Dense Axonal Arborizations in the Neostriatum.* J. Neurosci. ***2009,*** 29, *444–453.*

large number of branches and is associated with efferent forebrain connections.[76–79] This class of neurons is associated with antidromic signal generation at efferent connections and may be involved in the hypothesized switching mechanism in the LC.[73] However, because of the anatomy of the LC, it is more difficult to identify a class of neurons that would likely have the lowest axonic impedance as seen from the cell body, and the mechanisms of operation may be different in the LC than in the SNc.

2.6 Intracellular voltage measurements

In order to generate excitons, an energy source may be required to either cause a low-energy electron from the highest occupied molecular orbit of a neuromelanin complex to gain enough energy to escape, or to otherwise trap an electron in the bandgap of a ferritin complex. Intracellular voltages could provide a source of energy to both generate excitons from neuromelanin and potentially ferritin and to also drive electron localization. Intracellular electric fields in excess of 1.0×10^6 V/m magnitude have been reported,[80,81] although it is noted that Debye screening could result in significant variability depending on the source of the field.[82] These field strengths are sufficient by themselves to create a high probability (>10%) of electron disassociation from an exciton at the interface to a pi-conjugated polymer, even for materials other than ferritin and neuromelanin.[83] Pacemaker activity of SNc and LC neurons helps to coordinate the generation of these voltages between neurons and could further facilitate localization.[63–66,84] In particular, pacemaker activity is believed to involve the small to medium SNc dopamine neurons, but not the larger dopamine neurons with a large number of axonal branches.[60] However, it is also noted that chemically or thermally activated transport might also be possible with ferritin, such that intracellular voltages might not be required for formation of electron transport mechanisms.[85,86]

2.7 Generation of Fe^{2+} ions from transferred electrons

The 8-nm diameter internal cavity of ferritin can hold up to 4500 iron atoms,[14] although ferritin in the SNc typically stores between 1500 and 1800.[87] The protein shell structure of ferritin allows ferrous ions to diffuse in and out of the core in conjunction with iron homeostasis mechanisms, through eight hydrophilic channels located at a threefold symmetry axis.[56,88–92] Ferritin stores iron in a form similar ferrihydrite ($[FeO(OH)]_8[FeO(H_2PO_4)]$), which is water insoluble, by using ferroxidase of the ferritin heavy chain protein to remove excess Fe^{2+} by oxidizing it to water-insoluble Fe^{3+}. The stored Fe^{3+} is reduced to water-soluble Fe^{2+} iron by receipt of a free electron, as well as other components that would be necessary to create stable products during homeostasis. While Fe^{2+} is highly reactive, it has been shown that neuromelanin and dopamine are involved iron homeostasis, and the high levels of neuromelanin and dopamine in SNc dopamine neurons suggests that iron is also present, either as Fe^{2+}, Fe^{3+} or in both forms as part of a transition process involving iron storage and release from ferritin.

A quantitative molecular dynamics model of ferritin has been reported, which establishes that the storage and release of iron is accomplished by changes in pH in the ferritin channel mechanisms.[93] Based on this model, a qualitative analysis can be performed in regards to what might happen if a free electron tunnels into the ferritin core through the protein cage Coulomb barrier. That electron could reduce a stable Fe^{3+} ion to Fe^{2+}, which is water soluble, which should result in an increase in acidity if water is present and the Fe^{2+} disassociates 2 water molecules to form $Fe(OH)_2 + 2H^+$. While a single such reaction would not appreciably increase the pH inside of the ferritin core, a sufficient number of reactions could.

In vitro, such as the Bera device, there is no water to allow the Fe^{2+} to go into solution. Instead, the stored Fe^{3+} ion would become unstable when it is reduced to Fe^{2+}, and the Fe^{3+} state would eventually re-form after the electron is ejected to result in the stable $[FeO(OH)]_8[FeO(H_2PO_4)]$. This process would repeat until the electron tunnels back out, which would increase in probability as it loses energy and its De Broglie wavelength increases. The larger De Broglie wavelength would also assist with the formation of electron minibands. In the Bera device, an applied voltage of 3 V across $\sim$3300 adjacent ferritin cores results in substantial current, even though the voltage differential between each ferritin core would only be $\sim$0.001 V, which is too low to support hopping or sequential tunneling.

2.8 Calcium release due to Fe^{2+} ion release

Because Fe^{2+} is highly reactive, it is unlikely to be directly involved in action potential generation. However, it has been observed that neural mechanisms that use Ca^{2+} are also activated by Fe^{2+},[94] and Fe^{2+} may assist in the generation of action potentials in a manner similar to that of Fe^{2+} as part of iron homeostatis for Fe^{2+} release due to electron transport.[95] It has also been reported that Fe^{2+} generates Ca^{2+} signals through reactive oxygen species mediated ryanodine receptor stimulation.[96] The generation of Ca^{2+} may thus be part of the cellular Fe^{2+} homeostasis mechanism. This effect has also been proposed to function as a cellular redox sensor.[97] Labile Fe^{2+} can be detected in cells using a reaction-based fluorescent probe, although this does not appear to have been tried on dopamine neurons of the SNc or norepinephrine neurons of the LC.[98] Detection of Fe^{2+} using fluorescent probes in live *Caenorhabditis elegans* specimens has been reported, albeit not in the dopamine neurons.[99]

It is noted that both SNc and LC neurons have observed Ca^{2+} action potentials, as can be recognized by the long recovery time that is required for certain modes of neuron firing. For example, the LC neurons have two modes of firing: tonic and phasic.[100] The tonic activity is characterized by low frequency, sustained discharge, and is characteristic of Na^{+} action potentials, and the phasic activity is characterized by a small number of firings followed by longer recovery periods of hundreds of milliseconds duration, which is characteristic of Ca^{2+} action potentials. Likewise, the dopamine neurons of the SNc also exhibit the timing characteristics of both Na^{+} action potentials and Ca^{2+} action potentials.[101]

2.9 Iron in SNc and LC neurons

High levels of iron appear to either contribute or correlate to neuron damage in Parkinson's disease for dopamine neurons in the SNc and norepinephrine neurons of the LC to a greater extent than for other dopamine and norepinephrine neurons. These high levels of iron are also an indicator not only of higher levels of iron in those regions of the brain but also of an associated functional difference relating to iron between those areas and other areas with catecholaminergic neurons and neuromelanin.[102] This damage does not appear to be related only to iron content, since other regions of the brain with higher iron content, like the red nucleus, do not atrophy due to Parkinson's disease and may actually increase in size, possibly because of iron storage in lipofuscin instead of neuromelanin and ferritin.[62] As such, it is possible that the damage to SNc dopamine neurons and the loss of neuromelanin observed due to Parkinson's disease may interfere with iron homeostasis and result in generation of Fe^{2+} that is not properly contained by normal cellular processes.

2.10 Clinical and laboratory studies of damage to SNc and LC neurons

If ferritin and neuromelanin are responsible for electron transport associated with voluntary action generation in the SNc, then damage to the SNc would be expected to also result in a loss of such voluntary action, which is consistent with observations. For example, selective damage to the SNc dopamine neurons can be caused by MPTP and results in akinesia.[103,104] This damage is unlike the damage caused by Parkinson's disease, which also causes damage to norepinephrine neurons of the LC.[102,105] It is similar, though, in that such damage results in a loss of the ability to initiate voluntary action, a loss that can be relieved by levodopa.[106] Damage to dopamine neurons caused by MPTP does not immediately result in cell death, and replacement of dopamine by treatment with levodopa may mitigate disruption of the Ca^{2+}-mediated signaling pathways,[107] which may be involved with Fe^{2+} associated action potential generation, as discussed above. The MPTP damaged dopamine neurons are able to respond to dendritic inputs, such as to generate reflex actions.[108]

Clinical studies have shown that a decrement in LC function affects specific components of cognition in healthy older adults.[109] Laboratory studies on animals have also shown that loss of LC neurons contributes to motor dysfunction.[110] Damage to the LC and associated neural tissues resulted in paresis, hallucinations and other cognitive disorders in one case report.[111] However, the reported complete destruction of the LC does not result in a loss of cognitive function in some animals.[112]

2.11 Relationship between dopamine neurons and movement

The relationship between dopamine neuron function and movement is demonstrated at a simple level by *C. elegans*, which has eight dopamine neurons. It has been observed that normal specimens with functioning dopamine neurons are able to make small adjustments to their speed to maintain constant rates of locomotion. However, mutant specimens with a defective gene for generating tyrosine hydroxylase, which is needed to control dopamine synthesis, made larger adjustments to their speeds, resulting in large fluctuations in their rates of locomotion. These mutant specimens also frequently exhibited both abnormally high and abnormally low average speeds.[113] Providing dopamine was found to correct the movement abnormality in the mutant specimens, such that the correlation between more competitive movement and dopamine is clear. It remains to be determined whether the provision of dopamine itself resulted in the improvement, or if it aided in iron homeostasis to enable electron transport between ferritin cores. *C. elegans* also lacks neuromelanin, although ferritin is present,[114]

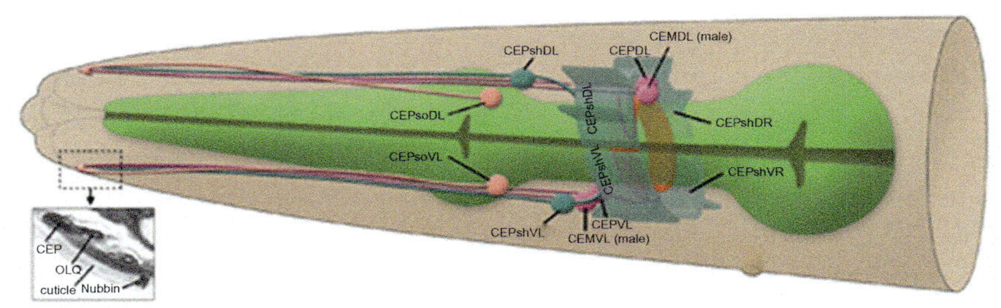

Fig. 9 Diagram showing placement of CEP dopamine neurons in *C. elegans.*[115]

and could provide the basis for coherent electron band formation in some of the eight dopamine neurons of *C. elegans*, with adequate spacing and an exciton generation mechanism. For example, consider the placement of the CEP dopamine neurons of the hermaphrodite *C. elegans* shown in Fig. 9.

The width of *C. elegans* at the location of the CEP neurons is approximately 50 μm, with a corresponding separation between CEP soma of less than 10 μm. In addition, the ftn-1 gene is expressed in the CEP neurons, which encodes the *C. elegans* equivalent of ferritin. The concentration of ferritin is unknown, but it appears to be used for iron homeostasis in *C. elegans*.

The lifespan of *C. elegans* is only several weeks, which might not be long enough for neuromelanin to develop. For example, it takes several years for neuromelanin to accumulate in human infants, who are not born with substantial levels of neuromelanin in the SNc or LC.[116] However, earthworms have longer lifespans of up to several years. They have been reported to have a neuromelanin-like substance in their dopamine neurons, which suggests that development of neuromelanin in dopamine neurons may have been guided by evolutionary vectors.[117]

2.12 Bereitschaftspotential, reaction time and the SNc and LC

Bereitschaftspotential, or "readiness potential," refers to the observation that neural activity occurs in the motor cortex, supplementary motor area and other areas of the brain before the conscious experience of the decision to initiate subsequent voluntary muscle movement.[118] This effect establishes that the conscious experience of making a decision is distinct from at least some of the neural processes that occur in areas of the brain that are determinative of those conscious experiences.

Both the SNc and LC are associated with reaction time. The SNc has been observed to perform a fundamental role in the temporal timing of tasks,[119] and Parkinson's Disease has been observed to cause a reduction in reaction time.[120] LC activation has also been observed to be closely linked in time to decision processes that link sensory inputs to motor outputs.[121] The reaction time to respond to simple visual stimuli, such as the display of a letter on screen (~ 0.20 s) is substantially faster than the reaction time to respond to complex visual stimuli that involve some level of analytic thinking, such as selecting a response based on the type of letter shown (~ 0.40 s).[122] This difference in reaction time suggests that there may be two different processes involved, one that uses the SNc and a second that uses the LC.

2.13 Evolution of the basal ganglia and relationship to voluntary action

The structure of the basal ganglia has remained largely unchanged over 500 million years of evolution.[39,40] While neuromelanin has largely been considered to be present only in mammals, ferritin has been demonstrated to be present in dopamine neurons of the basal ganglia across this evolutionary spectrum. The continuing presence of this structure indicates that it is of evolutionary importance.

The basal ganglia includes a number of regions that have distinct functions, which include the putamen, the caudate nucleus, the globus pallidus, the subthalamic nucleus and the substantia nigra.[123] These regions are functionally grouped into the "direct" pathway that is associated with voluntary action selection/movement initiation, and the "indirect" pathway, which is associated with movement inhibition, as shown in Fig. 10.

The importance of the SNc can be clearly seen in that it is the structure where the direct pathway initiates. Dopamine from the SNc acts to increase movement by stimulating the direct pathway and inhibiting the indirect pathway, and in this regard, the SNc is one of the most significant direct pathway components that is involved in action initiation.

The inputs to the SNc include the lateral hypothalamus and the paragigantocellular reticular nucleus, each of which also provide inputs to the LC, as well as the median Raphe and parabrachial nucleus, each of which also have high concentrations of neuromelanin.[124] As such, the neural tissues that have high concentrations of neuromelanin (and which likely also have associated high concentrations of ferritin, which is associated with iron homeostasis) either directly connect to each other or share

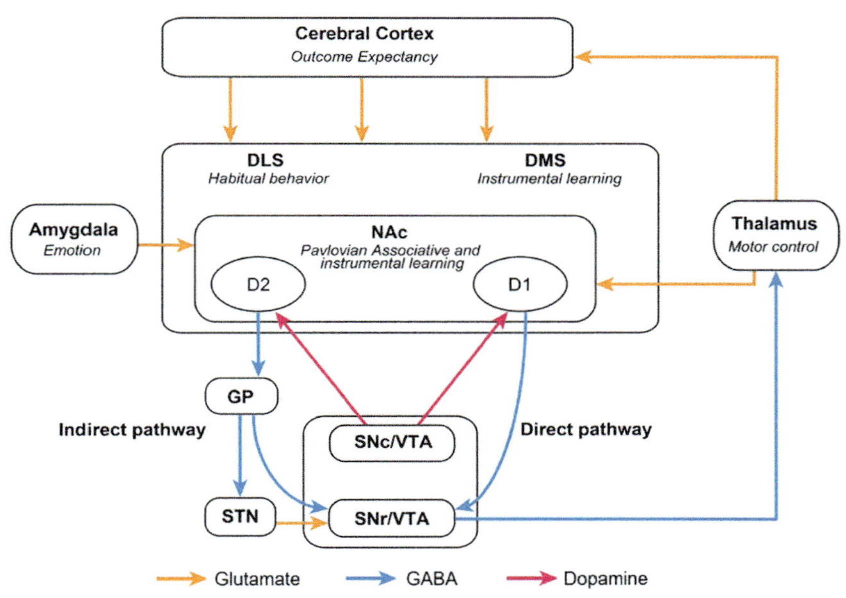

Fig. 10 Example of direct and indirect pathways. *From Macpherson, T.; Morita, M.; Hikida, T. Striatal Direct and Indirect Pathways Control Decision-Making Behavior.* Front. Psychol. ***2014***.

common afferent neurons. The inputs also include the zona incerta, the superior colliculus and other regions associated with the ability to recognize spatial locations.[125–127]

2.14 Quantum biology and neurology

Quantum mechanical effects have been shown to drive biological functions that were previously impossible to explain using non-quantum analysis,[128,129] including photosynthesis, photoreceptor cells, olfactory receptors, avian magnetoreception and enzymes, among others.[130] For example, the Fenna-Matthews-Olson complex and Light Harvesting Complex II have physical sizes that are on the same scale as quantum dots, and absorb energy at wavelengths similar to quantum dot structures to result in a high energy conversion efficiency.[13,131] "Quantum biology" has been proposed for explaining neurological effects, including the experience of consciousness, such as through "orchestrated objective reduction",[132] and nuclear spin entanglement.[133,134] However, direct experimental evidence of a quantum biological effect that explains a neural function has not yet been reported.[130]

3. QD electron transport

Electron transport arising from quantum mechanical effects is well known, and has been used in electronic devices such as tunneling diodes for decades. These quantum mechanical effects are also a design constraint or limitation for very small scale integrated circuits, where they must be taken into account to ensure proper operation of the circuit, but they are also emerging as a design tool for nanoscale circuits. It is no surprise, then, that such physical properties could be used by biological processes, and the presence of biostructures such as ferritin that are known to have QD properties in close proximity to materials like neuromelanin that are known to have pi-conjugated polymer structures with semiconducting material properties should flag the potential for quantum mechanical effects.

The quantum mechanical characteristics of QDs are physical characteristics and have been tested in materials that are similar to the in vivo environment, such as celluloid hydrogels.[135] Semiconductor devices that utilize QDs may be referred to as mesoscopic devices, which are designed to use the quantum mechanical properties of QDs for specific designed functions, but mesoscopic devices also include molecular transistors, quantum wires, nanodiodes, nanosensors and biomolecular devices.[136]

The quantum mechanical behavior of QDs has also been demonstrated to exist when they are placed within cells for specific purposes.[80] For example, it has been shown that QDs within a neuron can affect the action potential generation and firing of the neuron.[137] These effects were observed for an individual implanted QD, though, and did not address the possibility of naturally occurring structures within the cell possibly creating quantum confinement in a manner that is similar to that of a QD.

The effects of QDs formed from different materials and sizes on the quantum mechanical properties of assemblies of those QDs have been extensively studied, including electron transport in systems of multiple similar QDs and systems of QDs made of different materials.[138,139] One of these electron transport mechanisms is coherent electron bands, also referred to as minibands.[140,141] The effect of random variations and quasi-ordering of size and spacing of QDs has also been studied.[142–144] In particular, for a small disordered/quasi-ordered array of quantum dots disposed between two electrodes, the miniband conductivity was shown to be reduced due to the disorder, but also spread due to electron scattering, as shown in Fig. 11.

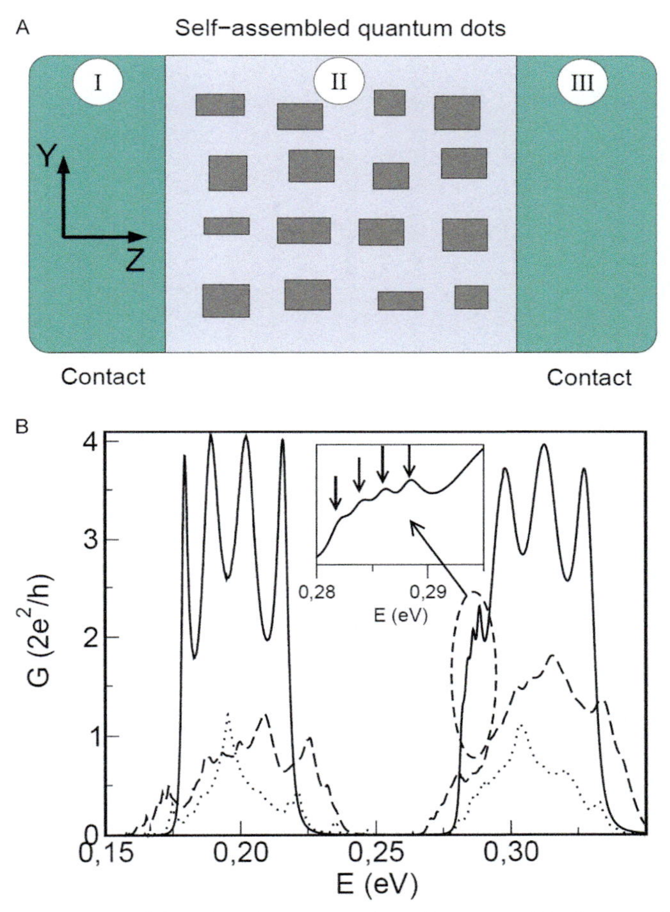

Fig. 11 (A) 4×4 array of disordered quantum dots; (B) conductivity in array. *From arXiv: cond-mat/0203255 [cond-mat.mes-hall].*

The effect of disorder in quantum dot arrays has also been shown to be reduced as the number of quantum dots in the array increases, but becomes less important as the voltage increases, as shown in Fig. 12.

While electron transport is more efficient in highly ordered arrays, these studies demonstrate that electron transport is also present in random/non-regimented/quasi-ordered QD arrays, at functional levels in such disordered or quasi-ordered arrays.

The creation of coherent electron bands was also demonstrated in a quantum well structure, which is a structure that constrains electron

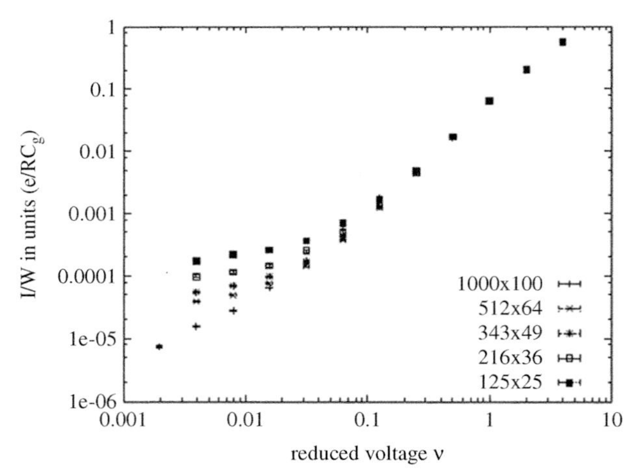

Fig. 12 Current v. Voltage for disordered arrays of quantum dots. *From arXiv:cond-mat/0203255 [cond-mat.mes-hall].*

movement in two dimensions, instead of three dimensions, like a QD. In that structure, a spatially extended, two-dimensional electron probability wave function exists when there are excitons formed by the applied electric field but when the applied electric field is too low to cause localization, which forms a coherent electron band.[145] As the electric field surrounding the quantum well structure is increased, the electron wave function reduces in extent (known as the Wannier-Stark regime) until it becomes fully localized into a single quantum well at a high electric field. This effect occurs when low-energy coherent electrons gain energy and decrease their wavelength. A diagram representing this effect is shown in Fig. 13.

The localization of electrons from such coherent electron bands at a specific location is a type of electron transport, and can be understood as occurring where the electric field characteristics cause the coherent electrons to increase their energy, such as where there is a larger field gradient relative to other areas. Assuming a completely balanced structure, then each quantum well has an equal probability of having an electron localizing in it, but when the structure is unbalanced (such as when one well has a lower impedance to ground), then the probability of electrons localizing in a quantum well could be a function of the electron field gradient in each well, a cascade effect due to electron crowding in the quantum well with the lowest impedance to ground or other imbalances that could modify the energy provided to each electron in that well, such as due to impurities or dimensional variations. The same effect will occur in three-dimensional coherent

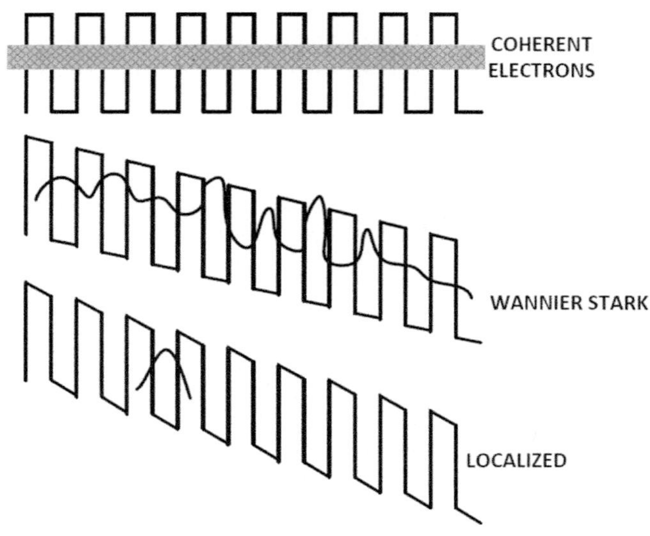

Fig. 13 Transition from coherent electrons to Wannier-Stark regime to localization.

electron structures, but is more difficult to show in a diagram. In addition, a combination of electron coherent tunneling, sequential tunneling, and hopping could potentially result in the same localization effect as coherent electron banding with the same localization drivers.

Coherent electron band formation has also been observed in ordered, quasi-ordered and random three-dimensional QD arrays.[146–148] The coupling between the QDs has been shown to modify the energy level spacing of one QD in such arrays as a function of the state of its neighbors. As a result, the state of neighboring QDs positively influences the formation of coherent electron bands.[149] Atomic force microscopy has been used to detect electron transport in quantum dot solids.[150,151]

Furthermore, the synergistic interaction of QDs (like ferritin) and pi-conjugated polymers (like neuromelanin) has been demonstrated to facilitate exciton diffusion and non-radiative energy transfer between those materials.[152–154] This transport occurs when the highest occupied molecular orbit (HOMO) or lowest unoccupied molecular orbit (LUMO) of the pi-conjugated polymer traps the electron of an exciton between the conduction band or valance band of the QD, respectively. These studies indicate that such materials can cooperate to create persistent free electrons by dissociation of electrons from excitons, such as those formed by exposure of ferritin or neuromelanin to electric fields or ions and the subsequent migration of the electrons between them.[155–157]

In regards to the interaction between ferritin and NMOs, as shown above, the ferritin appears to form a layer around the NMOs, which can be "unwrapped" as shown in Fig. 14.

As shown, the neuromelanin within the NMOs is approximately 30 nm in diameter, and has a variable separation (shown here as 30 nm) between the layer of ferritin cores, which are approximately 12 nm in diameter. Because neuromelanin is a pi-conjugated polymer, it is an electron donor material and could generate low-energy excitons in response to the electric field of an action potential or chemiexcitation. Such electrons would be attracted to both the hole remaining in the neuromelanin, as well as the reservoir of Fe^{3+} iron atoms in storage within each ferritin core, and would also be influenced by the material that separates the neuromelanin and the ferritin, as well as the H-chain and L-chain subunits of the ferritin. If an Fe^{3+} atom is reduced to an Fe^{2+} atom inside a ferritin core, it will be unable to be removed from the ferritin core immediately, and there will be a time period associated with that transition during which the Fe^{2+} atomic might be unable to retain the extra electron. The ability of the ferritin to store and release electrons will also be a function of the molecular dynamics of the ferritin, the electrons stored and external fields, as shown in Fig. 15.

With approximately 1500–1800 Fe^{3+} atoms,[87] ferritin in the SNc has the potential to store a large number of electrons, although the molecular dynamics of the ferrihydrite core will influence the dynamic behavior of ferritin.

The average distance between the electron and hole in an exciton is known as the Bohr radius, and can be expressed by the following equation:

$$R_{exc} = \varepsilon_r \left(m_e^0/\mu\right) a_o$$

where:

μ is the reduced mass of the exciton,
ε_r is the relative permittivity,
m_e^0 is the electron mass in a vacuum, and
a_o is the Bohr radius of a hydrogen atom

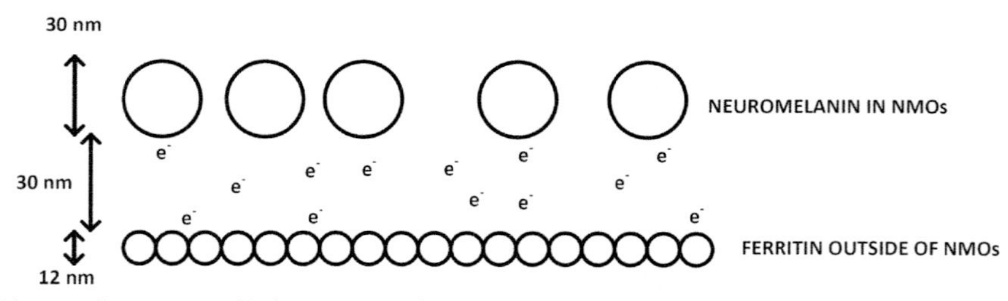

Fig. 14 "Unwrapped" disposition of neuromelanin and ferritin in SNc neurons.

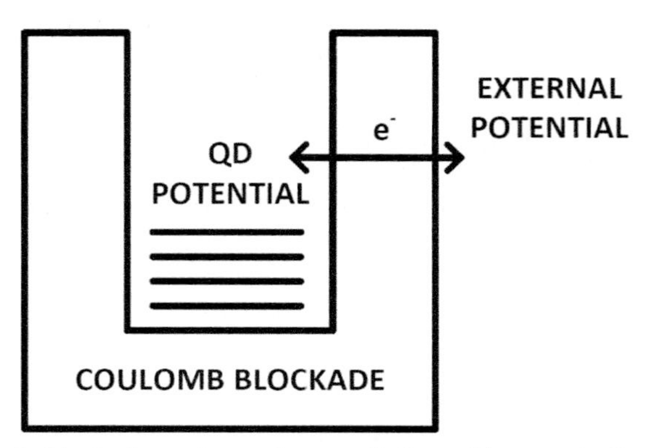

Fig. 15 Coulomb blockade and quantum well. *From Mesoscopic Electron Transport, Sohn, L.L., Ed.; Klewer Academic Publishers, 1997.*

.[158] Of note is that the Bohr radius is directly related to the relative permittivity. In neural tissue, the measured relative permittivity as a function of frequency ranges from approximately 30,000 at 10 kHz to approximately 5000 at 100 kHz,[159] but the relative permittivity of biological materials is highly variable.[160] The diffusion length and charge carrier lifetime are important physical characteristics of the processes that define the interaction between neuromelanin and ferritin, and can be difficult to determine.[161] While a comprehensive model of the specific materials and configurations would be necessary to analytically determine these parameters, it can be generally seen that the disassociated electrons would facilitate the formation of coherent electron bands and may have lifetimes of more than 10 ms.[156] Quantum gating effects at room temperature have also been shown to facilitate the formation of coherent electron bands and other quantum effects in random arrays.[139]

Combined force vectors, such as simultaneously applied electric and strain fields, can be used to control the creation of quantum effects in QDs of different size, such as to generate entangled photons at room temperature "on demand" in asymmetric QDs.[162] This behavior is also referred to as "strain-tunable".[163] The combination of such strain/stress field vectors has thus been demonstrated to result in controllable generation of quantum mechanical effects, even where the dimensional characteristics of the QDs are not matched. This suggests that such combined stress fields may be useful in the generation of other quantum effects at room temperature, such as formation of electron conduction bands and localization of electrons in those electron conduction bands.[164]

Other electron transport mechanisms have also been reported or hypothesized using quantum dot structures, such as sequential tunneling and hopping as discussed above using ferritin, and exciton polaron pairs,[165] phonon-electron coupling,[166] exciton diffusion,[167] and also electron transport in other biological structures such as pigment-protein complexes,[168] possibly due to resonant vibrations.[169] While these mechanisms are different from coherent electron bands, they may nonetheless result in electron transport in quantum dot structures or in biological structures that have similarities to quantum dots, such as the Fenna-Matthews-Olson complex.

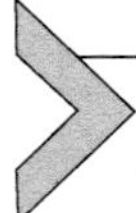

4. Theory and discussion

4.1 Overview

Classical models of SNc and LC behavior can account for many of the known functions of these neurons, either individually or in groups. However, there are functional aspects of these groups of neurons that are not fully understood using these classical models, in particular, functions related to voluntary action selection. The hypothesis discussed in this chapter suggests that ferritin and neuromelanin form a structure that operates in a manner similar to a disordered or quasi-ordered quantum dot array, to form an electron transport mechanism. The electron transport mechanism could be electron hopping, coherent tunneling, coherent electron bands or a combination of electron transport mechanisms that interconnect the soma of these neurons. While the specific function or functions performed by this electron transport mechanism cannot be determined solely from the published literature and without additional experimental work to confirm the theory, one possible function would be the generation of action potentials, under certain conditions and for certain types of actions. Further, synchronization and timing functions are performed by these neuron groups, and it is also possible that this electron transport mechanism is involved with those or other functions. In addition, a quantum mechanical electron transport mechanism that simultaneously connects soma of multiple neurons would be a unique physical structure that also instantaneously interconnects numerous efferent and afferent neural structures that are in communication with those multiple neurons in a manner that could be associated with the experience of consciousness, or the activation of the neurons at a level to be involved with the generation of coherent electron transport could also account for the number of different states associated with the human experience of consciousness.

4.2 Analysis

Numerous studies in the technical literature have established that ferritin is a QD and can generate excitons, either under normal conditions at room temperature or when exposed to a chemical and/or electrical environment that would be present within the cell bodies of dopamine and norepinephrine neurons. At low energies, the electrons associated with those excitons would be defined by a three-dimensional probability waveform that can have a wavelength of greater than 6 nm. In addition, numerous studies have shown that ferritin or other QDs having concentrations similar to those found in the SNc and LC can generate coherent electron bands, or otherwise support electron transport over substantial distances. Neuromelanin also has properties of QDs, and is at least a pi-conjugated polymer that can interact with QDs like ferritin to separate electrons from QD excitons and to extend the lifetime and coherence wavelength of those separated electrons. While the lowest unoccupied molecular orbital (LUMO) and highest occupied molecular orbit (HOMO) of neuromelanin is not known, the measurements of those values for melanin would potentially be synergistic with the conductance and valence energy bands of ferritin, which could function to trap electrons associated with excitons. This function would facilitate the creation of coherent electron bands in a disordered or quasi-ordered array of ferritin and neuromelanin molecules, or other electron transport mechanisms. Although the quantum mechanical effects of the specific combination of ferritin and neuromelanin does not appear to have been studied, there is no reason why these materials would not exhibit at least the quantum mechanical properties in combination that ferritin exhibits alone, and possibly may exhibit enhanced quantum mechanical properties. While other electron transport mechanisms such as exciton diffusion, coherent tunneling, sequential tunneling and electron hopping could also result in electron transport between ferritin-neuromelanin structures in SNc and LC neurons, this review will focus on coherent electron bands. Further, it is theoretically possible to create computer models of ferritin,[170] and to use finite difference Schrodinger-Poisson modeling to determine exactly whether electron transport is occurring and by what mechanism, but the spacing, dimensions, relative permittivity and other dimensional and material properties of the system would be difficult to determine to a sufficient level of accuracy. However, it may be possible to observe indirect effects of electron transport, as discussed further below.

The dopamine neurons of the SNc and the norepinephrine neurons of the LC contain a substantial number of NMOs. Ferritin forms a layer around

NMOs and is present in large quantities inside of those neurons, possibly due to iron homeostasis associated with the hypothesized electron transport mechanism. Ferritin is also present at a concentration in the intercellular fluid and glial cells between neurons that may support the formation of coherent electron bands (or other electron transport mechanisms) that extend through the associated neurons. The average spacing of these molecules within these neurons and in the intercellular fluid (100 nm or less), as shown in Fig. 16, may be within one order of magnitude of spacings that have been demonstrated to result in the formation of coherent electron bands for practical applications in semiconducting circuits (20 nm). This analysis omits the potential beneficial effects of other SNc and LC cells that contain ferritin and neuromelanin, such as astrocytes, microglia and oligodendrocytes, as well as observations of the spacing of SNc soma and iron concentration in the vicinity of those SNc soma, which demonstrate that the density and spacing assumptions used for this analysis are very conservative.

Ferritin should also operate in a synergistic manner with neuromelanin, at least because neuromelanin is a pi-conjugated polymer, but also because

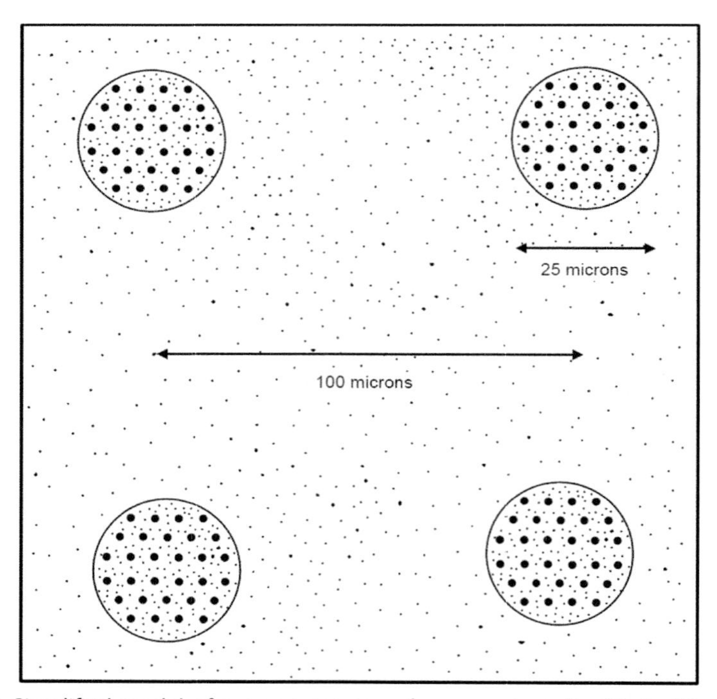

Fig. 16 Simplified model of average spacing of neuromelanin/ferritin in SNc and LC neurons.

neuromelanin itself might exhibit the quantum mechanical properties of a QD. Numerous studies indicate that room temperature quantum mechanical effects are possible, at least for periods of hundreds of femtoseconds, and possibly for periods longer than a millisecond. These include studies involving different material types of QDs operating as quantum gates, and studies involving QDs that incorporate magnetic materials. Based on these studies, the combination of ferritin and neuromelanin at densities found in the SNc and LC neurons could support the formation of coherent electron bands for electron transport, or other electron transport mechanisms (other regions of the brain also have NMOs at lower concentrations but could potentially use one or more of the hypothesized electron transport mechanisms).

The unusual structures of the SNc and LC neurons that provide the ferritin and neuromelanin disordered/quasi-ordered arrays also provide synchronized electric fields in a manner that could facilitate exciton formation and subsequent localization of electrons in one of those neurons at a time. As a preliminary matter, a substantial electric field might not even be required to generate excitons, because the measured electron band gaps of ferritin could allow thermal excitons to form without an associated electric field. It has also been shown that anionic components of the intracellular environment, such as chloride ions, can cause excitons to form through chemiexcitation. However, at least low-level fields are also present in the intracellular environment, and while peak intracellular field strengths of greater than 3.5×10^6 V/m magnitude have been measured and may be present, those peak fields would not be sustained indefinitely and would vary as a function of cellular dynamics. As such, while a specific mechanism for exciton generation has not been conclusively identified, many mechanisms that have been shown to generate excitons in QDs are present in SNc and LC neuron cell bodies.

If the coherent electron band mechanism or other electron transport mechanism is determined to exist, additional work will be needed to determine its function(s). SNc and LC neurons are similar to other neurons to the extent that they integrate dendritic voltages and generate an axonic impulse when an action potential is reached at the axon hillock as a result of dendritic excitation. This axonic impulse is transmitted to the synapses at the distal ends of the axon. However, SNc and LC neurons are different from most other neurons in several regards. In particular, these neurons generate the catecholamines dopamine (SNc) and norepinephrine (LC) (also known as noradrenalin), respectively, which are chemical neurotransmitters, and also have unusual axonic structures, as discussed further herein.

The conventional understanding of the function of these neurons is generally based on dendritic activation, although other classical (i.e., non-quantum) mechanisms of neuronal communication have also been proposed, such as electric field or ephaptic coupling. The electron transport mechanisms proposed in this chapter could likewise potentially provide an inter-neuron communication mechanism that is capable of complex information transfer and storage, unlike electric field or ephaptic coupling.

In addition to these functions, the SNc and LC regions include at least two different classes of catecholaminergic neurons, with a large number of neurons of a first class that have a small number of branches and a small number of neurons of a second class that have a very large number of branches (SNc) or antidromic signals (LC). Both of these classes of neurons express autonomous and synchronized pacemaker activity, but that activity is only sufficient to generate an action potential in the first class of neurons. While such pacemaker activity exists in other groups of neurons that do not develop neuromelanin, it is possible that the pacemaker activity in the SNc and LC is based on or assisted by these coherent electron bands (or other electron transport mechanism). Both the SNc and LC form part of the neural system for generating a physical reaction to external stimuli, which may require a higher level of synchronization than those other groups of neurons.

The axonal arbors associated with the second class of neurons in the SNc include hundreds of thousands or possibly upwards of 1 million synapses and have substantially greater energy requirements associated with both action potential generation and axon arbor energization. (A similar type of axonal arborization does not appear to be present in the LC neurons, but at least one mechanism (antidromic currents) could facilitate the impedance-based switching mechanism in a class of the norepinephrine neurons of the LC.) The amount of energy associated with firing of these neurons suggests that they do not fire as frequently as the first class of neurons. Although dendritic inputs can generate action potentials, such dendritic inputs might normally be associated with reflex-type reactions, or in response to coordinated dendritic stimuli from afferent neurons. Generation of non-reflex action potentials in the SNc and LC, such as when associated physical actions are not reflexive but are rather the result of a choice between a large number of options, could potentially involve a different physical mechanism in addition to dendritic inputs, such as a mechanism that would utilize electron transport facilitated by ferritin and neuromelanin. It is also noted that simultaneous generation of multiple action potentials in the largest of these neurons could result in seizures, if incompatible competing actions are activated, which does not happen under normal circumstances.

The extracellular electric fields associated with the dendrites of the efferent neurons of this second class of SNc and possibly LC neurons should result in an associated lower apparent impedance from the cell body for the associated axon. This lower impedance would cause electrons from neurons that are involved with the formation of coherent electron bands to localize in the associated neuron with a lower axonal impedance, because that neuron would be best situated to conduct those electrons to the ground potential of the intercellular fluid (it is noted that the same effect might occur from electron hopping or tunneling). Consider a simple four-neuron example, as shown in Fig. 17.

This structure has two "input" neurons that have afferent connections that are both stimulated, but where neither has reached the action potential. Each "input" neuron also has efferent connections to an "action" neuron. One "action" neuron is near its action potential due to dendritic potential, whereas the second "action" neuron is receiving no stimulation. If this organism has no electron transport function, no action will occur unless something else happens—either another stimulus to the "input" neurons or another stimulus to the "action" neuron. If the organism has an electron transport function for switching energy to the "input" neuron that is better

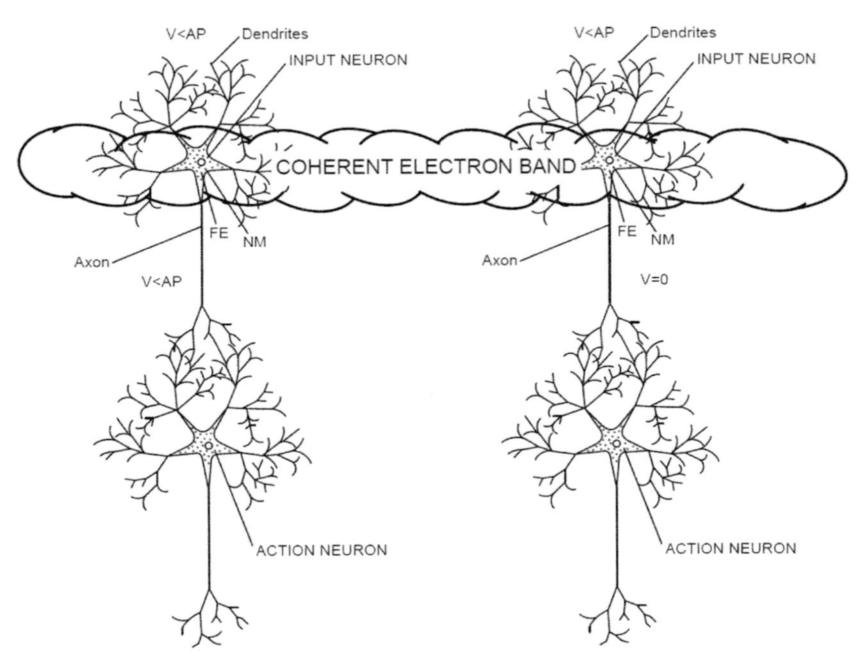

Fig. 17 Simplified model of input neurons with neuromelanin/ferritin electron coherence bands.

situated to actuate an "action" neuron, it will be able to act without any additional inputs. This mechanism could also potentially prevent the input neurons from firing simultaneously, first by holding the voltage to a common level that is lower than the action potential, and then by causing a reduction in the voltage in the non-firing neuron when the firing neuron has generated its action potential pulse and drops to a lower recovery voltage. Conversely, a similar organism without the electron transport function would not be able to act without additional inputs, and thus would be unable to obtain food, avoid danger, or otherwise improve its odds for survival, or might generate multiple concurrent action potentials that cause action conflicts and a seizure, or otherwise prevent movement.

As seen in Fig. 18 below, an axon with a large number of axonal branches has an effective impedance that is lower than an axon with a small number of axonal branches. This is because the impedances of each parallel branch (shown in Fig. 18 as a resistance for simplicity) generally add as the inverse of the sum of the inverses (1/[1/R1 + 1/R2 + ... 1/Rn]), in accordance with Ohms law (capacitances add directly, but capacitive impedances are small relative to resistance, at low frequencies).

The energy delivered by the electron transport function through the transfer of electrons to ferritin of a single neuron would not flow to ground. Instead, the electron transport function would operate to reduce the water-insoluble stored Fe^{3+} ions of the ferritin to water-soluble Fe^{2+} ions. These Fe^{2+} ions would enter into solution in the intracellular fluid, and could

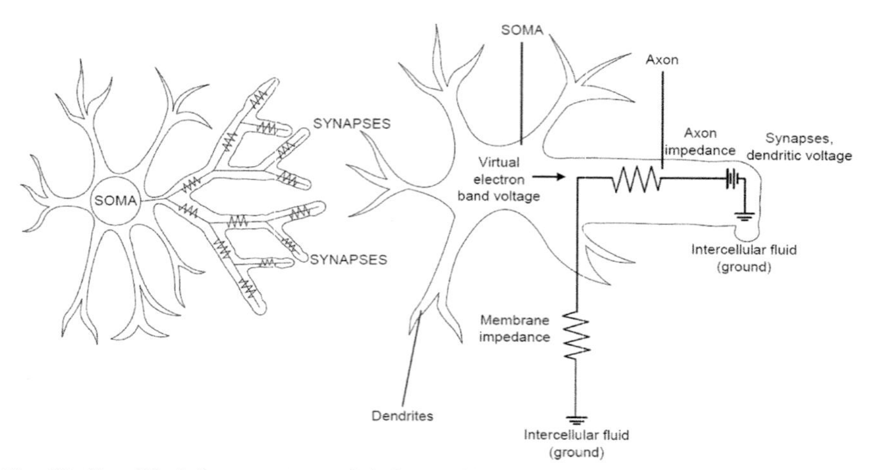

Fig. 18 Simplified electrotonic model of axonal branches and virtual electron band voltage, relative to intercellular fluid ground.

either directly increase the positive charge internal to the neuron, or trigger Ca^{2+} ion release through reactive oxygen species mediated ryanodine receptor stimulation, either of which would assist with the generation of an action potential.

Prior to electron transport, the electrons in coherent electron bands are probabilistically extended over a large number of axons. As shown in the simplified example of Fig. 19 below, as the electron transport function selects the axon path with the lowest impedance (which corresponds to the axon having the most synapses with dendrites that are at high dendritic voltages), it also electrically couples the parallel neural structures that are also connected to the associated efferent neurons. Unlike a conventional conductor, though, coherent electron waves are three-dimensional probabilistic distributions of individual electrons that are spatially and temporally coherent with the waveforms of other electrons, such that each electron is essentially simultaneously in contact with multiple different complex neural structures, either directly or through an overlapping connection to another electron. These coherent electron waveforms could be capable of storing state information associated with each of the physical structures that contain

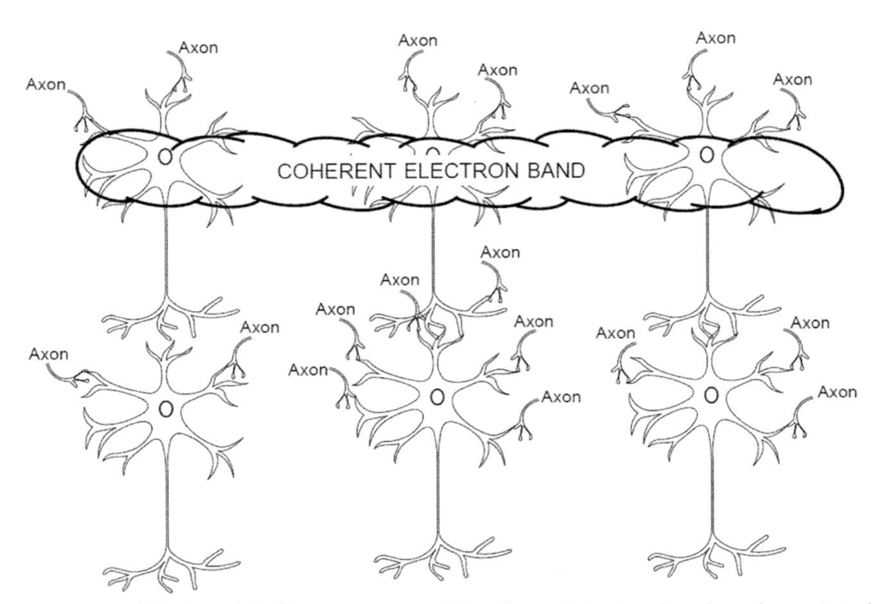

Fig. 19 Simplified model of input neurons with coherent electron bands and associated action neurons.

them, or other types of information. As discussed later in this chapter, the sequence of activation states of the SNc and LC neurons forms a state machine with a large number of possible states, and that structure would not require integration of the individual neural signals of synaptic inputs from afferent neurons or the synaptic outputs to efferent neurons to store large amounts of state information.

For an SNc neuron, most axonic connections are to the striatum; for LC neurons, axonic connections include several different areas, such as the cerebral cortex, the thalamus, and the spinal cord. However, LC axons with the lowest axonic impedance as seen from the cell body may be the ones that project to the cerebral cortex, which have a large number of branches and also have associated antidromic signals. The striatum controls movement, and the SNc is associated with neural signals that are associated with conscious movement, whereas the LC is associated with neural signals that are associated with conscious thought, as well as other functions. Based on this distinction, damage to SNc neurons that is sufficient to eliminate the coherent electron band formation, without associated damage to LC neurons, should result in akinesia or a condition similar to locked-in syndrome, if the electron conduction bands are associated with voluntary action selection. However, while non-reflexive voluntary action would be incapacitated, reflexive action based on dendritic stimulation alone should still be capable of generating an action potential, if it is present and if the associated neurons are still functional at a reduced level that supports the generation of action potentials solely from dendritic inputs. The observations of akinesia in MPTP-injured patients and ability of these patients to perform reflexive actions, and to recover the ability to perform voluntary action when treated with levodopa, appear to corroborate this theory of function. In particular, levodopa is believed to mitigate disruption of the Ca^{2+}-mediated signaling pathways that would be involved in action potential generation in response to Fe^{2+} generation and reactive oxygen species mediated ryanodine receptor stimulation. In contrast, a similar level of damage to the LC without associated damage to the SNc should not result in loss of consciousness. Instead, it should only impair higher level cognitive functions associated with the cerebral cortex. Observed loss of LC neurons appears to correlate with such impairment in reported studies. While these clinical studies are not conclusive evidence that ferritin and neuromelanin in the SNc and LC create an electron transport function that assists with the generation of action potentials, they are consistent with that hypothesis.

In regards to the experience of consciousness, the hypothesized electron transport would appear to be related to that experience to the extent that it creates a physical mechanism that integrates discrete neural structures associated with processing sensory and higher order thought. The experience of consciousness could be a function of the different complex neural structures that are activated at any specific time as well as the electron transport mechanism that couples those neural structures. However, it is noted that no direct physical or causal relationship between these different electron transport mechanisms and the other proposed quantum consciousness mechanisms (which generally involve entanglement in pyramidal cells or synapses) would appear to be present, based on the current understanding of quantum mechanical functions of coherence and entanglement. In addition, it is difficult to identify any specific time constraints for any of the proposed electron transport mechanisms as they might relate to the experience of consciousness, such as whether a persistent uniform coherence band structure or a network of electron tunneling or hopping mechanisms has any specific timing constraints. Research would need to be conducted to determine whether the persistence of any quantum mechanical effects has any bearing on the experience of consciousness, assuming that it can be shown that a quantum mechanical electron transport mechanism is present.

4.3 Proposed tests

Additional work could be performed to verify this theory using fluorescent probes that have been used to detect the presence of Fe^{2+} ions in cells. These probes might be able to detect the generation of Fe^{2+} ions in the dopamine neurons of *C. elegans* or earthworms in association with subsequent action potential generation. While testing on more complex neural structures of small mammals might ultimately be needed if the release of Fe^{2+} associated with action potentials in dopamine or norepinephrine neurons cannot be conclusively determined in simpler specimens, such testing would be significantly more difficult to perform. Detection of Fe^{2+} ions in association with action potential generation would provide compelling evidence of potential electron transport as proposed by this theory, because Fe^{2+} generation in association with an action potential would not normally occur.

Tests could also be performed to determine whether tissue specimens from the SNc and LC exhibit electrical characteristics that are sufficiently different from the electrical characteristics of other tissue samples to infer an electron transport mechanism, such as charge transport measurements, material characterization, atomic force microscopy or X-ray spectroscopy. While these tests would not necessarily prove that electron transport in association with action potential generation is present, they could establish whether a quantum dot solid is formed by the distribution of ferritin and neuromelanin in tissue samples. As discussed below, a small number of conductive atomic force microscopy (c-AFM) tests appear to indicate the existence of electron transport in SNc tissue, but further tests would be helpful to confirm this observation as well as to determine potential charge transport characteristics.

Reaction time tests could also be developed to try to determine whether certain classes of stimuli result in shorter reaction times, like those associated with simple visual stimuli (~0.20 s), and to identify the parameters associated with the level of task complexity that causes reaction time to increase (e.g., to ~0.40 s or greater), to determine whether the shorter reaction times could be associated with a single switching function of the SNc, if the longer reaction times might be associated with coordination of two switching functions associated with SNc actions and LC actions, or if other more complex neural information processing can be identified that corroborates the theory presented in this chapter. The hypothesized electron transport may be associated with a function that provides a basic evolutionary advantage, such as an improved ability to compete for food or provide defense, to process peripheral visual information or to coordinate audio and visual information processing. If so, then those functions might be associated with faster reaction times and primarily involve the SNC or LC, whereas reaction times associated with more complex neural functions that would not necessarily provide an evolutionary advantage with a faster response (e.g., those relating to quality discrimination, social status, societal mores) might involve other areas in addition to the SNc and LC. These tests could also be coordinated with functional magnetic resonance imaging to identify what regions of the brain are associated with the specific type of decision making.

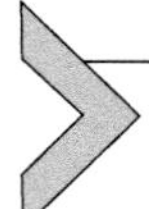

5. C-AFM test results

5.1 Introduction

In order to obtain experimental data that would provide additional evidence in regards to the hypothesized electron transport mechanism, c-AFM tests were performed on normal (non-diseased) human SNc tissue. C-AFM testing exposes a sample to electric fields that are sufficient to generate excitons, and has been used on materials other than neural tissue to generate excitons and electron transport. The equipment used to perform these c-AFM tests is capable of measuring current levels in the fA range, for sample areas at least as small as 2 μm × 2 μm.

The experimental procedures used to prepare the tissues for c-AFM testing and for performing the c-AFM testing are discussed below. Four tests were performed that yielded useful results. These tests indicate that electron transport may be occurring in these tissue samples at a measurable level. These experiments advance our understanding of whether it is possible for quantum mechanical electron transport to be present in neural tissue at room temperature by showing that levels of electron transport are present during c-AFM testing that would potentially be sufficient to create or contribute to the creation of an action potential.

5.2 Experimental procedures

The SNc tissue samples were obtained from AMSBIO of Cambridge, MA (catalog no. HP-253), and were fixed in 10% buffered neutral formalin and processed for embedding into low temperature melting paraffin. The tissue samples were then cut at a nominal thickness of 5 μm and mounted on (1) positively charged glass slides, and (2) glass slides wrapped in aluminum foil, to provide electrical contact for c-AFM testing. Each slide was cut in succession, which allowed areas on the foil-backed slides with dopamine neuron cell bodies to be identified by using the glass-backed slides to index those locations. The tissue samples were maintained at room temperature for short-term storage and at −4 °C for long-term storage in a sealed bag with desiccant, and were tested within 1 year of the date of purchase. Optical microscope data from the tissue samples on glass slides was compared to optical microscope data from the tissue samples on foil, and while the tissue samples on foil were more difficult to inspect for NMOs due to the absence of backlighting, the cellular structures associated with NMOs that were apparent from the tissue samples on foil also appeared in the tissue samples on the glass slide. These cellular structures were used to select areas for c-AFM testing. The correspondence between observable features on these two types of slides provides a high level of confidence that dopamine neuron cell body groups were identified.

Two tissue sample preparation procedures were used. For a first sample, the tissue was soaked in xylene twice for 5 min each time to remove wax, per AMSBIO recommendations. For a second sample, the tissue was soaked in xylene twice for 30 min after heating for 10 min at 65 °C, followed by successive ethanol soaks with increasing levels of water, to rehydrate the tissue. The second protocol was followed to ensure adequate wax removal after review of the first c-AFM test using the first protocol, and was used for the other three c-AFM tests.

The SNc tissue samples were shipped directly from AMSBIO to EAG Laboratories of Sunnyvale, CA for c-AFM testing using a Dimension Icon AFM instrument from Bruker of Santa Barbara, California. The tests were conducted by Dr. Sara Ostrowski of EAG using standard commercial testing protocols, and the test results were also reviewed by Bruker senior applications scientist Dr. Yueming Hua, who confirmed that the tests were conducted properly and that the data was accurate. The c-AFM data files are in a standard format that can be used by the Bruker Nanoscope software or other commercially available AFM data analysis software, as discussed further in,[171] and include all of the test parameters associated with the tests.

C-AFM testing is conducted by using a micron-scale cantilever probe to tap along the surface of a specimen (there is also a non-tapping mode of c-AFM, but that was not used for these tests). The probe has a pyramidal-shaped tip that extends downwards for several microns, which has dimensions of approximately 5–10 nm radius on a new tip. The probe is biased with a voltage, and three current data points are generated during tapping mode operation: (1) peak current, (2) cycle-averaged current, and (3) contact-averaged current. Peak current is the "instantaneous" current at the point of contact and coincides with the peak applied force. This peak current is not necessarily the maximum current, because the rise time is limited by the bandwidth of the current probe module and the impedance of the sample, which can cause a lag in the current response. The time period during which the peak current is measured is also not truly instantaneous, but depends on the time period during which peak force is applied. The cycle-averaged current is the average current over one full tapping cycle and includes both the current measured while tip is in contact with the surface, and while it is off the surface. The contact-averaged current is the average current measured only when the tip is in contact with the surface. For these tests, the probe made 512 measurements at a frequency of approximately 0.2 Hz for each linear scan (trace and retrace), and made 512 linear scans, for a total of 262,144 measurement locations per sample. In addition to the current measurements, the c-AFM probe also provides quantitative measurement of (1) height, (2) elasticity, (3) force, (4) deformation and (5) dissipation.

In addition to the measurements at each point, current-voltage (I—V) curves can be generated at selected points as the applied voltage is varied from a maximum of 10 V to a minimum of −10 V. I—V curves were generated for some of the tests, but only 10 points were measured as part of the standard c-AFM analysis conducted by EAG, and the measurement points were randomly selected. In addition, the measurement points within the image data have to be associated with the I—V curves, and it was not clear from the data that was provided exactly where each measurement was taken. As such, those results are not discussed here, as they did not provide evidence that was as useful as the large arrays of current measurement data at a fixed bias. It is also noted that two different current amplifiers were used to measure the contact, peak and average currents, a first one having a sensitivity of 1 nA/V (which was used for the 76 μm × 76 μm sample tests), and a second one having a sensitivity of 20 pA/V (which was used for the 2 μm × 2 μm sample test).

5.3 Samples

Four samples were tested:

(1) Test One—short wax removal protocol, 76 μm × 76 μm test area, cell-shaped area

(2) Test Two—long wax removal protocol, 76 μm × 76 μm area, cell-shaped area

(3) Test Three—long wax removal protocol, 2 μm × 2 μm detail within Test Two area

(4) Test Four—long wax removal protocol, 76 μm × 76 μm area, no cell-shapes (dark tissue appearance)

5.4 Test results

The Test One c-AFM results provided numerous but dispersed positive peak current, cycle-averaged current, and contact-averaged current indications at a bias voltage of −10 V, as well as numerous and concentrated negative peak current, cycle-averaged current and contact-averaged current indications, but the majority of readings were approximately zero. The data from that test is not presented here. Based on those results, the amount of wax removal was evaluated, and it was determined that a more thorough wax removal protocol than the one that was recommended by AMSBIO would likely improve the ability to perform current measurements. A suitable protocol was located and used for the remaining tests. Fig. 20 is an optical photograph of the Test Two SNc tissue sample disposed on foil.

Because of the absence of backlighting, neuromelanin cannot be seen, but the size (approximately 25 μm) and oval shape of five areas in the center of the image are similar to the size and shape of the dopamine neurons and nuclei from stained tissue photographs that can be seen in the literature, such as.[54] Darker regions can be seen that surround cell nuclei that contain

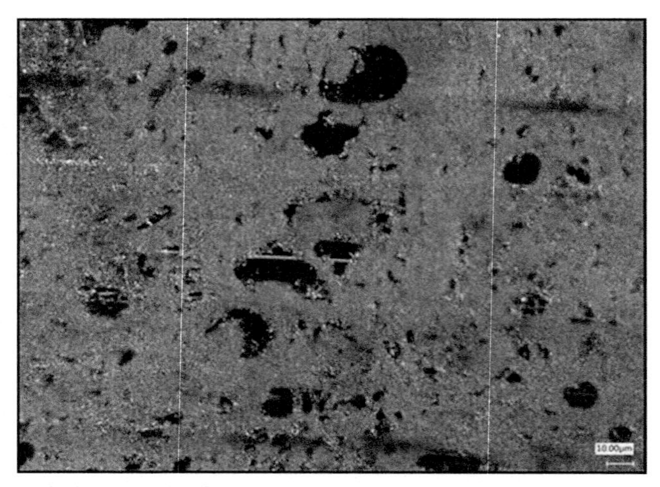

Fig. 20 Optical photograph of the Test Two SNc tissue sample disposed on foil.

neuromelanin and ferritin.[48] As noted, the location of these cell bodies corresponds to cell bodies identified on the glass-backed slides. It is also noted that the apparent spacing between cell bodies in this image is less than 10 μm, which is substantially closer than the assumed spacing of 100 μm in.[172]

A two-dimensional image of AFM height measurements for the area that is approximately contained within the square shown in Fig. 21. An autoscale generated by the Bruker data mapping software ranges from −3.1 to 3.1 μm, which correlates to the nominal 5 μm thickness of the SNc tissue samples. A histogram of this height data is shown in Fig. 22.

The darker, lower thickness areas generally have a variable height, when viewed at a higher resolution. Based on the thickness data and the assumptions that the foil would be relatively flat over the 76 μm × 76 μm sample and that the foil is at or below the lowest height measurement (approximately −2.4 μm), areas above that lowest height level are either (1) tissue structure, or (2) detritus that would have been suspended in intracellular or extracellular fluid but which remained after fluids were replaced by formalin and/or wax. The thinnest areas likely correspond to the detritus from material that would have been suspended in intracellular fluid, based on the spacing, shape and size of those areas. While these areas should contain neuromelanin and ferritin from the cell body, they would not exhibit the spacing that would be seen in a living cell, and may have a negative impact on electron transport current measurements relative to living cells, due to the importance of spacing in the generation of electron transport.

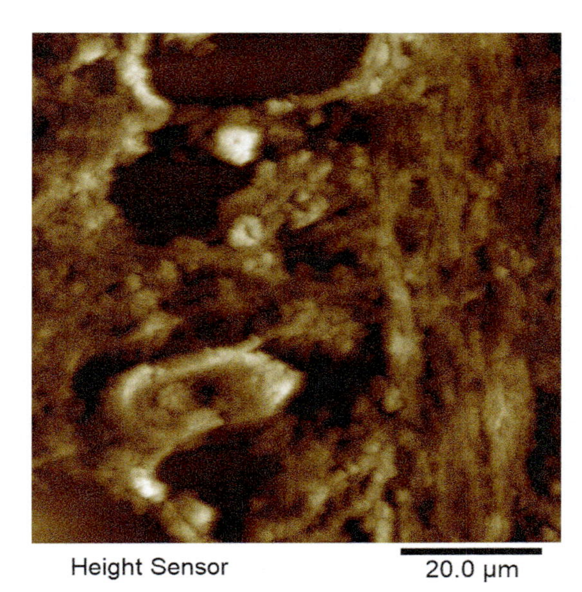

Fig. 21 AFM height measurement data map.

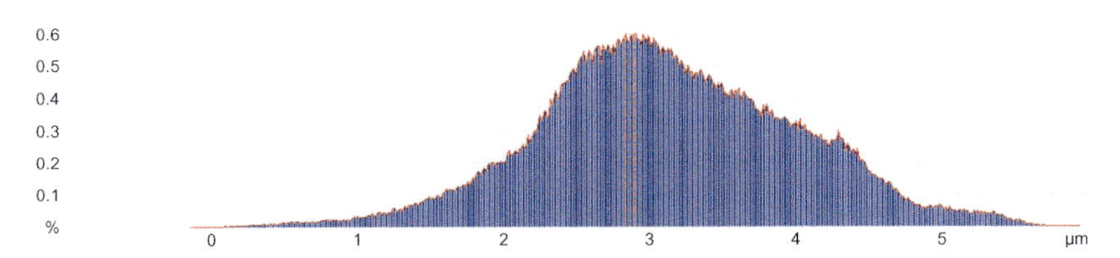

Fig. 22 AFM height measurement data histogram.

A two-dimensional image of the contact-averaged current is shown in Fig. 23. The auto-scale shows a range of −1.7 to 2.5 nA, but the histogram data for this image shown in Fig. 24 indicates that the contact-averaged current ranges from approximately −20 to 2 nA, with a small number of points that extend to −20 nA and that most data points are between −1 and 2 nA. The lowest contact-averaged current values are generally found in the thin areas that correspond to the intracellular detritus accumulations, which suggests that these areas have better electrical contact to the underlying foil layer.

A two-dimensional image of the cycle-averaged current is shown in Fig. 25. The auto-scale shows a range of −236 to 181 nA, but the associated histogram data in Fig. 26 shows that the cycle-averaged current includes a small number of data points that extend to approximately −20 nA.

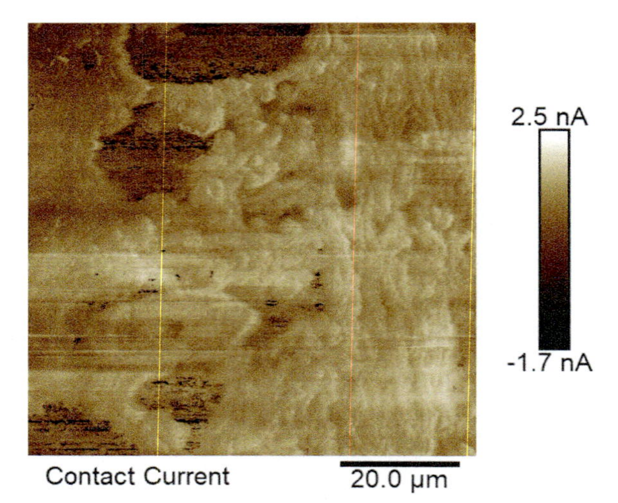

Fig. 23 C-AFM contact current measurement map.

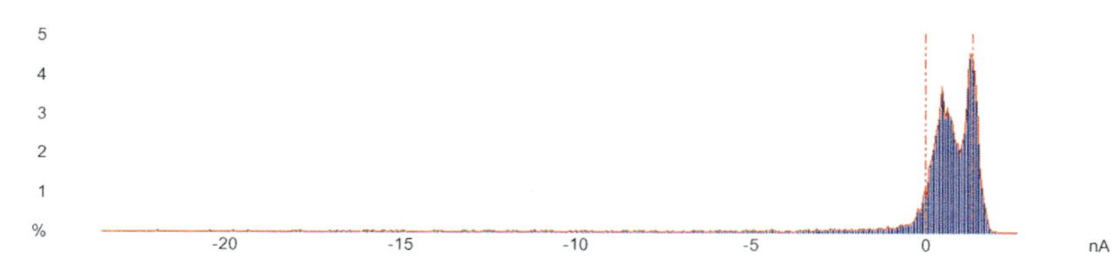

Fig. 24 C-AFM contact current histogram.

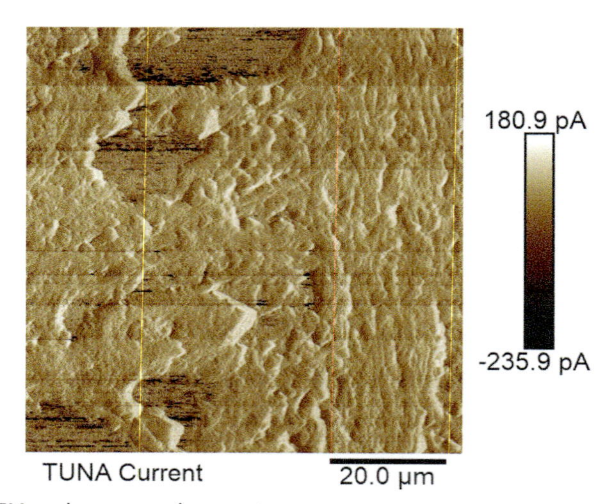

Fig. 25 C-AFM cycle-averaged current map.

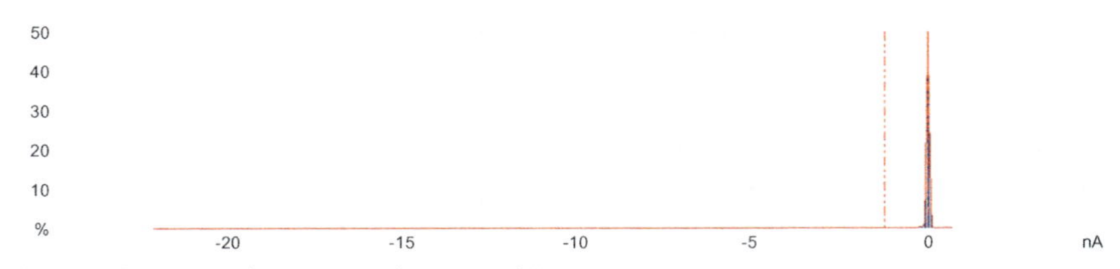

Fig. 26 C-AFM cycle-averaged current histogram.

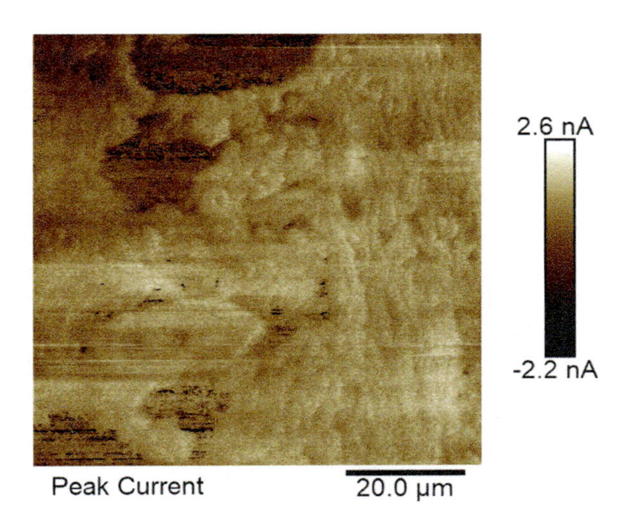

Fig. 27 C-AFM peak current map.

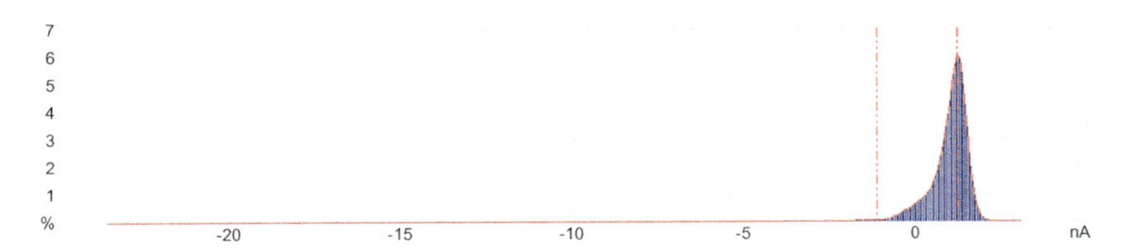

Fig. 28 C-AFM peak current histogram.

An image map of the peak contact current data for the Test Two sample is shown in Fig. 27. The associated histogram is shown in Fig. 28.

Several observations can be made from this data. First, the peak contact current data, the contact-averaged current data and the cycle-averaged current data vary significantly, and do not have the same distribution. This variation indicates that the current changes non-linearly and significantly over the measurement period for many contacts. Second, the peak contact current data, the contact-averaged current data and the cycle-averaged current data are not linearly related to the thickness, possibly even for negative current measurements, which would be expected if the impedance between the c-AFM probe tip and the foil substrate was linear or due to classical transport mechanisms. Third, the peak contact current data and the contact-averaged current data distributions have significant positive current components, which would not be expected from a −10 V probe tip bias voltage contacting a grounded sample, and it does not appear to be possible to explain these positive currents using classical electron transport mechanisms (while probe contamination could result in some positive current if the contamination was caused by negatively-charged material, homeostasis maintains the cellular environment at a slightly basic level but very close to neutral pH). Fourth, it is noted that the majority of height measurements are substantially greater than the minimum, which indicates that the negative currents that are measured are not caused by direct contact with the foil substrate, but rather that the conductive cellular detritus layer might not have impedance characteristics that vary linearly with thickness. Finally, there are a large number of positive peak contact current and contact-averaged current measurement readings in the detritus field, which indicates that the components that are responsible for that current behavior are distributed throughout those regions, as well as throughout the thicker tissues that appear to maintain at least some of their structure.

The arrows shown in the figures for height, contact current and averaged current show the approximate location of the second test sample, which was a 2 μm × 2 μm section from within the first sample region, at a point that corresponds to an area with one of the highest magnitude peak contact current readings. This location was chosen because the high levels of contact current should correspond to locations with a high concentration of ferritin and neuromelanin, if these complexes are functioning as quantum dots in a disordered (or semi-ordered) quantum dot array.

The source files for the image data of the tested samples, some additional optical images, the c-AFM data files and a link to the Bruker Nanoscope Analysis image data analysis software are provided in.[172] These c-AFM data files and the Nanoscope Analysis software can be used to reproduce the images shown in the other figures, as well as to analyze other aspects of the c-AFM data that may be of interest. While instructions for using the Nanoscope Analysis software are beyond the scope of this chapter, Bruker also provides a user manual for the software free of charge.

Fig. 29 is a three-dimensional image of the contact-averaged current readings of the Test Three high-resolution 2 μm × 2 μm section. Although the auto-scale shows a range of −3.4 to 8.2 pA, the histogram data in Fig. 30 shows that most contact-averaged current readings fall between −2 and 6 pA. Three round areas are identified with arrows that range from 200 to 400 nm in diameter, which corresponds to the size of NMOs that has been reported in the literature.[48] In addition, sheet-like regions of high contact-averaged current appear to be associated with different tissue structures, as shown in the height data shown below, but which do not necessarily correlate directly to those structures. The spatial resolution of the c-AFM height data is not accurate enough to determine whether these sheet-like structures include ferritin. The spacing between the NMOs and the sheet-like structures is within 400 nm, and between sheet-like structures is within 100 nm.

Fig. 31 is a three-dimensional projection of the contact-averaged current readings of the high-resolution 2 μm × 2 μm section. This figure more

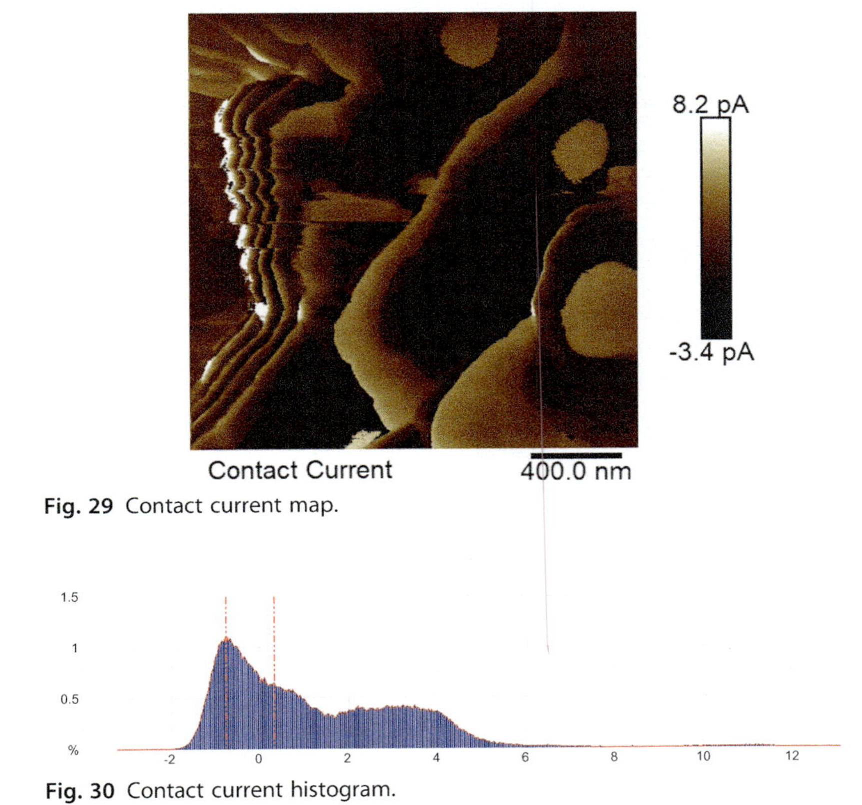

Fig. 29 Contact current map.

Fig. 30 Contact current histogram.

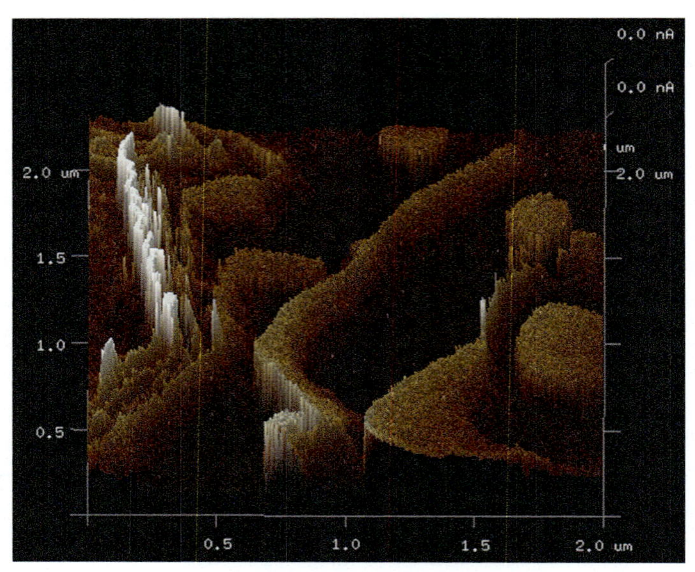

Fig. 31 Contact-averaged current map.

clearly shows the relative contact-averaged current differences between what may be NMOs and the sheet-like structures, and also shows that the highest contact-averaged currents are along the left-most side of the scan. These readings appear to be truncated at 8.2 pA, the maximum reading shown on the auto-scale. It is noted that the sheet-like regions of the highest measured currents correlate to the location of the ferritin sheets outside of the NMOs, and also correlate to the measured ability of planar arrays of ferritin to support electron transport. In particular, the c-AFM probe would contact the ferritin along an exposed edge of one of the sheets, and would apply an electric field parallel to the plane of the sheet, similar to the orientation of the applied field in the device disclosed in,[13] which discloses the ability of such structures to conduct current.

Fig. 32 is a three-dimensional projection of the height readings of the high-resolution 2 μm × 2 μm section. Fig. 33 is the associated histogram data. These figures more clearly show a lower thickness area on the left-most side of the scan, with areas of increasing thickness moving from left to right in the figure. The borders between regions where the thickness changes appear to be associated with the sheet-like contact-averaged current structures, but there is not a direct correlation between those two features at all points. It is also noted that what may be NMO structures in the contact current readings are not associated with any surface structures, and appear to be

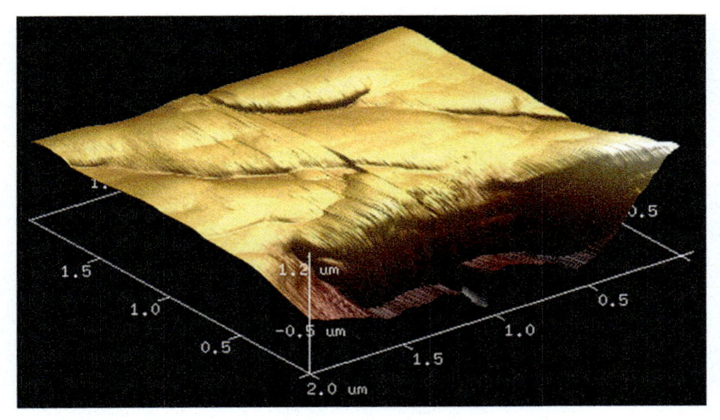

Fig. 32 AFM height map.

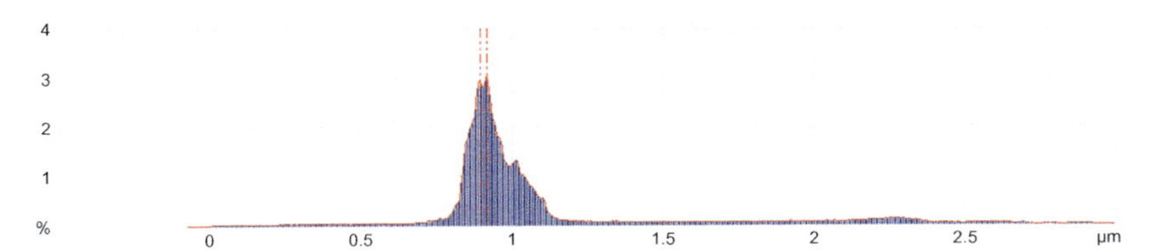

Fig. 33 AFM height histogram.

associated with NMOs that are covered by a layer of material that may include portions of the cytoskeleton, lipids or other materials.

Fig. 34 is a three-dimensional projection of adhesion readings of the high-resolution 2 μm × 2 μm section. This figure more clearly shows what may be spherical NMO structures based on the material property of adhesion, and also shows that the sheet-like areas of higher contact current appear to correlate more closely with areas having higher adhesion. While the interpretation of such adhesion readings is not addressed in this analysis, these adhesion readings appear to indicate the presence of different materials and the borders between those materials. It is also noted that neuromelanin and ferritin both have been observed to have measurable adhesion properties.[173,174]

Fig. 35 includes (1) a histogram of the peak contact current data (top), and (2) histogram of the cycle-averaged current data (bottom), respectively, for the Test Three sample. Several observations can be made. First, the peak contact current data, the contact-averaged current data and the cycle-averaged current data vary significantly, and do not have the same distribution, as was also the case with the Test Two data, also indicating that the

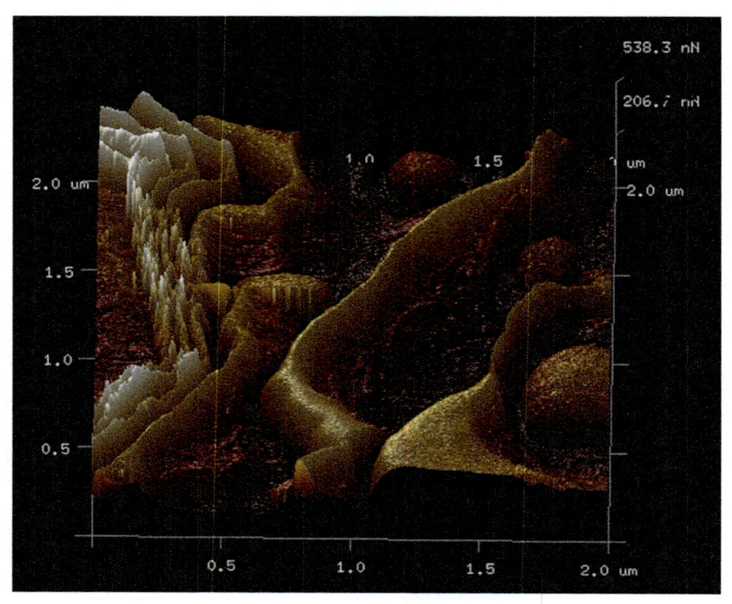

Fig. 34 AFM Adhesion map.

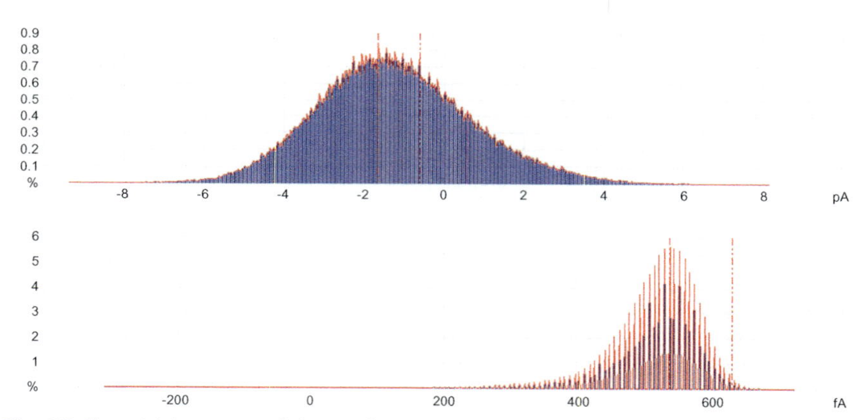

Fig. 35 Top: A histogram of the peak contact current data (top), and Bottom: a histogram of the cycle-averaged current data (bottom).

current changes non-linearly and significantly over the measurement period for each contact. Second, the peak contact current data, the contact-averaged current data and the cycle-averaged current data are not linearly related to the thickness, as was also the case with the Test Two data. Third, the peak contact current data and the contact-averaged current data distributions have significant positive current components, as was the case with the test Two data. Fourth, it is noted that the majority of height measurements are substantially greater than the minimum, which indicates that the negative currents that are measured are not caused by direct contact with the foil substrate, but rather that what may be the conductive cellular detritus layer does not have impedance characteristics that vary linearly with thickness. Finally, the peak contact current data and the contact-averaged current data include negative current values, but the cycle-averaged current data is predominantly positive. This is also the case for the Test Two data, but the histogram data covers a greater range, due to a number of negative cycle-averaged current data readings that comprise a very small percentage of the total number of readings. This data suggests that current is being transferred between the probe tip and the tissue without direct contact, possibly by electron tunneling, hopping or other quantum mechanical transport mechanisms that do not require a physical contact.

Fig. 36 is an optical photograph of the Test Four SN tissue sample disposed on foil. Because of the absence of backlighting, neuromelanin cannot be seen, but the darker area was investigated to determine whether it contained neuromelanin. There are no apparent dopamine neurons in the area that was tested, and this area does not correlate to dopamine neuron cell body groups on the glass slides.

Figs. 37 and 38 below are a three-dimensional projection image of contact-averaged current measurements for the area that is contained within the square shown in the photograph, and a histogram of that data.

Fig. 36 Test 4, optical image of SNc tissue on foil.

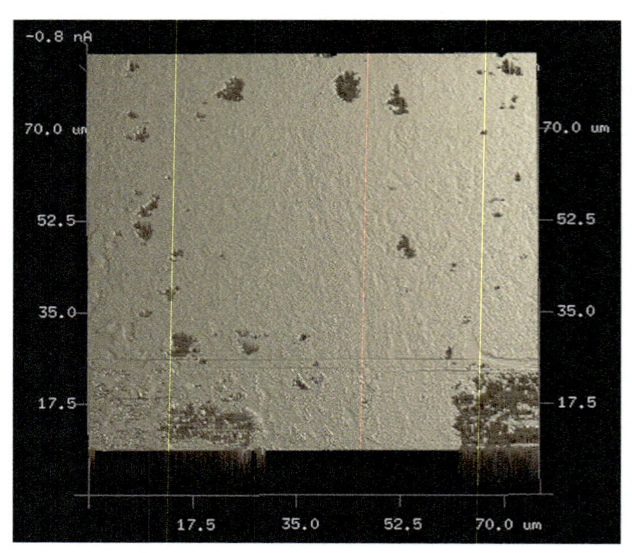

Fig. 37 Contact-averaged current map.

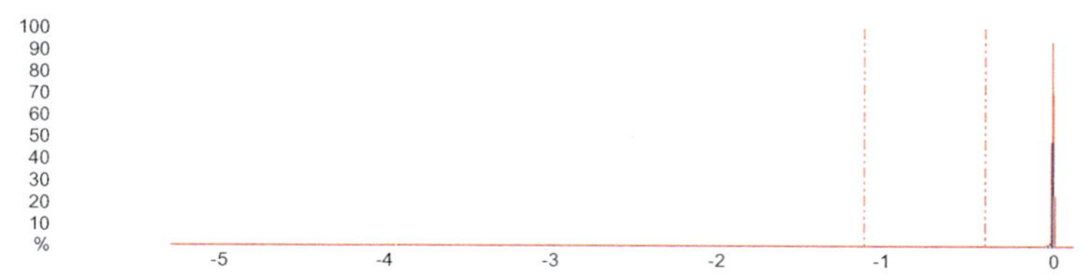

Fig. 38 Contact-averaged current histogram.

A histogram of this height data is shown below, which shows that these readings are very close to zero but the histogram data covers a greater range, due to a number of negative cycle-averaged current data readings that comprise a very small percentage of the total number of readings. A review of the data that was used to generate the histogram shows that greater than 90% of the readings are under 3 pA.

Fig. 39 is a two-dimensional image of the height data. The auto-scale shows a range of −2.6 to 2.4 μm relative to a middle point of 0, but the histogram data of Fig. 40 shows that the absolute height mostly ranges from approximately 3–7 μm. The height at the top of the foil layer is not known.

Fig. 41 below is a two-dimensional side image of the contact-averaged current. The auto-scale shows a range of −2.1 nA to 421 pA, but the histogram data shows that the contact-averaged current is primarily approximately zero.

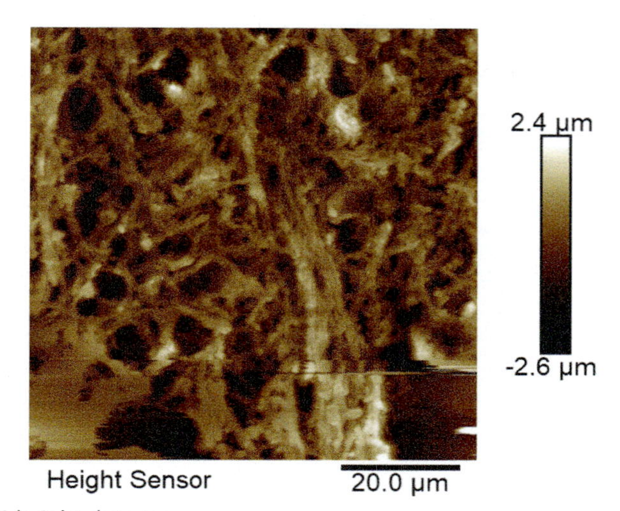

Fig. 39 AFM height data map.

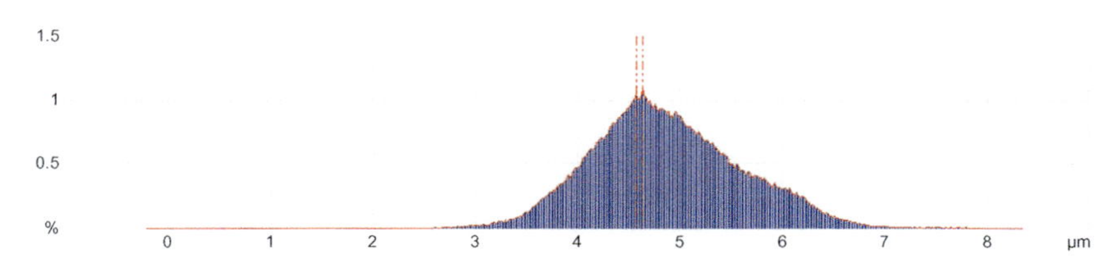

Fig. 40 AFM height data histogram.

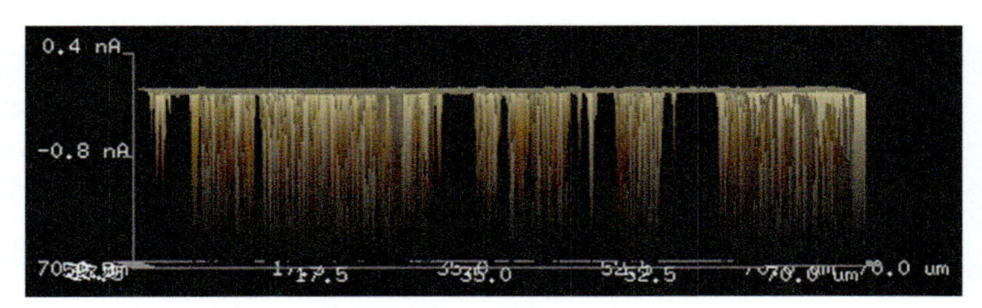

Fig. 41 Contact averaged current map.

Fig. 42 is (1) a histogram of the peak contact current data (top), and (2) a histogram of the cycle-averaged current data (bottom) for the Test Four sample, respectively. Several observations can be made. First, the peak contact current data, the contact-averaged current data and the cycle-averaged current data do not vary significantly, and have approximately the same distribution of almost 0 A. This indicates that the current does not change

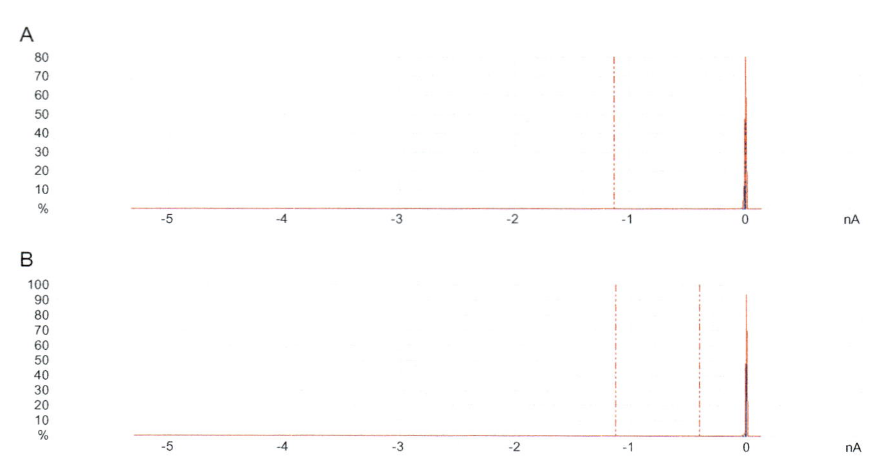

Fig. 42 (A) A histogram of the peak contact current data, and (B) A histogram of the cycle-averaged current data (bottom).

significantly over the measurement period for each contact. Second, the peak contact current data, the contact-averaged current data and the cycle-averaged current data are not linearly related to the thickness, which would be expected if the impedance between the c-AFM probe tip and the foil substrate was linear. Third, these current readings are unlike the current readings for the other three tested areas, and demonstrate that the typical current response for c-AFM testing of tissue samples is not the generation of measurable contact currents. These results suggest that the results observed in the other three areas are not typical of c-AFM testing of tissue samples that do not include high concentrations of ferritin and neuromelanin, although it is noted that there does not appear to be any reported data for c-AFM testing of other bulk organic materials (c-AFM testing has been performed for individual organic molecules). As such, it is not known if other tissues exhibit these properties.

5.5 Analysis

The most significant observation from these tests is the generation of positive peak contact current data, contact-averaged current data and cycle-averaged current data with a negative applied bias voltage. Results such as these, that apparently cannot be explained based on classical transport mechanisms, at least create an inference that quantum mechanical transport mechanisms may be present.[175] The applied bias voltage of -10 V and a distance between a planar electrode and a foil ground plane of 5 μm (which is the nominal thickness of the tested SN tissue samples disposed on foil substrate) would result in an electric field magnitude of 2 million volts per meter.

The relatively sharp tip of the c-AFM probe creates a substantial increase in the electric field gradient, and the tissue thickness was variable over the test regions and often less than 5 μm in thickness, which would further increase the electric field magnitude in areas near the tip. Electric fields at these magnitudes are high enough to cause ionization of molecular structures with low ionization energies. If such molecular structures exhibit quantum dot characteristics, the free electrons and the associated holes created by ionization could form excitons, and support the creation of coherent electrons and formation of electron transport in a disordered or semi-ordered array, as has been experimentally observed at room temperature in semiconductor devices.[140,141] A large number of coherent electrons forming electron minibands or other electron transport mechanisms (such as electron hopping or tunneling) could generate a transient effective negative voltage that is greater in magnitude than the −10 V bias of the c-AFM probe tip, and would explain the positive current flow into the c-AFM tip. In the absence of such electron transport, it would be expected that the electrons supplied from the −10 V bias of the c-AFM probe tip would only create negative currents, which is what was observed in locations where the thickness of the sample was much lower. It is noted that quantum mechanical electron transport could also be occurring in those locations, but would be much harder to distinguish from classical transport mechanisms.

While there are a number of references in the technical literature that discuss c-AFM measurements of quantum dots, these references are generally directed to measuring individual quantum dot currents, with a small number of additional individual quantum dots in the near vicinity that may have some impact on the current measurements, but which do not form large arrays of quantum dots that can generate electron transport currents.[176,177] Other studies involve measurements of conduction currents through monolayers of quantum dots.[178,179] While positive currents with applied negative bias voltage has been reported, none of these studies involved large-scale three-dimensional devices that included thousands of quantum dots or more, or measurement of tissues with quantum dots disposed in the tissue. As such, there does not appear to be sufficient pre-existing literature to assist with confirming the interpretation of these test results.

There was a notable difference between the positive currents measured for Test One, and the positive currents measured for Tests Two and Three (although negative current readings were generally similar for Tests One, Two and Four). While the Test One results yielded a relatively small number of positive current readings, Tests Two and Three yielded a substantially greater number of positive current measurements, which indicates that wax removal is important to improve the ability to measure c-AFM currents. The second wax removal protocol was identified from a resource for tissue staining, and is also beneficial for c-AFM testing of tissue, as it appears to remove additional wax layers that might be only several hundred nanometers in thickness. Due to the apparent absence in the literature of any reports of c-AFM testing on tissue, additional research in this area would be useful, not only on wax removal protocols for animal tissues for c-AFM testing, but also on c-AFM testing of other suspected biological quantum dot structures, such as the Fenna-Matthews-Olson (FMO) complex and Light Harvesting Complex II (LHCII). In addition, while Test Four did not result in any notable positive contact currents and thus acts as a control to some extent, additional testing of different tissues would be useful to rule out any classical causes for these positive currents, such as material-specific effects.

The Bruker Dimension c-AFM system compensates for capacitive charging current, so such currents would not be a factor in any measurements. Regardless of whether the tissue impedance is mostly resistive or includes capacitive or even inductive impedance components (which is unlikely), capacitive and inductive circuit elements would still draw charging current at contact and should not be able to provide positive contact current to a negative biased probe. Furthermore, the absence of such currents in the Test Four sample indicates that the effect of such resistive, capacitive or inductive contact and peak current components are likely negligible in the Test Two and Test Three results.

The ability of the Bruker Dimension Icon to measure peak contact current data, contact-averaged current data and cycle-averaged current data provides some insight into the transient nature of these currents. The scan rate for these c-AFM measurements was approximately 0.2 Hz, which results in an average time between contact measurements on the order of 200 microseconds, but the Bruker user manual (available from www.bruker.com) does not specify how much of that period is part of the measurement time for each of the types of currents measured, and instead suggests that the peak and contact-averaged measurement periods are likely a function of the material properties of the specimen being tested. It is assumed that the transit time of the probe tip between two successive contact points is the period for the cycle-averaged current data, and further assumed that the measurement period for the contact-averaged current is no more than 10% of the transit time and that the measurement period for the peak current is directly related to the ratio between the peak current and the contact averaged current. In other words, a constant current flow during the contact averaged current period is assumed. It is noted that there is a high-speed data capture function of the Bruker Dimension that can be used for limited periods of time, and that such high-speed data capture would address some of these uncertainties, but that function was not activated for these tests.

These time period assumptions are applied for an exemplary single point measurement from the Test Two data, which has a peak current of 1.09 nA, a contact averaged current of 1.25 nA and a cycle averaged current of 37.5 pA. One Coulomb represents the charge of 6.2415×10^{18} electrons, such that the measured peak current would correspond to a transfer of:

$$(1.09 \times 10^{-9}) \times (6.2415 \times 10^{18}) = 6.8 \times 10^{9} \text{ electrons}$$

if it were sustained for a second, and the measured contact averaged current would correspond to a transfer of:

$$(1.25 \times 10^{-9}) \times (6.2415 \times 10^{18}) = 7.8 \times 10^{9} \text{ electrons}$$

if it were sustained for a second. As such, the length of time during which the peak current is measured and the length of time during which the contact current is measured will have a bearing on the number of electrons that were transferred during those measurement periods. If the period is very short, such as 1×10^{-6} s, then the number of electrons transferred would be smaller, approximately 6.8×10^{3} and 7.8×10^{3} electrons, respectively. Likewise, if the period is longer, then the number of transferred electrons would be correspondingly larger. These estimates assume constant current flow during the measurement period.

However, it is acknowledged that the assumption of constant current flow (at least at all points) is unrealistic. The following table shows the data measured at 5 points for (1) height, (2) contact averaged current, (3) adhesion, (4) deformation, (5) cycle averaged current, (6) peak current, (7) peak to contact averaged current ratio and (8) peak to cycle-averaged current ratio.

Height (nm)	Contact current (nA)	Adhesion (nN)	Deformation (nm)	Cycle current (nA)	Peak current (nA)	Peak-contact ratio	Peak-cycle ratio
703	0.922	−14.1	21.1	0.00473	0.697	0.756	0.0513
−1030	1.15	−4.30	168	−0.00375	0.974	0.847	−0.00326
−793	1.21	−11.6	16.7	0.057	1.05	0.868	0.0471
−1770	1.33	26.8	8.67	−0.353	1.19	0.895	−0.2651
457	1.78	−15.6	16.5	0.00623	1.72	0.966	0.0350

A number of observations can be made from this limited data set. First, the contact averaged current does not appear to be directly related to thickness, which is the height value at the point of measurement plus the average thickness of the sample, approximately three microns. Because these contact averaged current measurements are all positive currents flowing into the probe, which had a -10 V bias, a linear relationship would not be expected in any event.

Second, the contact averaged current does not appear to be directly related to material properties such as adhesion or deformation. While those material property measurements are larger than nominal values and show a substantial amount of variability, it is noted that similar readings and levels of variability were observed in the Test Four data, and therefore those results do not by themselves appear to be the reason for the generation of the anomalous positive currents.

Third, the ratio of peak current to contact averaged current is somewhat related, and increases with increasing contact averaged current. This data suggests that the associated time periods during which these measurements are taken may have a similar relationship, and that estimates of the number of electrons that are transferred to the probe from the tissue during these periods may be accurate to at least some degree, with the primary uncertainties being the duration of each period and the changes in current magnitude during those periods.

Fourth, the ratio of contact averaged current to cycle averaged current exhibits a high degree of variability, including swings between positive and negative ratios. The differences between these measurements suggests that a high frequency current redistribution process may be occurring, which would correlate to known electron coherence times in the hundreds of femtosecond range at room temperature for other possible biomolecular quantum dot structures such as FMO or LHCII. If a large number of ferritin and neuromelanin molecules are involved in the generation of excitons and coherent electron transport mechanisms, a complex high frequency current pattern would be expected.

Assuming that each electron represents an exciton generated by a ferritin or neuromelanin molecule, these estimates would suggest that the measured currents may have involved electron transfer from multiple adjacent dopamine neurons in the tissue samples, which may contain on the order of 1×10^6 ferritin cores in a 25 μm diameter cell.[56] Because the tissue being analyzed in these tests (1) is not living tissue, (2) includes only a fraction of a cell body, (3) presumably excludes the ferritin and neuromelanin that is not maintained in an array, and (4) involves an exciton generation mechanism that is not related to living tissue exciton generation, the actual levels of electron transport could be greater or lesser than these numbers. However, further assuming that action potentials in an average neuron can be generated by an influx of approximately 2×10^7 sodium ions, these c-AFM measurements indicate that a sufficient level of electron transport might be able to occur in the SNc to be a factor in the generation of action potentials, although further research and a more rigorous analysis would be required to substantiate that hypothesis. No conclusions are made about these effects playing a role in ion channel activity, because the numbers of ions generating a flux in an action potential must be related to the time it takes for them to cross the membrane and hence to the amount of current associated with it. In addition, the interaction of this hypothesized effect with the complex and well-known conventional action potential current mechanisms is unknown.

The correlation between the NMO-sized regions and the positive contact currents is also notable, and further suggests that neuromelanin has quantum dot properties, similar to those that have been observed for ferritin. The adhesion (and elasticity) readings also correlate to the size and shape of NMOs. None of these regions appeared to have an observable surface feature based on the AFM height data, and the NMOs may have been underneath some thickness of intervening tissue structure, a lipid layer or other materials. The positive contact currents at the boundaries in the Test Three data could likewise be explained by ferritin accumulations in these locations. The highest contact current peaks correspond to the thinnest region, which could include exposed ferritin complexes that were able to make physical contact with the c-AFM probe, whereas lower contact-averaged current regions may include ferritin complexes that have covering tissue or material, which would limit currents measured by the probe to electron tunneling, hopping or other quantum mechanical electron transport effects.

Another observation that can be made from these test results is that the presence of a wax coating appears to have an impact on the results. While the thickness of the remaining wax coating after the first and second wax removal protocols is not known, the second wax removal protocol included a substantially longer exposure to the wax solvent xylene after heating, and successive ethanol soaks. It is also possible that tissue hydration increased tissue conductivity or other electrical effects, although that should not have had any direct influence on either quantum mechanical electron transport or the generation of positive contact-averaged current, peak contact current and cycle-averaged current in response to a negative probe bias. It would not be unexpected for a remaining thickness of several hundred nanometers of wax to be present after the first wax removal protocol, and for that layer to be removed by the second wax removal protocol. A thickness of that amount of an intervening space could be sufficient to prevent the tip of the c-AFM probe from interacting with a disordered array of neuromelanin and ferritin to receive electron transport currents, which could explain the difference in the number of positive current measurements between the Test One results (fewer positive current measurements and additional wax) and the Test Two and Three results (extensive positive current measurements and less wax).

The distribution of positive contact-averaged current, peak contact current and cycle-averaged current in the second test location is consistent with estimates of the density of ferritin in SNc tissue, and suggests that c-AFM probe measurements within several hundred nanometers of these structures are sufficient to measure currents due to electron transport. Positive contact-averaged current, peak contact current and cycle-averaged current were also measured at locations in the thinner material regions that appear to correspond to detritus from intracellular fluid for Test Two, but not Test One. The difference between the Test One and Two results in these areas may also indicate that a wax layer of several hundred nanometers thickness or more is sufficient to prevent electron transport current from passing from the ferritin and neuromelanin into the c-AFM probe tip in these regions.

It is hypothesized that the measured electron transport currents could be used by SNc neurons to coordinate retinotopically-mapped neural inputs to the SNc with other corresponding retinoptopically-mapped neural inputs from cortical regions associated with image data processing and object recognition.[180] For example, this electron transport mechanism could explain the ability to identify objects within image data and to understand what those objects are, by integrating neural inputs into the SNc from different regions of the cerebral cortex associated with image data processing and object recognition, which are also retinotopically organized.[181] The ability of coherent electrons to integrate these inputs into what is understood as the conscious experience suggests that coherent electrons could be capable of separating the neural inputs into a large number of states, and of sharing this state information with other coherent electrons in some manner, although additional work would be required to test that hypothesis. Controlled electrical or chemical stimulation of the SNc synaptic connections could potentially be used to determine whether these inputs are determinative of the conscious experience.

It is unlikely that these currents could be caused by an ionic or other electrochemical reaction for a number of reasons. First, neural tissue must provide a neutral (i.e., non-acidic and non-basic) environment, and many cellular mechanisms exist to remove potentially damaging acidic or basic compounds, such as reactive oxygen species. In order to force currents into a c-AFM probe that is biased at −10 V, a chemical reaction would be needed, as well as a circuit that is associated with that reaction, such as an electrochemical battery. Such batteries typically include electrodes that are involved in the associated chemical reaction, but the Bruker c-AFM probe is made of conductive diamond or other non-reactive materials. Some of the highest electrochemical cell voltages involve highly reactive materials like lithium, and in order to obtain a voltage of greater than −10 V typically requires a serial connection of electrochemical cells. It is also worth noting that the highest currents were measured at what appears to be the edge of a ferritin sheet, but ferritin stores $Fe3+$, which should absorb free electrons and which has no additional electrons to donate to an electrochemical process. And while this specific tissue has not been tested previously, other C-AFM tests of biological tissues as well as electrostatic force microscopy (EFM) tests, Kelvin probe force microscopy tests (KPFM) and other similar tests have measured no such anomalous currents.[182]

As a final note, while wax was removed from the neural tissue samples that these tests were performed on were removed on, and those samples were rehydrated for Tests Two Through Four, the tissue is not the same as living cells. The measurements presented may not reflect the behavior of living cells, and may only be due to the remaining tissues after the storage and preparation step. For example, the ionic concentration of cytosol is capable of screening and shielding electrical charge, which could possibly block electrical effects, although this would not necessarily directly impact the quantum mechanical electron transport mechanisms such as electron hopping, electron tunneling or the creation of electron minibands. In addition, neurons possess other lysosomes and vacuoles of similar size, shape and profile as ferritin and neuromelanin. Additional work investigating the electrical properties of microtubules,[183] and certain lipids, such as dolichols, which are present in neuromelanin[184] has shown that these biomolecules can increase conductivity in membranes,[185] although such conductivity would not explain the generation of positive contact currents with a negative probe bias voltage, nor has it been shown or even suggested that such biomolecules form disordered quantum dot arrays that are similar to engineered quantum dot arrays. In the absence of staining that can identify ferritin and neuromelanin, it is possible that some other biomolecule may contribute in whole or in part to the measured currents (although it is noted that no such currents were measured in Test Four).

5.6 C-AFM testing—Conclusion

These c-AFM tests appear to demonstrate highly non-linear behavior that correlates to electron transport in quantum dot arrays. The currents measured could be the result of electron transport levels that would be sufficient to influence or to potentially determine action potential generation. Additional testing of different types of tissue and at different bias voltages could be performed to confirm these test results.

The postulated electron transport mechanism might seem unlikely, but other electron conductance mechanisms are widely accepted to exist and to regulate heartbeat, breathing, or even individual neurons themselves.[186] Such transport mechanisms are among known practical applications for QD arrays in semiconductor device applications. Therefore, it is at least possible that the SNc, LC, or possibly other regions of the brain have evolved to use the disordered or quasi-ordered arrays of neuromelanin and ferritin as an electron transport mechanism, as hypothesized. Additional research to investigate this proposed physical mechanism could be conducted.

6. Electron transport and voluntary action selection

The neural electron transport mechanism described above could be related to the voluntary action selection mechanism by providing a global workspace function that is used to connect the cell bodies of a large number of neurons. This electron transport mechanism would provide a function, and from an evolutionary perspective, it would be expected that voluntary action selection provides a functional advantage to an organism that provides that organism with a competitive advantage over organisms that lack the function associated with voluntary action selection, such as the ability to move faster for obtaining food or shelter, the ability to mate more effectively and so forth. However, the experience of consciousness might not be the function, but just an artifact of the function. While it might appear that the experience of consciousness generates the function from the point of view of the observer of that experience, it may also be the case that the experience of consciousness is generated by voluntary action selection.

In that regard, there are a number of different functions that could be performed by the electron transport mechanism. One function would be to synchronize the firing of neurons in tissues with high concentrations of neuromelanin. While such synchronization is observed in neurons, it occurs in neural tissue that lacks ferritin and neuromelanin as well as neural tissue that contains ferritin and neuromelanin, and as such, might be redundant to functions provided by other mechanisms. Because the presence of an ROS like Fe^{2+} is damaging to cellular mechanisms, it does not appear that it would be favored as an evolutionary vector if a less damaging mechanism was available that performed the same function, such as electric fields or ephaptic coupling.

Another function that could be provided by the electron transport mechanism would be to suppress action potential voltages and to thus prevent neurons from firing as readily. This function was observed in the c-AFM tests, where the transient negative probe voltage appears to have generated a dynamic electron transport state that had a sufficiently negative voltage to be able to drive positive currents into the negatively-biased probe. However, even if this function is present in neural tissues, it would not prevent multiple neurons in those neural tissues from reaching an action potential at the same time. In a structure like the SNc that is associated with voluntary action selection, such simultaneous firings could result in a seizure. If the objective is to find a cellular mechanism that explains the ability of the SNc to actuate only a single voluntary action and to avoid seizures that would result from multiple concurrent actuations, a simple action potential voltage suppression mechanism would not provide an evolutionary advantage.

Another function that could be provided by the electron transport mechanism would be a voluntary action selection function that effectively switches an individual neuron in a group of neurons. As discussed above, this switching mechanism could be caused by the localization of electrons in the soma with the lowest impedance axonic path, which could cause generation of Fe^{2+} and associated cellular responses that result in the generation of an action potential. While that specific function has not yet been observed in solid state devices that utilize quantum dot structures, it is nevertheless theoretically possible, and would provide a function that has a substantial evolutionary advantage of selecting a best neural action from a number of potential neural actions and which prevents multiple voluntary actions from being simultaneously initiated.

Regardless of the specific cellular function provided by the electron transport mechanism, the important difference between the electron transport mechanism and any other known neural communications mechanism is its ability to simultaneously couple thousands of neurons with a singular mechanism in a functional manner that could be associated with the conscious experience. No other mechanism, including electric and magnetic fields, is capable of that function. For example, while both local and global electric and magnetic fields are generated by the brain, those fields couple neurons that are actively involved with the conscious experience (such as neurons associated with the experiences of hearing and eyesight) as well as neurons that are not (such as neurons that cause changes in blood pressure, blood sugar level and perspiration).[187] In contrast, the neural tissues that have sufficient ferritin and neuromelanin concentrations to support the hypothesized electron transport mechanism are in areas that receive highly processed neural signals that are directly and functionally associated with the conscious experience.[125,126]

The human conscious experience can be understood as a sequence of states. Each instantaneous state has a number of components—images, sounds and other inputs being received by the senses, a current recollection of the logical meaning of objects and sounds, the current recollection of the history of those objects and sounds and so forth. For example, consider the image in Fig. 43.

The significance of this image depends to some extent on state—this could be an image from a viewer who is walking to the left of the frame,

Fig. 43 Roadway scene.[188]

the right of the frame, or standing still; the cars could be moving or stopped; the car in the foreground could be similar to one that the viewer is waiting for, but the driver could be different; or numerous other state variables could be important. The amount and sequence of neural processing that is involved with the evaluation of each of these factors can be vast. For example, the neural signals associated with all image data originate at the retina, are transmitted through the lateral geniculate nucleus to the primary visual cortex (known as V1), then to a series of additional neural structures that identify borders of objects, color, motion and other primitives of objects. The neural signals are generated by each of these areas in various combinations of in sequence and in parallel, are then processed by neural structures that are capable of identifying logically distinct objects, such as people or buildings, and the objects identified by these neural signals are eventually provided to the mechanism that generates the conscious experience. None of the neural preprocessing that generates the experience associated with the state of the image objects forms part of the conscious experience, and instead occurs below the level of conscious perception, like the neural signals of the autonomic nervous system that are responsible for control of internal organs and smooth muscles. If the ubiquitous cranial electric and magnetic fields were responsible for the experience of consciousness, then these subconscious neural signals would form part of that experience. In addition, they would be capable of functional control, but they are not by their very definition.

Assuming that a sequence of combinations of instantaneous neural states forms the human conscious experience, it can be shown that the neural electron transport mechanism could support the sequence of states that comprise at least some aspects of the human conscious experience. In particular, the instantaneous combination of these neural states could be associated with the perception of reality based on the development of synaptic connections over time as the viewer learns to correlate the visual appearance of objects with the tactile experience of those objects, the sounds associated with those objects, the behavior of those objects and so forth. These neural states are thus correlated to functional and sensory interactions with real world objects, and are created by synaptic structures that are formed (learned) by interacting with those objects but which can be extended to objects that have not been directly experienced.

A simple hypothetical example is provided by *C. elegans*. The hermaphrodite variety of this animal has 8 dopamine neurons, of which 4 CEP neurons are located very close to each other in the head, as previously shown. These 4 CEP neurons have a single dendrite that originates from a mechanosensory structure at the front of the worm, and which essentially receives a single input that generates an action potential. The CEP neuron fires when it reaches a high enough level of stimulus from this single input. Each CEP neuron thus has at least 2 states—either an insufficient level of mechanosensory stimulation to cause the CEP neuron to fire (an "off" state), or a sufficient level to cause it to fire (an "on" state). These two states can also be referred to as "binary" states, and are similar to the digital logic used by computers to process data, in that regard. Almost all neurons exhibit this binary state behavior, and it is their primary mode of operation. To analogize, a neuron can be viewed as being similar to a light switch, with an "on" and an "off" position or state.

However, a neuron that is capable of being coupled to other neurons through an electron transport mechanism that is activated by electrical or chemical activity within the soma of the neuron has two other states—"active" and "inactive," where neurons in the active state are coupled to other neurons through the electron transport mechanism, and neurons in the inactive state are not. The active state neurons are capable of participating in the generation of an action potential as a result of the electron transport mechanism. The combination of active neurons are capable of functioning as a single binary switch, with the additional feature that only one of the combination of switches will be turned on as at any time. Which one of these switches is turned on would depend on the axonic impedance, or possibly other factors.

The CEP neurons also express the gene for ferritin, and may have sufficient ferritin to support electron transport, as hypothesized. At least two of the CEP neurons are almost in contact, and the other pair are at least several microns away, but they could also possibly be connected to the first pair by the electron transport mechanism if there is sufficient ferritin present. In this regard, it has been observed that dopamine in these neurons increases the efficiency by which *C. elegans* can integrate information in those neurons, which is consistent with the hypothesis that an electron transport function may exist in those neurons.[122] If that is the case, then those 4 neurons would form a 4 bit "state machine," which is a mechanism that has a combination of 4 different binary variables. These four different binary variables result in a total of 2^4 different potential states. This is shown in Fig. 44 below, where the 4 CEP neurons are represented by squares with an active state or inactive state, as indicated in each square.

These 16 states represent all of the possible "experiences" that *C. elegans* can have with these 4 CEP neurons and the hypothesized electron transport mechanism. These 16 different states would be states of that electron

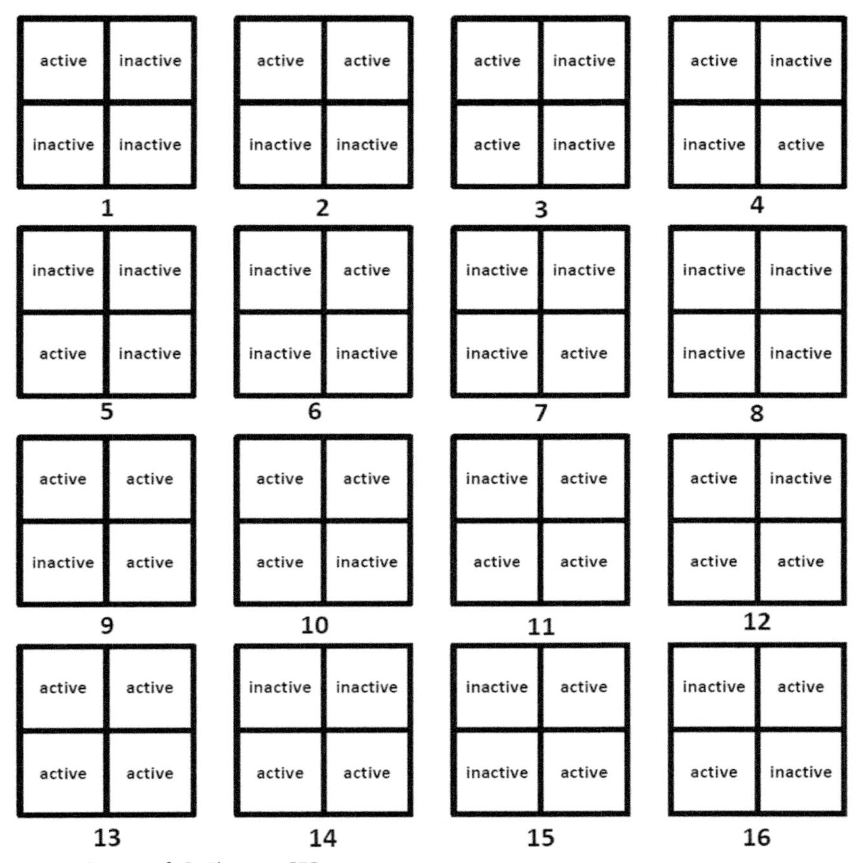

Fig. 44 States of *C. Elegans* CEP neurons.

transport mechanism in combination with these 4 CEP neurons. They would form a single structure, part of which would be a real physical structure, and part of which would be an electron transport structure that has a physical effect on the 4 CEP neurons. While each CEP neuron would have its own individual function, it would also have a function that is a result of the function of the other 3 CEP neurons that it is connected to by the electron transport mechanism. In addition, while localization of electrons due to the axonic impedance of each neuron has not been proven, if that effect is present, then only one of the 4 CEP neurons might be able to fire at a time from that mechanism. That additional function would increase the number of potential states. The electron transport mechanism could also provide some other function, such as to depress the voltage in these CEP neurons in the same manner that it does in the c-AFM measurements, to further limit the ability of more than one CEP neuron from firing at a time.

It is noted that the structure of the axonic connections of the CEP neurons is very different from that of the human SNc dopamine neuron axons, and the CEP neurons do not function in the same manner as the SNc neurons. If they did, then the number of states would be increased by an additional factor related to the number of active neurons and the number of possible low-impedance axonic connections. In addition, a single active CEP neuron would not receive any additional energy from another inactive CEP neuron (e.g., states 1 and 5–7), so those states would not contribute to the generation of an action potential. Whether or not electron transport between the ferritin structures within a single active CEP neuron would create a "conscious" experience is not known, but if it did, then that instantaneous conscious experience would change as other CEP neurons become active and interact using the electron transport mechanism.

This hypothetical is not presented to suggest that *C. elegans* experiences consciousness like the human experience of consciousness on a smaller scale, but rather only to provide a simple example of how these 4 CEP neurons could interact using the hypothesized electron transport mechanism to form a singular neural mechanism (or switch) with a large number of different states. If this neural mechanism existed, then these states would be nothing like the human experience of consciousness, as will be discussed below. However, one aspect of the human experience of consciousness that can be understood by these 16 states can be seen by considering "state transitions."

A state transition is when a state changes from a first state to a second state. In the human experience of consciousness, awareness of state transitions is widespread and is reflected in the changes in the state of what is seen and heard, among other things. The order of spoken words and the sequence at which cars arrive at an intersection are examples of how such state transitions form part of the human experience of consciousness, and specific neural structures are required for an awareness or memory of such state transitions. For these 4 CEP neurons of *C. elegans*, a state transition could occur from any one of these 16 states to any other one of these 16 states, and a series of state transitions could occur in any order, such as "1- 15 – 3 – 8 – 1" or "7 – 6 – 16 – 2 – 4 - 9." The ability to *remember* a series of such state transitions would require additional neural structures that do not appear to be present in *C. elegans*, but the ability to *process* state changes would be inherent in the functions of *C. elegans*, such as to go from a fast movement state to a slow movement state when the mechanosensory input to the CEP neurons changes from all being "off," or receiving no inputs, to one or more being "on," or receiving an input. As such, while *C. elegans* might be able to instantaneously experience a series of state transitions, such as "1 – 2 – 1 – 14 – 12 – 9," it would not remember that series unless it had some additional neural mechanism that is capable of memory.

The CEP neurons of *C. elegans* are not homologous to SNc neurons, but this state hypothesis of consciousness can be extended to the human experience of consciousness. For example, when choosing a voluntary action, a number of different factors need to be balanced and compared to select the optimal apparent action. To understand this problem in a conventional context, consider a common situation—navigating an intersection, as shown in the photograph on the top, which has been abstracted to key elements in the diagram on the bottom in Fig. 45.

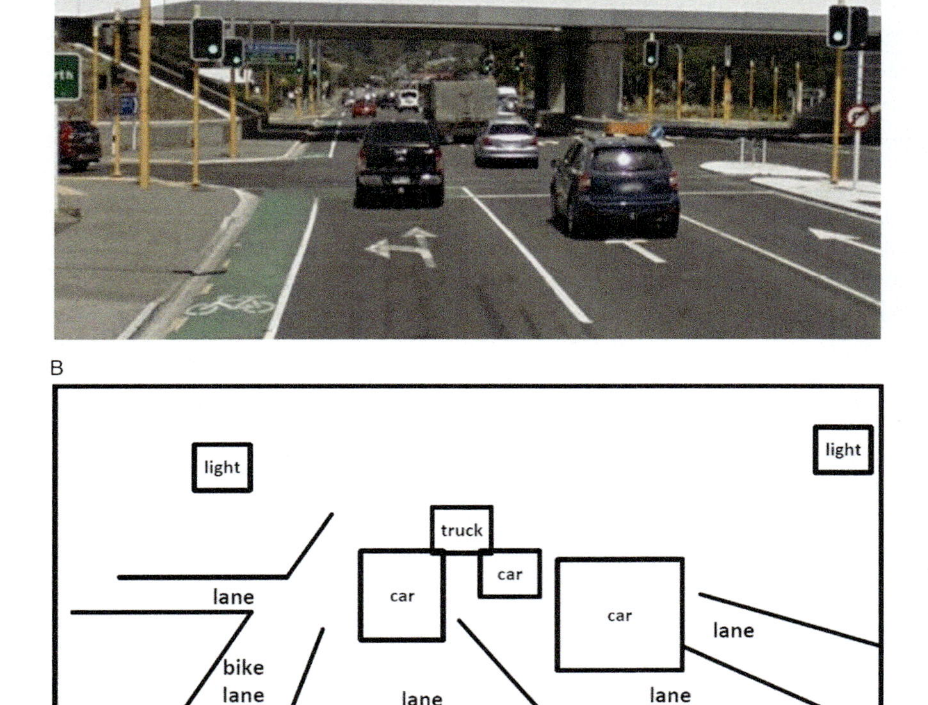

Fig. 45 Left: Photo of intersection. Right: Key elements.

If the photograph represented the field of view (FOV) of a person, the image data that creates the field of view is initially received by over 100 million rod and cone cells, which generate neural signals on approximately 1 million retinal ganglion cell axons. These neural signals are conveyed "retinotopically"—in the same arrangement as the image data is received at the retina—to the visual cortex. As discussed above, these neural signals are subsequently highly processed after leaving the visual cortex to identify the objects in the FOV, such as cars and overpasses.

As shown in the abstraction of the photograph, there are at least 11 different primary objects in the FOV that would be of immediate interest for the purposes of any voluntary action selection of a person arriving at the intersection in a car—three cars, one truck, four lanes, one bike lane and two traffic lights. However, numerous additional other factors are also being tracked by the different neural structures of the brain. For example, there might be a number of different objects outside of the FOV that are nonetheless mapped to neural structures that are part of the voluntary action selection process, such as vehicles that are next to but outside of the FOV, the relative speed of each vehicle, the driving history of each vehicle (e.g., whether they have been swerving or driving erratically), the color of the traffic lights, the length of time that the lights have been the current color, important sounds such as sirens and so forth. The neural structures that are responsible for detecting and tracking these variables are complex, and many also receive retinotopically-mapped neural signals from the visual cortex that are processed to identify and track shapes. Those neural signals travel along parallel paths through the parietal and temporal lobes to associate the shapes with known objects and to effectively select the object that most likely corresponds to the shape, as well as numerous other neural structures. All of these neural signals are integrated to create the singular experience of consciousness, and drive the voluntary action selection process.

For example, if the FOV is associated with an 11×23 array of quasi-retinotopically-mapped dopamine neurons (actual retinotopic mapping has a substantially higher density in the center of the field of vision) that are coupled by the hypothesized electron transport mechanism, those dopamine neurons can be allocated to an 11×23 state diagram, as shown in Fig. 46 with dashed lines superimposed on the abstracted FOV.

This state diagram is similar to the 4 state diagram of the *C. elegans* CEP neurons, except that it is capable of $\sim 2^{253}$ different states ($\sim 10^{84}$). Thus, even with a relatively small number of dopamine neurons that are coupled by the hypothesized electron transport mechanism, the number of different

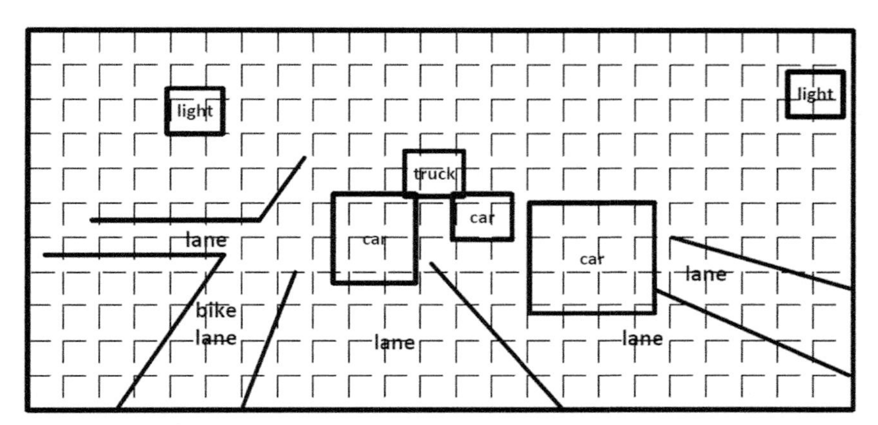

Fig. 46 States of intersection image.

states that can be represented by those neurons is very large. When it is considered that the associated neural signals might not be limited to just two states but could instead represent different energy levels that are associated with state information (such as movement and memory), or that these state signals might be provided to dopamine neurons other than those that are receiving retinotopically-mapped image data, the number of different states could be even larger. These states do not represent images, abstract ideas or other such information, but rather correspond to the combination of sensory inputs as a function of time as those states have been correlated by synaptic structures and different sensory mechanisms.

For example, one of the first activities typically engaged in by newborns includes synchronizing the tactile shapes of their own bodies and objects with the way their bodies and those objects appear.[189] Through trial and error, synaptic connections are formed and reinforced that accurately predict these correlations, and which allow the child to recognize objects visually, to associate those visual objects with physical objects and to distinguish different objects from each other. These neural structures can be understood as developing to provide a set of signals to the relevant set of SNc neurons that provide successful motor responses to sensory stimuli.[135, 136, 151] The active and inactive SNc neurons that are associated with those sensory signals and the associated signals from these additional neural structures can create the state of experience of seeing the objects and understanding what they are using the hypothesized neural electron transport mechanism.

While a number of different neural structures appear to include the hypothesized neural electron transport mechanism, based on neuromelanin content (e.g., the SNc, the LC, the ventral tegmental area (VTA) and the raphe nucleus (RN)), the SNc appears to be most closely associated with the voluntary action selection mechanism, and receives both retinotopically-mapped image neural signals as well as neural signals from a large number of other neural structures. If the SNc neurons that support the electron transport mechanism are viewed as a state machine, then it can be seen that even a small number of such neurons would be capable of representing a large enough number of states to explain the range of human experience in all of its intricate details and variations, as discussed above. Thus, while the human experience of consciousness seems to be capable of an almost infinite variety of different experiences, even a small number of SNc neurons (253 in the example above) are capable of representing a number of states that would seem to be large enough to encompass all of those experiences. However, as discussed above, there are 10,000 or more of the largest volume SNc neurons that would be associated with the electron transport mechanism, such that the actual number of potential states is extremely large. In addition, the number of associated neurons in the LC, the VTA, the RN and other neural structures that are possibly involved, while relatively small compared to the ~80 billion neurons of the brain, are still sufficient to provide a large number of potential states that could account for the wide variety of experiences.

The human experience of consciousness is more than sensory inputs, and also includes memories, the symbolic meanings of what is being experienced, abstract ideas such as love and hate, and numerous other aspects. The neural signals that are input into the SNc and other neural structures that could support electron transport could potentially combine to generate these different aspects of the human experience of consciousness, even if they use just a small number of neurons in each neural structure. For example, it is noted that cognitive impairment in Parkinson's Disease patients impacts many aspects of the human experience of consciousness, such as attention and frontal executive functions, memory, visuospatial skills and language.[190] The neural structures associated with these cognitive functions are found in many diverse areas of the brain, including the prefrontal cortex, amygdala, hippocampus, basal ganglia, motor cortex, cerebellum, occipital lobe, parietal lobe and temporal lobe,[135, 136, 151] but many of these areas do not show any signs of neurological damage in pathological analysis of Parkinson's Disease injury. If the SNc, LC or other catecholaminergic neurons are damaged and are also responsible for integrating the associated cognitive functions into the electron transport function, then it could be the loss of those catecholaminergic neurons that results in the associated cognitive impairment, as opposed to damage to the neurons that are associated with the cognitive function. While this evidence does not establish that it is damage to the catecholaminergic neurons from Parkinson's Disease that causes this cognitive impairment, it is consistent with that hypothesis.

Although some of these neural structures are separated by neural tissue that does not appear to be able to physically support electron transport, such as the SNc and the LC, they are integrated by virtue of receiving neural inputs from the same afferent neural structures. In that regard, it is noted that the human experience of consciousness can be separated into at least two components that appear to be associated with the SNc and the LC, with the SNc being associated with sensory processing and the LC being associated with the internal dialog. These components interact with each other at various levels, but are capable of functioning independently of each other.

If the hypothesized neural electron transport mechanism is present and functions in this manner, then a prosthetic neuron that has a similar ferritin and neuromelanin structure may be able to interact with it in some manner. Whether that interaction would enable conscious states to be monitored, recorded or modified might only be determined by experiment, and could end up having no effect at all or only effects that are unpleasant or painful. However, even a negative effect would demonstrate that the hypothesized neural electron transport mechanism relates to the human experience of consciousness in some manner.

7. Conclusion

The hypothesized neural electron transport mechanism is potentially capable of integrating a large number of neural input signals into a single physical mechanism that could correspond to the singular experience of human consciousness. If that is the case, then the experience of consciousness might be a function of the differential activation states of these neurons relative to the activation state of every other neuron. The state of this physical structure has no analog, but is loosely similar to the way in which different pictures can be generated on a display by using different combinations of active and inactive pixels. That display must be viewed by an observer for the picture to be perceived, whereas it would be the net effect of the complex integration of neural signals by the hypothesized electron transport mechanism that generates the experience of consciousness. In any event, the presence of a structure in neural tissue that is capable of supporting electron transport between soma of SNc dopamine neurons and LC norepinephrine neurons is evidence that a function may be associated with that structure, even if it is different from the hypothesis presented in this chapter.

References

1. Hopfield, J. J. Electron transfer between biological molecules by thermally activated tunneling. *Proc. Natl. Acad. Sci. U. S. A.* **1974**, *71*, 3640.
2. Stuchebrukhov, A. A Long Distance Electron Tunnelling in Proteins. *Laser Phys.* **2010**, *1*, 125–138.
3. Earhart, R. The Sparking Distances Between Plates for Small Distances. *Philosoph. Mag. J. Sci.* **1901**, *1* (6th Series), 147–159.
4. Born, M. Theory of Nuclear Disintegration. *Z. f. Phys.* **1929**, *58*, 306.
5. Gorter, C. *Physica* **1951**, *17*, 777.
6. Ekimov, A. I.; Onushchenko, A. A. Quantum Size Effect in Three-Dimensional Microscopic Semiconductor Crystals. *JETP Lett.* **1981**, *34*, 345–349.
7. Hennequin, B. *Aqueous Near Infrared Fluorescent Composites Based on Apoferritinencapsulated PbS Quantum Dots*; PhD thesis, University of Nottingham, 2008. http://eprints.nottingham.ac.uk/11071/1/Thesis_B._Hennequin.pdf. (Accessed 18 April 2018).
8. Reed, M.; Randall, J.; Aggarwall, R.; Matyi, R.; Moore, T.; Wetsel, A. Observation of discrete electronic states in a zero-dimensional semiconductor nanostructure. *Phys. Rev. Lett.* **1988**, *60*, 535.
9. McFadden, J.; Al-Khalili, J. The Origins of Quantum Biology. *Proc. R. Soc. A* **2018**, *474*, 2220.
10. Löwdin, P. Some Aspects of Quantum Biology. *Biopolym. Symp.* **1964**, *1*, 293–311.
11. Fleming, G.; Scholes, G. Quantum Mechanics for Plants. *Nature* **2004**, *431*, 256–257.
12. Olšina, J.; Dijkstra, A.; Wang, C.; Cao, J. *Can Natural Sunlight Induce Coherent Exciton Dynamics?*; 2014. arXiv:1408.5385 [physics.chem-ph].
13. Bera, S.; Kolay, J.; Pallabi Pramanik, P.; Bhattacharyyab, A.; Mukhopadhyay, R. Long-Range Solid-State Electron Transport Through Ferritin Multilayers. *J. Mater. Chem. C* **2019**, 7, 9038–9048.
14. Kell, D. B.; Pretorius, E. Serum Ferritin is an Important Inflammatory Disease Marker, as it is Mainly a Leakage Product From Damaged Cells. *Metallomics* **2014**, *6*, 748–773.
15. Carmona, U.; Li, L.; Zhang, L. B.; Knez, M. Ferritin Light-Chain Subunits: Key Elements for the Electron Transfer across the Protein Cage. *Chem. Commun.* **2014**, *50*, 15358–15361.
16. Awschalom, D. D.; Smyth, J. F.; Grinstein, G.; DiVincenzo, D. P.; Loss, D. Macroscopic Quantum Tunneling in Magnetic Proteins. *Phys. Rev. Lett.* **1992**, *68*, 3092–3095.
17. Tejada, J.; Zhang, X. X.; del Barco, E.; Hernández, J. M.; Chudnovsky, E. M. Macroscopic Resonant Tunneling of Magnetization in Ferritin. *Phys. Rev. Lett.* **1997**, *79*, 1754–1757.
18. Schäfer-Nolte, E.; Schlipf, L.; Ternes, M.; Reinhard, F.; Kern, K.; Wrachtrup, J. Tracking Temperature-Dependent Relaxation Times of Ferritin Nanomagnets With a Wideband Quantum Spectrometer. *Phys. Rev. Lett.* **2014**, *113*, 217204.
19. Kumar, K. S.; Pasula, R. R.; Lim, S.; Nijhuis, C. A. Long Range Tunneling Processes Across Ferritin-Based Junctions. *Adv. Mater.* **2016**, *28*, 1824–1830.
20. Axford, D.; Davis, J. J. Electron Flux Through Apo-And Holoferritin. *Nanotechnology* **2007**, *18*, 145502.

21. Rakshit, T.; Mukhopadhyay, R. Solid-State Electron Transport in Mn-, Co-, holo-, and Cu-Ferritins: Force-Induced Modulation is Inversely Linked to the Protein Conductivity. *J. Colloid Interface Sci.* **2012**, *388*, 282–292.
22. Bostick, C. D.; Mukhopadhyay, S.; Pecht, I.; Sheves, M.; Cahen, D.; Lederman, D. Protein Bioelectronics: A Review of What We Do and Do Not Know. *Rep. Prog. Phys.* **2018**, *81*, 026601.
23. Fittipaldi, M.; Innocenti, P.; Ceci, P.; Sangregorio, C.; Castelli, L.; Sorace, L.; Gatteschi, D. Looking for Quantum Effects in Magnetic Nanoparticles Using the Molecular Nanomagnet Approach. *Phys. Rev. B* **2011**, *83*, 104409.
24. Gossuin, Y.; Muller, R.; Gillis, P. Relaxation Induced by Ferritin: A Better Understanding for an Improved MRI Iron Quantification. *NMR Biomed.* **2004**, *17*, 427–432.
25. Tackeuchi, A.; Kuroda, T.; Yamaguchi, K.; Nakata, Y.; Yokoyama, N.; Takagahara, T. Spin Relaxation and Antiferromagnetic Coupling in Semiconductor Quantum Dots. *Physica E* **2006**, *32*, 354–358.
26. Papaefthymiou, G. C. The Mössbauer and Magnetic Properties of Ferritin Cores. *Biochim. Biophys. Acta* **2010**, *1800*, 886–897.
27. Cole, J. H.; Hollenberg, L. C. Scanning Quantum Decoherence Microscopy. *Nanotechnology* **2009**, *20*, 495401.
28. Moro, F.; Turyanska, L.; Wilman, J.; Fielding, A. J.; Fay, M. W.; Granwehr, J.; Patane, A. Electron Spin Coherence Near Room Temperature in Magnetic Quantum Dots. *Sci. Rep.* **2015**, *5*, 10855.
29. Caram, J. R.; Zheng, H.; Dahlberg, P. D.; Rolczynski, B. S.; Griffin, G. B.; Fidler, A. F.; Dolzhnikov, D. S.; Talapin, D. V.; et al. Persistent Interexcitonic Quantum Coherence in CdSe Quantum Dots. *J. Phys. Chem. Lett.* **2014**, *5*, 196–204.
30. Colton, J. S.; Erickson, S. D.; Smith, T. J.; Watt, R. K. Sensitive Detection of Surface- and Size-Dependent Direct and Indirect Band Gap Transitions in Ferritin. *Nanotechnology* **2014**, *25*, 135703.
31. Smith, T. J. *The Synthesis and Characterization Of Ferritin Bio Minerals for Photovoltaic, Nanobattery, and Bio-Nano Propellant Applications*; Brigham Young University, 2015. https://scholarsarchive.byu.edu/etd/6045. (Accessed 17 April 2018).
32. Xu, D.; Watt, G.; Harb, J.; Davis, R. Electrical Conductivity of Ferritin Proteins by Conductive AFM. *Nano Lett.* **2005**, *5* (4), 571–577.
33. Kolay, J.; Bera, S.; Rakshit, T.; Mukhopadhyay, R. Negative Differential Resistance Behavior of the Iron Storage Protein Ferritin. *Langmuir* **2018**, *34* (9), 3126–3135.
34. Schwartz, J. R.; Roth, T. Neurophysiology of Sleep and Wakefulness: Basic Science and Clinical Implications. *Curr. Neuropharmacol.* **2008**, *6*, 367–378.
35. Oades, R. D.; Halliday, G. M. Ventral Tegmental (A10) System: Neurobiology. 1. Anatomy and Connectivity. *Brain Res.* **1987**, *12*, 117–165.
36. Margolis, E. B.; Lock, H.; Hjelmstad, G. O.; Fields, H. L. The Ventral Tegmental Area Revisited: Is There an Electrophysiological Marker for Dopaminergic Neurons? *J. Physiol.* **2006**, *577* (Pt 3), 907–924.
37. Damier, P.; Hirsh, E. C.; Agid, Y.; Graybiel, A. M. The Substantia Nigra of the Human Brain. I. Nigrosomes and the Nigral Matrix, a Compartmental Organization Based on Calbindin D28K Immunohistochemistry. *Brain* **1999**, *122*, 1421–1436.
38. Haining, R. L.; Achat-Mendes, C. Neuromelanin, One of the Most Overlooked Molecules in Modern Medicine, is not a Spectator. *Neural Regen. Res.* **2017**, *12*, 372–375.
39. Grillner, S.; Robertson, B. The Basal Ganglia Over 500 Million Years. *Curr. Biol.* **2016**, *26* (20), R1088.
40. Reiner, A. The Conservative Evolution of the Vertebrate Basal Ganglia. In *Handbook of Behavioral Neuroscience, The Conservative Evolution of the Vertebrate Basal Ganglia*; 2010.

41. Zecca, L.; Bellei, C.; Costi, P.; Albertini, A.; Monzani, E.; Casella, L.; Gallorini, M.; Bergamaschi, L.; et al. New Melanic Pigments in the Human Brain That Accumulate in Aging and Block Environmental Toxic Metals. *Proc. Natl. Acad. Sci. U. S. A.* **2008**, *105*, 17567–17572.
42. Crippa, P. R.; Cristofoletti, V.; Romeo, N. A Band Model for Melanin Deduced From Optical Absorption and Photoconductivity Experiments. *Biochim. Biophys. Acta* **1978**, *538*, 164–170.
43. Obeid, M. T.; Hussain, W. A. Optical Properties of Pure Synthetic Melanin and Melanin Doped With Iodine and Sodium-Borohydride. *Arch. Phys. Res.* **2013**, *4*, 40–46.
44. Mostert, A. B.; Powell, B. J.; Pratt, F. L.; Hanson, G. R.; Sama, T.; Gentle, I. R.; Meredith, P. Role of Semiconductivity and Ion Transport in the Electrical Conduction of Melanin. *Proc. Natl. Acad. Sci. U. S. A.* **2012**, *109*, 8943–8947.
45. Ito, S. Encapsulation of a Reactive Core in Neuromelanin. *Proc. Natl. Acad. Sci. U. S. A.* **2006**, *103*, 14647–14648.
46. Said, M.; Galadanci, G.; Babaji, G. *Study of the Electronic and Transport Properties of Melanin Using Density Functional Theory (DFT), EuroSciCon Conference on Chemistry and Green Chemistry Research*; 2018.
47. Brash, D.; Goncalves, L. Bechara E (2018), Chemiexcitation and Its Implications for Disease. *Trends Mol. Med.* **2018**, *24* (6), 527–541.
48. Tribl, F.; Asan, E.; et al. Identification of L-Ferritin in Neuromelanin Granules of the Human Substantia Nigra: A Targeted Proteomics Approach. *Mol. Cellular Proteomics MCP.* **2009**, *8*, 1832–1838.
49. Sulzer, D.; Cassidy, C.; Horga, G.; et al. Neuromelanin Detection by Magnetic Resonance Imaging (MRI) and Its Promise as a Biomarker for Parkinson's Disease. *Parkinson's Disease* **2018**, *4*, 11.
50. Zecca, L.; Tampellini, D.; Gerlach, M.; Riederer, P.; Fariello, R. G.; Sulzer, D. Substantia Nigra Neuromelanin: Structure, Synthesis, and Molecular Behaviour. *Mol. Pathol.* **2001**, *54*, 414–418.
51. Bolzoni, F.; Giraudo, S.; Lopiano, L.; Bergamasco, B.; Fasano, M.; Crippa, P. R. Magnetic Investigations of Human Mesencephalic Neuromelanin. *Biochim. Biophys. Acta, Mol. Basis Dis.* **2002**, *1586* (2), 210–218.
52. Berezin, A. Two- and Three-Dimensional Kronig-Penney Model With Δ-Function-Potential Wells of Zero Binding Energy. *Phys. Rev. B* **1986**, *33*, 2122.
53. Chang, S.; Chien, C.; Jeng, B. An Efficient Algorithm for the Schrodinger-Poisson Eigenvalue Problem. *J. Comput. Appl. Math.* **2007**, *205* (1), 509–532.
54. Halliday, G.; Ophof, A.; et al. A-Synuclein Redistributes To Neuromelanin Lipid In The Substantia Nigra Early In Parkinson's Disease. *Brain* **2005**, *128* (11), 2654–2664.
55. Gibb, W. R.; Lees, A. J. Anatomy, Pigmentation, Ventral and Dorsal Subpopulations of the Substantia Nigra, and differential cell death in Parkinson's Disease. *J. Neurol. Neurosurg. Psychiatry* **1991**, *54*, 388–396.
56. Bertini, I.; Lalli, D.; Mangani, S.; Pozzi, C.; Rosa, C.; Theil, E. C.; Turano, P. Structural Insights Into the Ferroxidase Site of Ferritins From Higher Eukaryotes. *J. Am. Chem. Soc.* **2012**, *134*, 6169–6176.
57. Choi, S. H.; Kim, J.-W.; Chu, S.-H.; Park, Y.; King, G. C.; Lillehei, P. T.; Kim, S.-J.; Elliot, J. R. Ferritin-Templated Quantum-Dots for Quantum Logic Gates. *Proc. SPIE* **2005**, *5763*, 213–232.
58. Reinert, A.; Morawski, M.; Seeger, J.; et al. Iron Concentrations in Neurons and Glial Cells With Estimates on Ferritin Concentrations. *BMC Neurosci.* **2019**, *20*, 25.
59. El Hady, A.; Machta, B. B. Mechanical Surface Waves Accompany Action Potential Propagation. *Nat. Commun.* **2015**, *6*, 6697.
60. Pissadaki, E. K.; Bolam, J. P. The Energy Cost of Action Potential Propagation in Dopamine Neurons: Clues to Susceptibility in Parkinson's Disease. *Front. Comput. Neurosci.* **2013**, 7, 13.

61. Matsuda, W.; Furuta, T.; Nakamura, K. C.; Hioki, H.; Fujiyama, F.; Arai, R.; Kaneko, T. Single Nigrostriatal Dopaminergic Neurons Form Widely Spread and Highly Dense Axonal Arborizations in the Neostriatum. *J. Neurosci.* **2009**, *29*, 444–453.
62. Colpan, M. E.; Slavin, K. V. Subthalamic and Red Nucleus Volumes in Patients With Parkinson's Disease: Do They Change With Disease Progression? *Parkinsonism Relat. Disord.* **2010**, *16*, 398–403.
63. Joensson, M.; Thomsen, K. R.; Andersen, L. M.; Gross, J.; Mouridsen, K.; Sandberg, K.; Ostergaard, L.; Lou, H. C. Making Sense: Dopamine Activates Conscious Self-Monitoring Through Medial Prefrontal Cortex. *Hum. Brain Mapp.* **2015**, *36*, 1866–1877.
64. Parvizi, J.; Damasio, A. Consciousness and the Brainstem. *Cognition* **2001**, *79*, 135–160.
65. T1, S.; Ishida, Y.; Isobe, K. I. Age-Dependent Changes in Axonal Branching of Single Locus Coeruleus Neurons Projecting to Two Different Terminal Fields. *J. Neurophysiol.* **2000**, *84*, 1120–1122.
66. Alreja, M.; Aghajanian, G. K. Pacemaker Activity of Locus Coeruleus Neurons: Whole-Cell Recordings in Brain Slices Show Dependence on cAMP and Protein Kinase A. *Brain Res.* **1991**, *556*, 339–343.
67. Rudow, G.; O'Brien, R.; Savonenko, A.; et al. Morphometry of the Human Substantia Nigra in Ageing and Parkinson's Disease. *Acta Neuropathol.* **2008**, *115* (4), 461–470.
68. Schomburg, E.; Anastassiou, C.; Buzsáki, G.; Koch, C. The Spiking Component of Oscillatory Extracellular Potentials in the Rat Hippocampus. *J. Neurosci.* **2012**, *32* (34), 11798–11811.
69. Auerbach, A. A.; Bennett, M. V. A Rectifying Electrotonic Synapse in the Central Nervous System of a Vertebrate. *J. Gen. Physiol.* **1969**, *53*, 211–237.
70. Freed, K. F. A Self-Consistent Field Theory of Electron Localization in Disordered Systems: The Anderson Transition. *J. Phys. C Solid State Phys.* **1971**, *4*, L331.
71. Svirskis, G.; Gutman, A.; Hounsgaard, J. Electrotonic Structure of Motoneurons in the Spinal Cord of the Turtle: Inferences for the Mechanisms of Bistability. *J. Neurophysiol.* **2001**, *85*, 391–398.
72. Schwindt, P.; Crill, W. A Persistent Negative Resistance in Cat Lumbar Motoneurons. *Brain Res.* **1977**, *120*, 173–178.
73. Marzo, A.; Totah, N. K.; Neves, R. M.; Logothetis, N. K.; Eschenko, O. Unilateral Electrical Stimulation of Rat Locus Coeruleus Elicits Bilateral Response of Norepinephrine Neurons and Sustained Activation of Medial Prefrontal Cortex. *J. Neurophysiol.* **2014**, *111*, 2570–2588.
74. Blythe, S. N.; Wokosin, D.; Atherton, J. F.; Bevan, M. D. Cellular Mechanisms Underlying Burst Firing in Substantia Nigra Dopamine Neurons. *J. Neurosci.* **2009**, *29* (49), 15531–15541. 2009.
75. Menegas, W.; Akiti, K.; Amo, R.; Uchida, N.; Watabe-Uchida, M. Dopamine Neurons Projecting to the Posterior Striatum Reinforce Avoidance of Threatening Stimuli. *Nat. Neurosci.* **2018**, *21*, 1421–1430.
76. Aston-Jones, G.; Cohen, J. An Integrative Theory of Locus Coeruleus-Norepinephrine Function: Adaptive Gain and Optimal Performance. *Annu. Rev. Neurosci.* **2005**, *28*, 403–450.
77. Schwarz, L.; Luo, L. Organization of the Locus Coeruleus-Norepinephrine System. *Curr. Biol.* **2015**, *25*, R1051–R1056.
78. Samuels, E.; Szabadi, E. Functional Neuroanatomy of the Noradrenergic Locus Coeruleus: Its Roles in the Regulation of Arousal and Autonomic Function Part II: Physiological and Pharmacological Manipulations and Pathological Alterations of Locus Coeruleus Activity in Humans. *Curr. Neuropharmacol.* **2008**, *6*, 254–285.
79. Brightwell, J.; Taylor, B. Noradrenergic Neurons in the Locus Coeruleus Contribute to Neuropathic Pain. *Neuroscience* **2009**, *160*, 174–185.

80. Tyner, K. M.; Kopelman, R.; Philbert, M. A. "Nanosized Voltmeter" Enables Cellular-Wide Electric Field Mapping. *Biophys. J.* **2007**, *93*, 1163–1174.
81. Gatenby, R.; Frieden, B. Coulomb Interactions Between Cytoplasmic Electric Fields and Phosphorylated Messenger Proteins Optimize Information Flow in Cells. *PLoS One* **2010**, *5* (8), e12084.
82. Tuszynski, J.; Wenger, C.; Friesen, D.; Preto, J. An Overview of Sub-Cellular Mechanisms Involved in the Action of TTFields. *Int. J. Environ. Res. Public Health* **2016**, *13* (11), 1128.
83. Nenashev, A. V.; Wiemer, M.; Jansson, F.; Baranovskii, S. D. Theory of Exciton Dissociation at the Interface Between a Conjugated Polymer and an Electron Acceptor. *J. Non Cryst. Solids* **2012**, *358*, 2508–2511.
84. Guzman, J. N.; Sanchez-Padilla, J.; Chan, C. S.; Surmeier, D. J. Robust Pacemaking in Substantia Nigra Dopaminergic Neurons. *J. Neurosci.* **2009**, *29*, 11011–11019.
85. Yan, H.; Bergren, A.; McCreery, R.; Della Rocca, M. L.; Martin, P.; Lafarge, P.; Lacroix, J. Activationless Charge Transport Across 4.5 to 22 nm in Molecular Electronic Junctions. *PNAS* **2013**, *110* (14), 5326–5330.
86. Garrigues A, Wang L, del Barco E, Nijhuis C (2016), Electrostatic Control Over Temperature-Dependent Tunnelling Across a Single-Molecule Junction. Nat. Commun. 7, 11595.
87. Galazka-Friedman, J.; Bauminger, E.; Szlachta, K.; Friedman, A. The Role of Iron in Neurodegeneration—Mössbauer Spectroscopy, Electron Microscopy, Enzyme-Linked Immunosorbent Assay and Neuroimaging Studies. *J. Phys. Condens. Matter* **2012**, *24*, 24.
88. Theil, E. C. Ferritin Protein Nanocages Use Ion Channels, Catalytic Sites, and Nucleation Channels to Manage Iron/ Oxygen Chemistry. *Curr. Opin. Chem. Biol.* **2011**, *15*, 304–311.
89. Theil, E. C. Ferritin: The Protein Nanocage and Iron Biomineral in Health and in Disease. *Inorg. Chem.* **2013**, *52*, 12223–12233.
90. Bou-Abdallah, F.; Zhao, G.; Biasiotto, G.; Poli, M.; Arosio, P.; Chasteen, N. D. Facilitated Diffusion of Iron(II) and Dioxygen Substrates Into Human H-Chain Ferritin. A Fluorescence and Absorbance Study Employing the Ferroxidase Center Substitution Y34W. *J. Am. Chem. Soc.* **2008**, *130*, 17801–17811.
91. Tosha, T.; Ng, H. L.; Bhattasali, O.; Alber, T.; Theil, E. C. Moving Metal Ions Through Ferritin-Protein Nanocages From Three-Fold Pores to Catalytic Sites. *J. Am. Chem. Soc.* **2010**, *132*, 14562–14569.
92. Turano, P.; Lalli, D.; Felli, I. C.; Theil, E. C.; Bertini, I. NMR Reveals Pathway for Ferric Mineral Precursors to the Central Cavity of Ferritin. *Proc. Natl. Acad. Sci. U. S. A.* **2010**, *107*, 545–550.
93. Sala, D.; Ciambellotti, S.; Giachetti, A.; Turano, P.; Rosato, A. Investigation of the Iron(II) Release Mechanism of Human H-Ferritin as a Function of pH. *J. Chem. Inf. Model.* **2017**, *57* (9), 2112–2118.
94. Lopin, K. V.; Gray, P.; Obejero-Paz, C. A.; Thévenod, F.; Jones, S. W. Fe2+ Block and Permeation of CaV3.1 (α1G) T-Type Calcium Channels: Candidate Mechanism for Non-Transferrin-Mediated Fe2+ Influx. *Mol. Pharmacol.* **2012**, *82*, 1194–1204.
95. Riegel, A. C.; Williams, J. T. CRF Facilitates Calcium Release From Intracellular Stores in Midbrain Dopamine Neurons. *Neuron* **2008**, *57*, 559–570.
96. Hidalgo, C.; Donoso, P.; Carrasco, M. A. The Ryanodine Receptors Ca2+ Release Channels: Cellular Redox Sensors? *IUBMB Life* **2005**, *57*, 315–322.
97. Hidalgo, C.; Núñez, M. Calcium, Iron and Neuronal Function. *IUBMB Life* **2007**, *59*, 280–285.
98. Xuan, W.; Pan, R.; Wei, Y.; Cao, Y.; Li, H.; Liang, F.; Liu, F.; Wang, W. Reaction-Based "Off–On" Fluorescent Probe Enabling Detection of Endogenous Labile Fe2+ and Imaging of Zn2+-induced Fe2+ Flux in Living Cells and Elevated Fe2+ in Ischemic Stroke. *Bioconjug. Chem.* **2016**, *27* (2), 302–308.

99. Hare, D.; New, E.; de Jonge, M.; McColl, G. Imaging Metals in Biology: Balancing Sensitivity, Selectivity and Spatial Resolution. *Chem. Soc. Rev.* **2015**, *44* (17), 5941–5958.
100. Berridge, C.; Waterhouse, B. The Locus Coeruleus-Noradrenergic System: Modulation of Behavioral State and State-Dependent Cognitive Processes. *Brain Res. Brain Res. Rev.* **2003**, *42* (1), 33–84.
101. Yung, W.; Häusser, M.; Jack, J. Electrophysiology of Dopaminergic and Non-Dopaminergic Neurones of the Guinea-Pig Substantia Nigra Pars Compacta In Vitro. *J. Physiol.* **1991**, *436*, 643–667.
102. Zucca, F. A.; Bellei, C.; Giannelli, S.; Terreni, M. R.; Gallorini, M.; Rizzio, E.; Pezzoli, G.; Albertini, A.; et al. Neuromelanin and Iron in Human Locus Coeruleus and Substantia Nigra During Aging: Consequences for Neuronal Vulnerability. *J. Neural Transm. (Vienna)* **2006**, *113*, 757–767.
103. Davis, G. C.; Williams, A. C.; Markey, S. P.; Ebert, M. H.; Caine, E. D.; Reichert, C. M.; Kopin, I. J. Chronic Parkinsonism Secondary to Intravenous Injection of Meperidine Analogues. *Psychiatry Res.* **1979**, *1*, 249–254.
104. Langston, J. W. The MPTP Story. *J. Parkinsons Dis.* **2017**, 7, S11–S22.
105. Gesi, M.; Soldani, P.; Giorgi, F. S.; Santinami, A.; Bonaccorsi, I.; Fornai, F. The Role of the Locus Coeruleus in the Development of Parkinson's Disease. *Neurosci. Biobehav. Rev.* **2000**, *24*, 655–668.
106. Langston, J. W.; Ballard, P. Parkinsonism Induced by 1-Methyll-4-Phenyl-1,2,3,6-Tetrahydropyridine (MPTP): Implications for Treatment and the Pathogenesis of Parkinson's Disease. *Can. J. Neurol. Sci.* **1984**, *11* (1 Suppl), 160–165.
107. Mattson, M. P. Parkinson's Disease: Don't Mess With Calcium. *J. Clin. Invest.* **2012**, *122* (4), 1195–1198.
108. Morris, M. E. Movement Disorders in People With Parkinson Disease: A Model for Physical Therapy. *Phys. Ther.* **2000**, *80* (6), 578–597.
109. Hämmerer, D.; Callaghan, M. F.; Hopkins, A.; Kosciessa, J.; Betts, M.; Cardenas-Blanco, A.; Kanowski, M.; Weiskopf, N.; et al. Locus Coeruleus Integrity Predicts Memory in Aging. *Proc. Natl. Acad. Sci. U. S. A.* **2018**, *115*, 2228–2233.
110. Rommelfanger, K. S.; Edwards, G. L.; Freeman, K. G.; Liles, L. C.; Miller, G. W.; Weinshenker, D. Norepinephrine Loss Produces More Profound Motor Deficits Than MPTP Treatment in Mice. *Proc. Natl. Acad. Sci. U. S. A.* **2007**, *104*, 13804–13809.
111. Kronenburg, A.; Spliet, W.; Broekman, M.; Robe, P. Locus Coeruleus Syndrome as a Complication of Tectal Surgery. *BMJ Case Rep.* **2015**, Apr 22.
112. Korf, J.; Aghajanian, G. K.; Roth, R. H. Stimulation and Destruction of the Locus Coeruleus: Opposite Effects on 3-methoxy-4-hydroxyphenylglycol Sulfate Levels in the Rat Cerebral Cortex. *Eur. J. Pharmacol.* **1973**, *21*, 305–310.
113. Omura, D. T.; Clark, D. A.; Samuel, A. D.; Horvitz, H. R. Dopamine Signaling is Essential for Precise Rates of Locomotion by C. elegans. *PLoS One* **2012**, 7, e38649.
114. Anderson, C. P.; Leibold, E. A. Mechanisms of Iron Metabolism in CAENORHABDITIS ELEGANS. *Front. Pharmacol.* **2014**, *5*, 113.
115. Altun, Z.; Herndon, L.; Wolkow, C.; Crocker, C.; Lints, R.; Hall, D. H. *WormAtlas*; 2002–2019. http://www.wormatlas.org.
116. Itzev, D. E.; Ovtscharoff, W.; Marani, E.; Usunoff, K. G. Neuromelanin-Containing, Catecholaminergic Neurons in the Human Brain: Ontogenetic Aspects, Development and Aging. *Biomed. Rev.* **2002**, *13*, 39–47.
117. Fyffe, W. E.; Kronz, J. D.; Edmonds, P. A.; Donndelinger, T. M. Effect of High-level Oxygen Exposure on the Peroxidase Activity and the Neuromelanin-Like Pigment Content of the Nerve Net in the Earthworm, Lumbricus terrestris. *Cell Tissue Res.* **1999**, *295*, 349–354.

118. Kornhuber, H. H.; Deecke, L. Brain Potential Changes in Voluntary and Passive Movements in Humans: Readiness Potential and Reafferent Potentials. *Pflügers Archiv: Eur. J. Physiol.* **2016**, *468* (7), 1115–1124.
119. Jahanshahi, M.; Jones, C.; Dirnberger, G.; Frith, C. The Substantia Nigra Pars Compacta and Temporal Processing. *J. Neurosci.* **2006**, *26* (47), 12266–12273.
120. Vaishali, P.; Shastri, M.; Baig, N. Effect of Parkinson's Disease on audiovisual Reaction Time in Indian Population. *Int. J. Biol. Med. Res.* **2012**, *3* (1), 1392–1396.
121. Clayton, E.; Rajkowski, J.; Cohen, J. Aston-Jones G (2004), Phasic Activation of Monkey *locus ceruleus* Neurons by Simple Decisions in a Forced-Choice Task. *J. Neurosci.* **2004**, *24* (44), 9914–9920.
122. Logan, G.; Cowan, W.; Davis, K. On the Ability to Inhibit Simple and Choice Reaction Time Responses: A Model and a Method. *J. Exp. Psychol. Hum. Percept. Perform.* **1984**, *10* (2), 276–291.
123. Nolte, J. *The Human Brain: An Introduction to Its Functional Anatomy*, 5th Ed; Mosby, 2002.
124. Menegas, W.; Bergan, J.; Ogawa, S.; Isogai, Y.; Umadevi Venkataraju, K.; Osten, P.; Uchida, N.; Watabe-Uchida, M. Dopamine Neurons Projecting to the Posterior Striatum Form an Anatomically Distinct Subclass. *Elife* **2015**, *31*, 4.
125. Merker B (2007), Consciousness Without a Cerebral Cortex: A Challenge for Neuroscience and Medicine. Behav. Brain Sci. 30(1):63-81; discussion 81-134.
126. Merker, B. The Efference Cascade, Consciousness, and Its Self: Naturalizing the First Person Pivot of Action Control. *Front. Psychol.* **2013**, *4*, 501.
127. Penfield, W.; Jasper, H. H. *Epilepsy and the Functional Anatomy of the Human Brain*; Little, Brown & Co: Boston, 1954.
128. Brookes, J. C. Quantum Effects in Biology: Golden Rule in Enzymes, Olfaction, Photosynthesis and Magnetodetection. *Proc. Math. Phys. Eng. Sci.* **2017**, *473*, 20160822.
129. Bordonaro M, Ogryzko V, Quantum Biology at the Cellular Level—Elements of the Research Program. Biosystems, 112:11–30.
130. Jedlicka, P. Revisiting the Quantum Brain Hypothesis: Toward Quantum (Neuro) biology? *Front. Mol. Neurosci.* **2017**, *10*, 366.
131. Fleming, G.; Scholes, G.; Cheng, Y. Quantum Effects in Biology. *Procedia Chem.* **2011**, *3*, 38–57.
132. Hameroff, S.; Penrose, R. Consciousness in the Universe: A Review of the 'Orch OR' Theory. *Phys. Life Rev.* **2014**, *11* (1), 39–78.
133. Fisher, M. Quantum Cognition: The Possibility of Processing With Nuclear Spins in the Brain. *Ann. Phys. (N Y)* **2015**, *362*, 593–602.
134. Weingarten, C.; Doraiswamy, M.; Fisher, M. A New Spin on Neural Processing: Quantum Cognition. *Front. Hum. Neurosci.* **2016**, *10*, 541.
135. Khabibullin, A.; Alizadehgiashi, M.; Khuu, N.; Prince, E.; Tebbe, M.; Kumacheva, E. Injectable Shear-Thinning Fluorescent Hydrogel Formed by Cellulose Nanocrystals and Graphene Quantum Dots. *Langmuir* **2017**, *33*, 12344–12350.
136. Das, M. P. A. Mesoscopic Systems in the Quantum Realm: Fundamental Science and Applications. *Adv. Nat. Sci. Nanosci. Nanotechnol.* **2010**, *1*, 4.
137. Meshik, X. Quantum Dot- and Aptamer-Based Nanostructures for Biological Applications, thesis. In *University of Illinois at Chicago*; 2015.
138. Schliwa, A.; Hönig, G.; Bimberg, D. Electronic Properties of III-V Quantum Dots. *Lect. Notes Comput. Sci. Eng.* **2014**, *94*, 57–85.
139. Burkard, G.; Loss, D.; DiVincenzo, D. P. Coupled Qua Ntum Dots as Quantum Gates. *Phys. Rev. B* **1999**, *59*, 2070–2078.
140. Sun, K.; Li, Y.; Stroscio, M. A.; Dutta, M. Miniband Formation in Superlattices of Colloidal Quantum Dots and Conductive Polymers. *ECS Trans.* **2008**, *6*, 1–12.

141. Lazarenkova, O. L.; Balandin, A. A. Electron and Phonon Energy Spectra in a Three-Dimensional Regimented Quantum Dot Superlattice. *Phys. Rev. B* **2002**, *66*, 245319-1–245319-9.
142. Mahler, G.; Wawer, R. Quantum Networks: Dynamics of Open Nanostructures. *VLSI Design* **1998**, *8*, 191–196.
143. Gomez, I.; Domínguez-Adame, F.; Diez, E.; Orellana, P. Transport in Random Quantum Dot Superlattices. *J. Appl. Phys.* **2002**, *92*, 4486–4489.
144. Nozik, A. J.; Beard, M. C.; Luther, J. M.; Law, M.; Ellingson, R. J.; Johnson, J. C. Semiconductor Quantum Dots and Quantum Dot Arrays and Applications of Multiple Exciton Generation to Third-Generation Photovoltaic Solar Cells. *Chem. Rev.* **2010**, *110*, 6873–6890.
145. Bradshaw, L.; Leavitt, R. *The Spatial Coherence of Electron Wavefunctions and the Transition From Miniband to Stark-Ladder Electric Field Regimes in InGaAs/lnAIAs-on-lnP Superlattices*; Army Research Laboratory. ARL-TR-1536, 1998. http://www.dtic.mil/get-tr-doc/pdf?AD=ADA337751. (Accessed 17 April 2018).
146. Khituna, A.; Balandinb, A.; Liuc, J. L.; Wang, K. L. The Effect of the Long-Range Order in a Quantum Dot Array on the In-Plane Lattice Thermal Conductivity. *Superlattice. Microst.* **2001**, *30*, 1–8.
147. Jongen, M. *Quantum Field Theory on a Random Lattice: The Propagator on Percolation Lattices*; Radboud University Nijmegen, 2013. http://www.ru.nl/publish/pages/760962/master_thesis_martijn_jongen.pdf. (Accessed 17 April 2018).
148. Park, J.; Prabhakaran, P.; et al. Photopatternable Quantum Dots Forming Quasi-Ordered Arrays. *Nano Lett.* **2010**, *10* (7), 2310–2317.
149. Jaskolski, W.; Zielinski, M.; Bryant, G. W. Coupling and Strain Effects in Vertically Stacked Double InAs/GaAs Quantum Dots: Tight-Binding Approach. *Acta Phys. Pol., A* **2004**, *106*, 193–205.
150. Cockins, L.; Miyahara, Y.; Bennett, S.; Clerk, A.; Studenikin, S.; Poole, P.; Sachrajda, A.; Grutter, P. Energy Levels of Few-Electron Quantum Dots Imaged and Characterized by Atomic Force Microscopy. *Proc. Natl. Acad. Sci.* **2010**, *107* (21), 9496–9501.
151. Rourk, C. Indication of Quantum Mechanical Electron Transport in Human Substantia Nigra Tissue From Conductive Atomic Force Microscopy Analysis. *Biosystems* **2019**, *179*, 30–38.
152. Guzelturk, B.; Hernandez-Martinez, P. L.; Sharma, V. K.; Coskun, Y.; Ibrahimova, V.; Tuncel, D.; Govorov, A. O.; Sun, X. W.; et al. Study of Exciton Transfer in Dense Quantum Dot Nanocomposites. *Nanoscale* **2014**, *6*, 11387–11394.
153. Su, Y. W.; Lin, W. H.; Hsu, Y. J.; Wei, K. H. Conjugated Polymer/Nanocrystal Nanocomposites for Renewable Energy Applications in Photovoltaics and Photocatalysis. *Small* **2014**, *10*, 4427–4442.
154. Kisslinger, R.; Hua, W.; Shankar, K. Bulk Heterojunction Solar Cells Based on Blends of Conjugated Polymers with II—VI and IV—VI Inorganic Semiconductor Quantum Dots. *Polymers* **2017**, *9* (2), 35.
155. Lattante, S. Electron and Hole Transport Layers: Their Use in Inverted Bulk Heterojunction Polymer Solar Cells. *Electronics* **2014**, *3*, 132–164.
156. Ruizhi, W.; Yan, X.; Wang, Y.; Li, H.; Sheng, C. Long Lived Photoexcitation Dynamics in π-Conjugated Polymer/PbS Quantum Dot Blended Films for Photovoltaic Application. *Polymers* **2017**, *9*, 352.
157. Konstantatos, G.; Sargent, E. H. Solution-Processed Quantum Dot Photodetectors. *Proc. IEEE* **2009**, *97*, 1666–1683.
158. Barbagiovanni, E.; Lockwood, D.; Simpson, P.; Goncharova, L. Quantum Confinement in Si and Ge Nanostructures: Theory and Experiment. *Appl. Phys. Rev.* **2014**, *1*, 011302.

159. Gun, L.; Ning, D.; Liang, Z. Effective Permittivity of Biological Tissue: Comparison of Theoretical Model and Experiment. *Math. Probl. Eng.* **2017**, *2017*, 7249672.
160. Li, C.; Zhang, Z.; Alexov, E. On the Dielectric "Constant" of Proteins: Smooth Dielectric Function for Macromolecular Modeling and Its Implementation in DelPhi. *J. Chem. Theory Comput.* **2013**, *9* (4), 2126–2136.
161. Proppe, A. H.; et al. Picosecond Charge Transfer and Long Carrier Diffusion Lengths in Colloidal Quantum Dot Solids. *Nano Lett.* **2018**, *18*, 7052–7059.
162. Zallo, E.; Trotta, R.; Krapek, V.; Huo, Y. H.; Atkinson, P.; Ding, F.; Sikola, T.; Rastelli, A.; Schmidt, O. G. Strain-Induced Active Tuning of the Coherent Tunneling in Quantum Dot Molecules. *Phys. Rev. B* **2014**, *89*, 241303.
163. Zhang, J.; Wildmann, J. S.; Ding, F.; Trotta, R.; Huo, Y.; Zallo, E.; Huber, D.; Rastelli, A.; et al. High Yield and Ultrafast Sources of Electrically Triggered Entangled-Photon Pairs Based on Strain-Tunable Quantum Dots. *Nat. Commun.* **2015**, *6*, 10067.
164. Wilmer, B. L.; Webber, D.; Ashley, J. M.; Hall, K. C.; Bristow, A. D. Role of Strain on the Coherent Properties of GaAs Excitons and Biexcitons. *Phys. Rev. B* **2016**, *94*, 075207.
165. De Sio, A.; Troiani, F.; Maiuri, M.; et al. Tracking the Coherent Generation of Polaron Pairs in Conjugated Polymers. *Nat. Commun.* **2016**, 7. https://doi.org/10.1038/ncomms13742.
166. Lacy, M. Phonon-Electron Coupling as a Possible Transducing Mechanism in Bioelectronic Processes Involving Neuromelanin. *J. Theor. Biol.* **1984**, *111* (1), 201–204.
167. Kholmicheva, N.; Moroz, P.; Eckard, H.; Jensen, G.; Zamkov, M. Energy Transfer in Quantum Dot Solids. *ACS Energy Lett.* **2017**, *2* (1), 154–160.
168. Zhang, Z.; Wang, J. Origin of Long-Lived Quantum Coherence and Excitation Dynamics in Pigment-Protein Complexes. *Sci. Rep.* **2016**, *6*, 37629.
169. Malý, P.; Somsen, O. J. G.; Novoderezhkin, V. I.; Mančal, T.; van Grondelle, R. The Role of Resonant Vibrations in Electronic Energy Transfer. *Chemphyschem* **2016**, *17* (9), 1356–1368.
170. Torras, J.; Alemán, C. Massive Quantum Regions for Simulations on Bio-Nanomaterials: Synthetic Ferritin Nanocages. *Chem. Commun.* **2018**, *54* (17), 2118–2121.
171. Rourk, C. Ferritin and Neuromelanin "Quantum Dot" Array Structures in Dopamine Neurons of the Substantia Nigra Pars Compacta and Norepinephrine Neurons of the Locus Coeruleus. *Biosystems* **2018**, *171*, 48–58.
172. Rourk, C. Conductive Atomic Force Microscopy Data From Substantia Nigra Tissue. *Data Brief.* **2019**, *27*, 103986.
173. Solano, F. Melanins: Skin Pigments and Much More—Types, Structural Models, Biological Functions, and Formation Routes. *New J. Sci.* **2014**, *2014*, 498276.
174. Tesfay, L.; Huhn, A. J.; Hatcher, H.; Torti, F. M.; Torti, S. V. Ferritin Blocks Inhibitory Effects of Two-Chain High Molecular Weight Kininogen (HKa) on Adhesion and Survival Signaling in Endothelial Cells. *PLoS One* **2012**, 7 (7), e40030.
175. Lambert, N.; Emary, C.; Chen, Y.; Nori, F. Distinguishing Quantum and Classical Transport through Nanostructures. *Phys. Rev. Lett.* **2010**, *105*, 176801.
176. Zhang, S.; Xue, F.; Wu, R.; Cui, J.; Zm, J.; Yang, X. Conductive Atomic Force Microscopy Studies on the Transformation of GeSi Quantum Dots to Quantum Rings. *Nanotechnology* **2009**, *1* (13), 20.
177. Tanaka, I.; Kamiya, I.; Sakaki, H. Imaging and Probing Electronic Properties of Self-Assembled InAs Quantum Dots by Atomic Force Microscopy With Conductive Tip. *Appl. Phys. Lett.* **1999**, *74*, 844.
178. Wold, D.; Haag, R.; Rampi, M. A.; Frisbie, C. D. Distance Dependence of electron tunneling through self-assembled monolayers measured by Conducting Probe Atomic

Force Microscopy: Unsaturated Versus Saturated Molecular Junctions. *J. Phys. Chem. B* **2002**, *106* (11), 2813–2816.
179. Casuso, I.; Fumagalli, L.; Samitier, J.; Padrós, E.; Reggiani, L.; Akimov, V.; Gomila, G. Electron Transport Through Supported Biomembranes at the Nanoscale by Conductive Atomic Force Microscopy. *Nanotechnology* **2007**, *18*, 46.
180. Desimone, K.; Viviano, J.; Schneider, K. Population Receptive Field Estimation Reveals New Retinotopic Maps In Human Subcortex. *J. Neurosci.* **2015**, *35* (27), 9836–9847.
181. Erlikhman, G.; Gurariy, G.; Mruczek, R. E. B.; Caplovitz, G. P. The Neural Representation of Objects Formed Through the Spatiotemporal Integration of Visual Transients. *Neuroimage* **2016**, *42*, 67–78.
182. Cheong, L.; Zhao, W.; Song, S.; Shen, C. Lab on a tip: Applications of Functional Atomic Force Microscopy for the Study of Electrical Properties in Biology. *Acta Biomater.* **2019**, *99*, 33–52.
183. Friesen, D. E.; Craddock, T. J.; Kalra, A. P.; Tuszynski, J. A. Biological Wires, Communication Systems, and Implications for Disease. *Biosystems* **2015**, *127*, 14–27.
184. Engelen, M.; Vanna, R.; Bellei, C.; et al. Neuromelanins of Human Brain have Soluble and Insoluble Components With Dolichols Attached to the Melanic Structure. *PLoS One* **2012**, 7 (11), e48490.
185. Korolev, N. P.; Gus' kova, R. A.; Mil'gram, V. D.; Fedorov, G. E.; Osipov, V. I. Effect of the C35 Analog of Dolichol on the Electrical Properties of Planar Black Bilayers. In *Nauchnye doklady vysshei shkoly. Biologicheskie nauki(No. 7)*; 1987; pp. 37–40.
186. Tyson, J. J.; Albert, R.; Goldbeter, A.; Ruoff, P.; Sible, J. Biological Switches and Clocks. *J. R. Soc. Interface* **2008**, *5* (Suppl 1), S1–S8.
187. Rinkel, M.; Greenblatt, M.; Coon, G.; Solomon, H. Relation of the Frontal Lobe to the Autonomic Nervous System in Man. *Arch. NeurPsych.* **1947**, *58* (5), 570–581.
188. Photo via <a href="https://www.goodfreephotos.com/">Good Free Photos</a>.
189. Bahrick, L. Body Perception: Intersensory Origins of Self and Other Perception in Newborns. *Curr. Biol.* **2013**, *23* (23), R1039–R1041.
190. Watson, G.; Leverenz, J. Profile of Cognitive Impairment in Parkinson's Disease. *Brain Pathol.* **2010**, *20* (3), 640–645.

CHAPTER FOUR

Ion plasmon collective oscillations underlying saltatory conduction in myelinated axons and topological-homotopy concept of memory

Witold A. Jacak and Janusz E. Jacak*
Department of Quantum Technology, Wrocław University of Science and Technology, Wrocław, Poland
*Corresponding author: e-mail address: janusz.jacak@pwr.edu.pl

Contents

Abstract

We demonstrate a new wave-type model of saltatory conduction in myelinated axons. A poor conductivity of the neuron cytosol limits electrical current signal velocity upon the cable theory, to 1–3 m/s, whereas the saltatory conduction undergoes with the velocity of signal transduction 100–300 m/s. We propose the wave-type mechanism of the saltatory conduction in the form of kinetics of ion plasmon-polariton being the hybrid of the electromagnetic wave and of the ion plasma oscillations, which meets the observations. Plasmons oscillations have a quantum character to some extent as the coherent oscillations of all ions are possible due to their repulsion which can be understood upon the quantum approach like random phase approximation by Pines and

Advances in Quantum Chemistry, Volume 82
ISSN 0065-3276
https://doi.org/10.1016/bs.aiq.2020.04.005

Bohm. Basing on the difference of electricity of myelinated white matter and non-myelinated gray matter in cortex we outline the topological concept of the memory functioning.

1. Introduction

The brain and neurons operate at temperatures ca. 36°C. At such temperatures ions are charge carriers and electrons change their locations due to chemical reactions mostly. A full dissociation of electrons is energetically inaccessible as requires larger energy scale. A crystalline structure is also not a state of brain matter at such temperatures. Thus all the electricity of neurons is governed by ions. For neurons we deal with electrolytes of various composition both inside neurons and in the intercellular liquid. The total charge is balanced though locally some polarizations occur conditioning living functions. Several factors determine charge distribution, Coulomb forces tending to equalize charge distribution, diffusion of various type ions due to their concentration gradients, active transfer of ions against concentration gradient, electrochemical mechanisms at synapses. All these constitute electro-signaling in neurons, with ions as charge and information carriers. Moreover, concentrations of ions in neuron cytosol and in intercellular liquid is small, typically on mM scale which gives pH of order of 7.5. Viscosity of such electrolytes is roughly similar as of water, though the diffusion of small particles is ca. fourfold slower than in pure water due to scattering on other components including macromolecules in the solution. The ion composition of the cytosol and outer intercellular liquid is different, Na^+ ca. 10 mM inside a cell and ca. 140 mM outside, whereas K^+ ca. 140 mM inside and 4 mM outside. Such a counter gradient distribution is achieved by energy cost of an active transfer ions across the membrane via ADPase controlled channels. This difference in various ion concentrations is critical for osmoregulation, since if the ion levels were the same inside a cell as outside, water would enter constantly by osmosis—since the levels of macromolecules inside cells are higher than their levels outside. Instead, sodium ions are expelled and potassium ions taken up by the Na^+/K^+–ATPase, potassium ions then flow down their concentration gradient through potassium-selection ion channels, this loss of positive charge creates a negative membrane potential. To balance this potential difference, negative chloride ions also exit the cell, through selective chloride channels.

The loss of sodium and chloride ions compensates for the osmotic effect of the higher concentration of organic molecules inside the cell. The Na^+/K^+–ATPase enzyme is active (i.e., it uses energy from ATP). For every ATP molecule that the pump uses, three sodium ions are exported and two potassium ions are imported; there is hence a net export of a single positive charge per pump cycle. Alcohol inhibits sodium–potassium pumps in the cerebellum and this is likely how it corrupts cerebellar computation and body coordination. Na^+/K^+–ATPase plays also a central role in mechanism of the action potential (AP) transduction along myelinated axons. Moreover sodium and potassium selective pores that span cell membranes function to conduct sodium and potassium ions down their electrochemical gradient, doing so both rapidly (up to the diffusion rate of these ions in bulk water) and selectively (despite the subangstroms difference in ion radius). Biologically, these channels act to set or reset the resting potential in many cells. In neurons, the sodium channels are responsible for the rising phase of AP, whereas the delayed counterflow of potassium ions shapes the AP. Before an AP occurs, the axonal membrane is at its normal resting potential, and Na^+ channels are in their deactivated state, blocked on the extracellular side by their activation gates. In response to an increase of the membrane potential to about −55 mV, the activation gates open, allowing sodium ions to flow into the neuron through the channels, and causing the voltage across the neuronal membrane to increase to +30 mV (in human neurons). Because the voltage across the membrane is initially negative, as its voltage increases from −70 mV at rest to a maximum of +30 mV, it is said to depolarize. This increase in voltage constitutes the rising phase of an AP. When enough Na^+ ions entered the neuron and the membrane's potential has become high enough, the sodium channels inactivate by closing their inactivation gates. With the sodium channel no longer contributing to the membrane potential, the potential decreases back to its resting potential as the neuron repolarises and subsequently hyperpolarises itself. This decrease in voltage constitutes the falling phase of the AP. When the membrane's voltage becomes low enough, the inactivation gate reopens and the activation gate closes in a process called deinactivation. With the activation gate closed and the inactivation gate open, the Na^+ channel is once again in its deactivated state, and is ready to participate in another AP. The temporal behavior of Na^+ channels can be modeled by a Markovian scheme (M) or by the Hodgkin–Huxley-type (HH) formalism. In the former scheme, each channel occupies a distinct state with differential equations describing transitions between states; in the latter, the channels are treated as a population that are affected by three independent gating variables. Each of these variables can attain a value between 1 (fully permeant to ions) and 0 (fully nonpermeant), the product of these variables yielding the percentage of conducting channels. The HH model can be shown to be equivalent to the M model. The pore of sodium channels contains a selectivity filter made of negatively charged amino acid residues, which attract the positive Na^+ ion and keep out negatively charged ions such as chloride. The cations flow into a more constricted part of the pore that is 0.3 by 0.5 nm wide, which is just large enough to allow a single Na^+ ion with a water molecule associated to pass through. The larger K^+ ion cannot fit through this area. Ions of different sizes also cannot interact as well with the negatively charged glutamic acid residues that line the pore. Na^+ channels both open and close more quickly than K^+ channels, producing an influx of positive charge (Na^+) toward the beginning of the AP and an efflux (K^+) toward the end. Both channels can be described by the system of two retarded coupled nonlinear differential equations perfectly deriving the shape and size of the AP spike in the model of HH or M on the time scale of 1 ms, whereas Na^+/K^+–ATPase restores in much longer time the steady state of a node of Ranvier. Such a long time for restoring of the node of Ranvier to a next activity conveniently prevents a back shift of the signal.

Despite the realistic model of AP spike formation at nodes of Ranvier,[1] the transduction of the igniting signal between consecutive nodes of Ranvier is not well described as of yet. The cable model of ion diffusion through the myelinated segment is ineffective because the upper limit of the velocity of this diffusion is by 1–2 orders of the magnitude lower than observed velocity. The cable model well describes the slow charge kinetics in dendrites[2,3] and in nonmyelinated axons,[4–7] being typically of order lower than 1 m/s, whereas the signal velocity in myelinated axons reaches 100–300 m/s, and reducing of it by even only of 10% ceases functionality of an organism. The mechanism of this kinetics is apparently beyond the diffusion transport upon the cable theory and is only called as the saltatory conduction, because transduction of the ignition signal between nodes of Ranvier resembles almost instant saltation of the voltage, being absolutely unreachable for ordinary diffusion of ions in the cable theory. Moreover, there are evidences that the saltatory conduction has a wave-type character. This conduction maintains even if several nodes of Ranvier are damaged or the axon is broken into two parts with ends separated by a small break. This absolutely precludes the diffusion type current conduction impossible to be maintained despite the break and evidences that the saltatory conduction must be of wave-type nature, which allows for jumping across small breaks and continuing of the traveling despite damage and inactivation of several nodes of Ranvier.

The issue of the chapter meets with a recent large increase of interest in plasmonics of metallic nano-systems and astonishing opportunities for sub-diffraction light manipulations by plasmon excitations in nanoscale metallic components,[8] which caused the related rapid development of the new field of nanoplasmonics that overlaps with the many application prospects of nanophotonics and subdiffraction opto-electronics.[8–13]

Plasmons correspond to oscillations of local charge density of electrons with respect to positive jellium. This effect is quantum in nature because the coherent movement of all charge carriers is conditioned by their mutual repulsion and cannot be understood classically. The successful quantum description of this phenomenon has been provided by Pines and Bohm[14,15] in the form of quantum random phase approximation (RPA) approach. Similar plasmon effect concerns also electrolytes with charge carriers in the form of ions instead of electrons.[16] Ca. 10^4 times larger masses of ions in comparison to electrons and the concentration of ions in electrolytes much lower than of electrons in metals highly reduce energy of ion plasmons to the scale which fits to energy scale of bio systems. Similarly the plasmonic size featured by a strong radiation of plasmons is shifted from the nanoscale for electrons to micrometer scale for ions, again just as for the scale of bio-cell organization. The RPA model of ion plasmons has been developed[16] including the application to synchronized plasmon oscillations in linear periodic alignments of electrolytes confined by membranes[17] in analogy to plasmon-polaritons in metallic nanochains.[9,12,18–23]

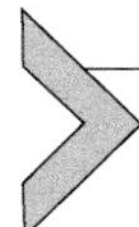

2. Insufficiency of the cable theory to explain the saltatory conduction

Signal kinetics in dendrites or unmyelinated axons is well described by the cable theory.[2–6,24–26] Upon this theory the velocity of an AP signal is characterized by $v_c = \frac{\lambda}{\tau} = \frac{\sqrt{G}}{C\sqrt{R}}$, where C is the capacity across the neuron cell membrane per unit of the axon length, G is the conductance across the membrane and R is the longitudinal resistance of the inner cytosol, both per unit of the neuron length, and λ and τ are space and time diffusion ranges defined in the cable model, respectively.[24–29] The velocity v_c scales as $\sqrt{d}$ with the dendrite (axon) diameter, d, because $C \simeq \frac{\varepsilon_0 \varepsilon \pi d}{\delta}$, $1/G \simeq \frac{\rho_1 \delta}{\pi d}$ and

$R \simeq \frac{4\rho}{\pi d^2}$, where δ is the cell membrane thickness, ρ is the longitudinal resistivity of the inner cytosol and ρ_1 is the resistivity across the membrane. For exemplary values of the membrane capacity per surface unit, $c_m = 1 \mu F/cm^2$, $\rho_1 \delta = 20{,}000\ \Omega cm^2$, $\rho = 100\ \Omega cm$ and the diameter of the cable $d = 2\ \mu m$, one gets $v_c \simeq 5$ cm/s.[24]

Due to a myelin sheath in myelinated axons both the capacity and conductance across the myelinated membrane are reduced, roughly inversely proportional to the myelin layer thickness. Thus, the cable theory velocity, $\sim \frac{\sqrt{G}}{C\sqrt{R}}$, grows ca. 10 times if the capacitance and conductance lower ca. 100 times. This is, however, still too low to match observations of kinetics in myelinated axons.[5–7,24–27,30–36] Moreover, the activity of nodes of Ranvier slows down the overall velocity of the discrete diffusion in the myelinated internodal segments to the level of factor 6 instead of 10.[24] For realistic axons with $d \sim 1\ \mu m$ the assessed cable theory velocity gives ca. 3 m/s instead of 100 m/s observed in such-size myelinated axons in peripheral human nervous system (PNS). Similar inconsistency in the cable model estimations is encountered in human central nervous system (CNS) with thinner myelinated axons of $d \in (0.2, 1)\ \mu m$.[30]

To model larger velocities upon the cable theory accommodated to the myelinated axons, much thicker axons are assumed with longer internodal distances, because upon the discrete diffusion model,[24] the velocity in a myelinated axon, $v_m \simeq \sqrt{\frac{l}{d_0}} v_c$, where l is the length of internodal segments and d_0 is the length of the node of Ranvier. For instance, in Refs. 30, 32 $d =$ 10 μm (and greater) and $l = 1150\ \mu m$ (and greater) have been assumed to gain the velocity of the AP transduction, $v_m \sim 40$ m/s (cf. also Refs. 33–36), i.e., there are ca. 10 times larger in dimension than the actual myelinated axons of a human. It is thus clear that the cable model is not effective in modeling of the observed quick saltatory conduction in myelinated axons.[5–7,25–27,30–37] The velocity predicted upon the discrete cable model for realistic axon parameters is by at least one order of the magnitude smaller than observed.

Interfering of the Huxley-Hodgkin (HH) mechanism at nodes of Ranvier with the cable model diffusion at internodal myelinated segments[1] results in the estimation of the AP propagation velocity[33,35] only ca. 6 time greater than in unmyelinated axons with the same geometry,[24–26] despite the reducing of the intercellular capacity and conductivity by the myelin sheath. The mentioned above simplified formula for this velocity (for iterative discrete diffusion model[24]) gives $v_m \sim \sqrt{\frac{l}{d_0}}$, and since d_0 is often

1μm and l is around 100μm, the increase in velocity of myelinated axons can be almost 10 times that of unmyelinated axons (more precisely the factor is closer to 6[24]).

Velocity of the AP transduction must keep the high value because the deviation by 10% ceases body functioning.[30] Continued mathematical attempts to optimize a model of HH mechanism mixed with the cable theory[24–26,33,35] in order to obtain a sufficiently high AP velocity in myelinated axons gave not a success for a long time, which strongly suggest that the way to understand the saltatory conduction must be linked with a different physical mechanism rather than any lifting of the cable theory of ion diffusion.

The cable theory is in fact the conventional model of a transmission line widely applied in the electronics and communication. The long line (like a coaxial line, particularly well corresponding to a bare neuron cell cord separated by a thin insulator membrane from the outer electrolyte of intercellular cytosol) is assumed to be a series of infinitely short segments of length dx, represented by the scheme as in Fig. 1. R is the longitudinal resistance of the inner line per its length unit, L is the distributed inductance of the inner wire per its length unit, C is the capacitance between two line components across the separating dielectric layer also per unit length of the line, G is the electrical conductivity across the barrier per unit length of the line.

Due to the Ohm law and the definition of the capacitance and the inductance one gets the relations between the voltage gated the line segment and the current along the inner wire,

$$\begin{aligned} \frac{\partial U(x,t)}{\partial x} &= -L\frac{\partial I(x,t)}{\partial t} - RI(x,t), \\ \frac{\partial I(x,t)}{\partial x} &= -C\frac{\partial U(x,t)}{\partial t} - GU(x,t), \end{aligned} \tag{1}$$

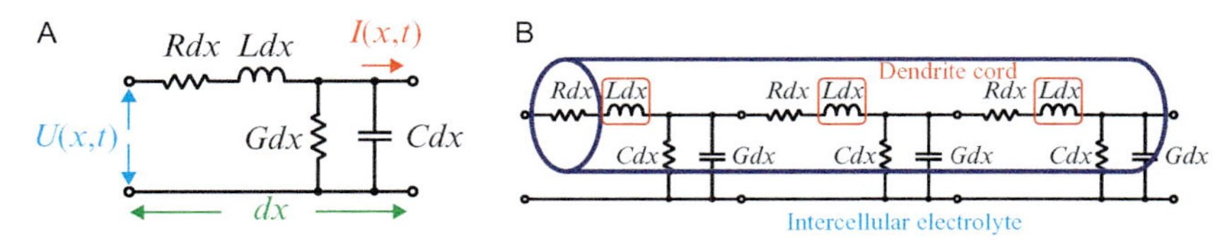

Fig. 1 (A) Elementary segment of the transmission line of length *dx*. *R* and *L* are resistance and inductance of the inner wire (the *upper line* in the figure) per length unit, *C* and *G* is the capacity and conductivity across the insulating barrier per unit of the length of the line. The electrical current flows only along the inner wire. (B) Transmission line applied to a dendrite—the inner wire is the dendrite cord (with $L = 0$) and outer coaxial screen is the intercellular electrolyte separated by the dendrite (or unmyelinated axon) cell membrane.

which are two coupled differential equations for the complex voltage V and current I. Taking the derivative $\frac{\partial}{\partial x}$ of both equations and substituting one into another, we can arrive in the second order differential equations,

$$\begin{aligned}\frac{\partial^2 U(x,t)}{\partial x^2} - LC\frac{\partial^2 U(x,t)}{\partial t^2} &= (RC + GL)\frac{\partial U(x,t)}{\partial t} + GRU(x,t),\\ \frac{\partial^2 I(x,t)}{\partial x^2} - LC\frac{\partial^2 I(x,t)}{\partial t^2} &= (RC + GL)\frac{\partial I(x,t)}{\partial t} + GRI(x,t).\end{aligned} \tag{2}$$

One can notice that both equations ate identical (they are conventionally called as telegrapher's equations). For the lossless case, i.e., when $R = G = 0$, Eq. (2) gains the form of the wave equations both for U and I, describing the ideal wave-type transmission along x direction with the velocity $v = \frac{1}{\sqrt{LC}}$ and without any losses. This corresponds to an ideal coaxial line. This is, however, not the case of a neuron, for which the assumption $L = 0$ is appropriate. Then Eq. (2) attains the form of 1D diffusion equation,

$$\begin{aligned}\frac{\partial^2 U(x,t)}{\partial x^2} &= RC\frac{\partial U(x,t)}{\partial t} + GRU(x,t),\\ \frac{\partial^2 I(x,t)}{\partial x^2} &= RC\frac{\partial I(x,t)}{\partial t} + GRI(x,t).\end{aligned} \tag{3}$$

Defining the parameters $\lambda = \frac{1}{\sqrt{GR}}$ and $\tau = \frac{C}{G}$, one can rewrite the above equation (for U) in a conventional form,

$$\lambda^2 \frac{\partial^2 U(x,t)}{\partial x^2} = \tau \frac{\partial U(x,t)}{\partial t} + U(x,t). \tag{4}$$

The parameter λ defines the spatial scale of the diffusion, whereas τ its time scale. The velocity of the diffusion of the signal, i.e., of the diffusive current along the dendrites (or unmyelinated axon) is assumed as $v_c = \frac{\lambda}{\tau} = \frac{\sqrt{G}}{C\sqrt{R}}$. This velocity is the larger the smaller C and R and the larger G are. The range of the diffusion, $\sim \lambda$, lowers, however, with growing G (due to the shunt escape of the current). Larger G results in larger velocity but severely limits the range. The overall behavior of the $U(x, t)$ (or I) diffusion defined by Eq. (4) is illustrated in Fig. 2, which presents the solution of Eq. (4) for initial condition in the form of periodic excitation in $x = 0$ point, $U(0, t) = U_0 cos(t)$. One can notice that the cable theory (i.e., Eq. 4) gives the nonwave-type propagation related to an ordinary current (and voltage signal) of diffusion type, thus on a relatively short distance with a quickly lowering amplitude. For the realistic values of R, C, G and d in axons, the

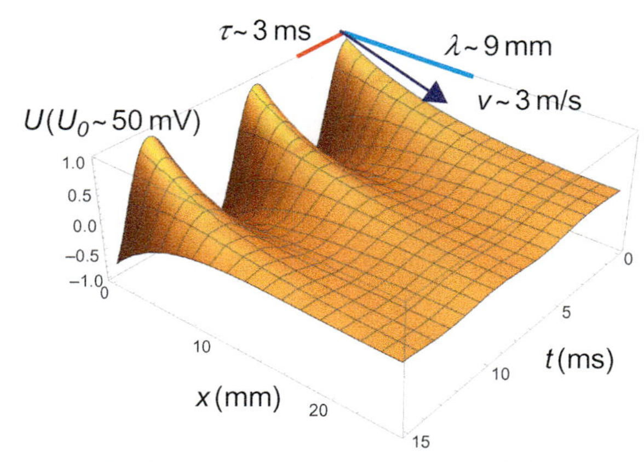

Fig. 2 The solution of Eq. (4) for $\lambda = 9$ mm and $\tau = 3$ ms and the ignition signal at $x = 0$ point, $U(0, t) = 50cos(t)$ mV. The unit proportions in the figure are conserved.

estimation of v_c gives 0.05–1 m/s. It is too low for explanation of the saltatory conduction in myelinated axons.

The observed conduction velocity of a myelinated axon is by factor 100–200 greater than that of a geometrically similar unmyelinated axon. Myelinated neurons make up a large proportion of all neurons in the human body, more so in the CNS. Specifically, all PNS neurons with diameters greater than around 1 μm and all CNS neurons with diameters greater than around 0.2 μm are myelinated.[33–35] In a study of myelinated axons in the human brain, most were found to have a diameter less than 1 μm. The cable theory applied to myelinated axons gives at most 6 times larger velocity and to consider velocity ~40 m/s the axon diameter and internodal length are assumed extremely large[30] in contrary to observations.[30,33,34] Data indicate that the v_m is roughly proportional to fiber diameter, d, but this scale factor does vary. The complex factors impacting the v_m in myelinated axons can be found in Refs. 33 and 35 upon the cable theory variants, including, for instance, myelin thickness and capacitance (taking into account that the cable theory speed v_c scales as $\sqrt{d}$). In the plasmon-polariton model we have got $v_m \sim d$.

It is evident that the diffusion of ions according the cable theory is too slow to explain the rapid saltatory conduction along myelinated segments of axons. Apparently different mechanism of this conduction is required beyond local diffusion. We propose such a new approach by synchronized oscillations of local ion density, which could travel along the periodically myelinated axon in the form of a wave plasmon-polariton, very well known form similar phenomenon in metallic periodic linear systems. The mathematical model of plasmon-polariton will be described in the following paragraph basing on the analysis of local oscillations of ion density, called as ion plasmons. The mathematical model of ion plasmons is presented in Ref. 16.

In discrete diffusion model[24] or in other modifications of the cable theory[25,26] the HH cycle is included as the integrative element of local electricity in the closest adjacent myelinated segments. In the plasmon-polariton model the HH cycles at consecutive nodes of Ranvier are decoupled of the synchronic oscillations of ion density in myelinated segments. The HH cycles are triggered by the plasmon-polariton wave packet traveling along the axon with high speed. HH cycles play, however, a role of the control over plasmon-polariton via the resonance selection of particular modes which are supplemented in energy and thus traversing arbitrary long distances with the constant amplitude despite Ohmic losses at ion oscillations.

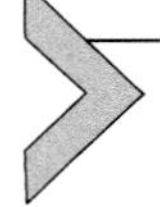

3. Plasmon-polaritons in a chain of finite ion systems—Model of the saltatory conduction in myelinated axons

In order to solve the problem of the saltatory conduction, we develop a new and original model for this mysterious conduction basing on the kinetic properties of collective plasmon-polariton modes propagating along linear and periodically modified electrolyte systems, which for an axon is the thin cord of the nerve cell periodically wrapped by Schwann cells creating periodic and relatively thick myelin sheath (Schwann cells myelinate axons in the PNS, whereas in the CNS axons are myelinated by oligodendrocyte cells[27]). The plasmon-polaritons were investigated and understood based on the well-developed domain of plasmonics,[8,12] especially nanoplasmonics applied to long-range low-damped propagation of plasmon-polaritons along metallic nanochains.[18] The main properties of these collective excitations occurring on the conductor/insulator interface[11,38] due to the hybridization of the surface plasmons (i.e., the charge density fluctuations on the conductor surface) with the electromagnetic wave are as follows: (1) much lower velocity of plasmon-polaritons than the velocity of light, which yields plasmon-polaritons with wavelengths much shorter than the wavelength of light at the same frequency, (2) the related strong discrepancy between the momenta of plasmon-polaritons and photons with the same energies causes that the external electromagnetic waves do not interact with plasmon-polaritons, i.e., photons cannot be excited or absorbed by plasmon-polaritons due to momentum conservation constraints, (3) all the electromagnetic field associated with propagation of plasmon-polaritons is compressed to the tunnel-volume of the chain, (4) all the radiative losses are quenched, and plasmon-polariton attenuation only occurs due to the Ohmic losses of oscillating charged carriers, which makes periodically corrugated conductors almost perfect waveguides for plasmon-polaritons, and (5) long-range and practically undamped propagation of plasmon-polaritons is experimentally observed in metallic nanochains.[9,18] The plasmon-polaritons in metallic nanostructures are likely to be exploited for future applications in optoelectronic where conversion of light signals into plasmon-polariton signals circumvents diffraction constraints that greatly limit the miniaturization of conventional optoelectronic devices (as the nanoscale of electron confinement inconveniently conflicts with the several-orders-of-magnitude larger scale of the wavelength of light at an energy similar to the nanoconfined electrons).[11,39]

All properties of plasmon-polaritons can be repeated in periodic linear arrangements of electrolyte systems with ions instead of electrons as charge carriers.[16,17] According to the larger mass of ions compared with that of electrons and the lower concentration of ions in electrolytes compared with that of electrons in metals, the plasmon resonances in finite electrolyte systems (e.g., liquid electrolyte confined to a finite volume by appropriately formed membranes, frequently found in biological cell structures) occur on the scale of micrometers rather than nanometers such as for metals and at frequencies several orders of magnitude lower (depending on the ion concentration).

For a spherical electrolyte system, the surface and volume plasmons are handled analogously to those of the metallic nanosphere as developed in Ref. 16. The ion surface plasmon frequencies are given for the multipole lth mode by the formula $\omega_l = \omega_p \sqrt{\frac{l}{\varepsilon(2l+1)}}$ with the bulk plasmon frequency $\omega_p = \sqrt{\frac{4\pi q^2 n}{m}}$ (n is the ion concentration, q and m are the ion charge and mass, respectively, and ε is the dielectric relative permittivity of the surroundings). For dipole surface plasmons ($l = 1$), this equation resolves to the Mie-type formula[40,41] $\omega_1 = \frac{\omega_p}{\sqrt{3\varepsilon}}$. The plasmon oscillations intensively radiate their own energy and are quickly damped due to Lorentz friction losses (i.e., due to radiation of e-m waves by oscillating charges[42,43]), which for large systems with a large number of ions participating in the plasmon oscillations (thus strengthening the Lorentz friction) are much greater than the

Joule-heat dissipation caused by Ohmic losses due to carrier scattering (scattering of ions on other ions, solvent and admixture atoms and the boundary of the system).[16,44]

Surprisingly, for a linear chain of spherical electrolyte systems, the radiation losses are completely reduced to zero in exactly the same manner as in metallic chains.[22,45,46] The radiation energy losses expressed by the Lorentz friction[42,43] are ideally compensated by the income of the energy due to the radiation from all the other spheres in the chain. As a result, the radiative losses are ideally balanced, and only the relatively small irreversible Ohmic energy dissipation remains due to ion scattering. Thus, the collective surface dipole plasmon-polaritons can propagate in the chain with strongly reduced damping, and if the energy is permanently supplemented to balance the small Ohmic losses, this propagation can occur over arbitrarily long distances without any damping.

3.1 Plasmon-polariton propagation in linear periodic electrolyte systems

We propose to apply the model of dipole plasmon-polariton excitations in a linear chain of electrolyte spheres to an axon cord periodically wrapped with myelin sheaths, as schematically depicted in Figs. 3 and 5. The periodicity makes the chain similar to a 1D crystal. Despite the cord of an axon being a continuous ion tube, the modeling of an axon by a chain of segments defined by the periodic myelin sheath meets well with plasmon-polariton kinetics maintaining the same character in discrete chains and in continuous but periodically corrugated wires. The interaction between the chain elements (or segments defined by the periodic myelin sheath) can be regarded as dipole-type coupling. For chains of electrolyte spheres, the results of the corresponding analysis for metallic chains can be adopted, which supports a dipole model of interaction.[39,47,48]

The dipole interaction resolves itself to the electric and magnetic fields created at any distant point by an oscillating dipole $\mathbf{D}(\mathbf{r}, t)$ pinned in $\mathbf{r}$. If the distant point is represented by the vector $\mathbf{r}_0$ (with the beginning fixed at the end of $\mathbf{r}$, where the dipole is placed), then the electric field produced by the dipole $\mathbf{D}(\mathbf{r}, t)$ takes the following form, including the relativistic retardation,[42,43]

$$\begin{aligned}\mathbf{E}(\mathbf{r},\mathbf{r}_0,t) &= \frac{1}{\varepsilon}\left(-\frac{\partial^2}{v^2\partial t^2}\frac{1}{r_0} - \frac{\partial}{v\partial t}\frac{1}{r_0^2} - \frac{1}{r_0^3}\right)\mathbf{D}(\mathbf{r},t-r_0/v) \\ &+ \frac{1}{\varepsilon}\left(\frac{\partial^2}{v^2\partial t^2}\frac{1}{r_0} + \frac{\partial}{v\partial t}\frac{3}{r_0^2} + \frac{3}{r_0^3}\right)\mathbf{n}_0(\mathbf{n}_0\cdot\mathbf{D}(\mathbf{r},t-r_0/v)),\end{aligned} \quad (5)$$

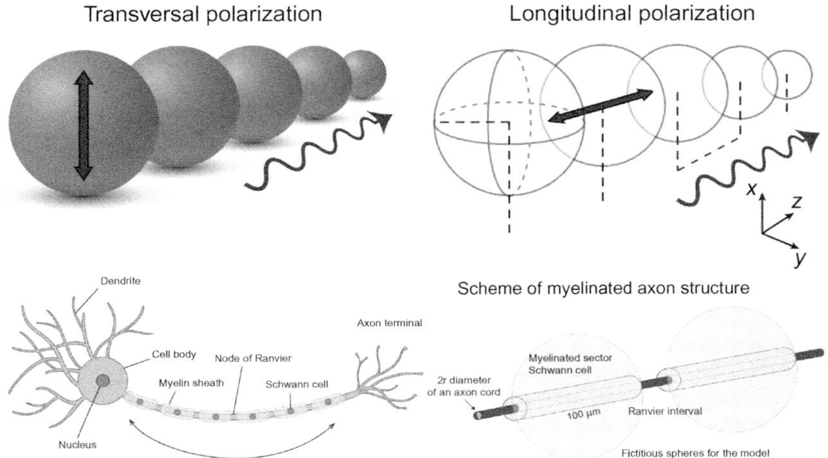

Fig. 3 In analogy to metallic nanochains, microchains of finite electrolytic segments ranged with dielectric membranes can be considered; in these electrolytic microchains, the ion plasmon-polariton modes can propagate similar to in metallic nanochains; we propose to model a periodically myelinated axon as a chain of electrolyte segments because the plasmon-polaritons can traverse equally discrete and periodically corrugated continuous linear plasmonic system alignments. *Reproduced from Jacak, W. Propagation of Collective Surface Plasmons in Linear Periodic Ionic Structures: Plasmon Polariton Mechanism of Saltatory Conduction in Axons. J. Phys. Chem. C 2015, 119 (18), 10015 with permission.*

with $\mathbf{n}_0 = \frac{\mathbf{r}_0}{r_0}$ and $v = \frac{c}{\sqrt{\varepsilon}}$, c is the light velocity. The terms with denominators of r_0^3, r_0^2, and r_0 are usually referred to as the near-field, medium-field, and far-field components of the interaction, respectively. The formula above will serve to describe the mutual interaction of the plasmon dipoles at each sphere in the chain. The spheres in the chain are numbered by integers l, and the equation for the surface plasmon oscillation of the lth sphere can be written as follows (where d denotes the separation between the centers of the spheres),

$$\begin{aligned}
&\left[\frac{\partial^2}{\partial t^2} + \frac{2}{\tau_0}\frac{\partial}{\partial t} + \omega_1^2\right] D_\alpha(ld, t) \\
&= \varepsilon\omega_1^2 a^3 \sum_{m=-\infty,\ m\neq l}^{m=\infty} E_\alpha\left(md, t - \frac{|l-m|d}{v}\right) \\
&+\varepsilon\omega_1^2 a^3 E_{L\alpha}(ld, t) + \varepsilon\omega_1^2 a^3 E_\alpha(ld, t),
\end{aligned} \tag{6}$$

$\alpha = z$ indicates the longitudinal polarization, whereas $\alpha = x(y)$ the transverse polarization (the chain orientation is assumed to be along the z direction, as illustrated in Fig. 3). The first term on the right-hand side of Eq. (6) describes the dipole coupling between the spheres, and the other two terms correspond to the plasmon attenuation due to Lorentz friction irradiation losses and the force field arising from an external electric field, respectively; $\omega_1 = \frac{\omega_p}{\sqrt{3\varepsilon}}$ is the frequency of the dipole surface plasmons. Ohmic losses are included via the term $\frac{2}{\tau_0}$ similar to what is applied to metals[9] but with the Fermi velocity of electrons in metals substituted by the mean velocity of ions for nondegenerated classical Boltzmann distribution regardless of the quantum statistics of ions, i.e.,

$$\frac{1}{\tau_0} = \frac{\mathsf{v}}{2\lambda_B} + \frac{C\mathsf{v}}{2a}, \tag{7}$$

where λ_B is the mean free path of the carriers (ions) in the bulk electrolyte, v is the mean velocity of the carriers at temperature T, $\mathsf{v} = \sqrt{\frac{3k_BT}{m}}$, m is the mass of the ion, k_B is the Boltzmann constant, C is a constant on the order of unity (to account for the type of scattering of carriers by the system boundary), and a is the radius of a sphere. The first term in the expression for $\frac{1}{\tau_0}$ approximates ion scattering losses such as those occurring in the bulk electrolyte (collisions with other ions, solvent and admixture atoms), whereas the second term describes losses due to the scattering of ions on the boundary of a sphere of radius a. According to Eq. (5), we can write the following quantities that appear in Eq. (6),

$$\begin{aligned} E_z(md,t) &= \frac{2}{\varepsilon d^3}\left(\frac{1}{|m-l|^3} + \frac{d}{v|m-l|^2}\frac{\partial}{\partial t}\right) \\ &\quad \times D_z(md, t - |m-l|d/v), \\ E_{x(y)}(md,t) &= -\frac{1}{\varepsilon d^3}\left(\frac{1}{|m-l|^3} + \frac{d}{v|m-l|^2}\frac{\partial}{\partial t} + \frac{d^2}{v^2|l-d|}\frac{\partial^2}{\partial t^2}\right) \\ &\quad \times D_{x(y)}(md, t - |m-l|d/v). \end{aligned} \tag{8}$$

Because of the periodicity of the chain, a wave-type collective solution of the dynamical equation in the form of Fourier component can be assumed (6),

$$\begin{aligned} D_\alpha(ld,t) &= D_\alpha(k,t)e^{-ikld}, \\ 0 \le k &\le \frac{2\pi}{d}. \end{aligned} \tag{9}$$

In the Fourier picture of Eq. (6) (the discrete Fourier transform (DFT) with respect to the positions and the continuous Fourier transform (CFT) with respect to time) this solution takes a form similar to that of the solution for phonons in 1D crystals. Note that DFT is defined for a finite set of numbers; therefore, we consider a chain with $2N+1$ spheres, i.e., a chain of finite length $L=2Nd$. Then, for any discrete characteristic $f(l)$, $l=-N,\ldots,0,\ldots,N$ of the chain, such as a selected polarization of the dipole distribution, we must consider the DFT picture $f(k)=\sum_{l=-N}^{N} f(l)e^{ikld}$, where $k=\frac{2\pi}{2Nd}n$, $n=0,\ldots,2N$. This means that $kd\in[0,2\pi)$ because of the periodicity of the equidistant chain. The Born-Karman periodic boundary condition, $f(l+L)=f(l)$, is imposed on the entire system, resulting in the form of k given above. For a chain of infinite length, we can take the limit $N\rightarrow\infty$, which causes the variable k to become quasi-continuous, although $kd\in[0,2\pi)$ still holds.

The Fourier representation of Eq. (6) takes the following form,

$$\begin{aligned}&\left(-\omega^2-i\frac{2}{\tau_0}\omega+\omega_1^2\right)D_\alpha(k,\omega)\\&=\omega_1^2\frac{a^3}{d^3}F_\alpha(k,\omega)D_\alpha(k,\omega)+\varepsilon a^3\omega_1^2E_{0\alpha}(k,\omega),\end{aligned}\tag{10}$$

with

$$\begin{aligned}F_z(k,\omega)=&4\sum_{m=1}^{\infty}\left(\frac{cos(mkd)}{m^3}cos(m\omega d/v)+\omega d/v\frac{cos(mkd)}{m^2}sin(m\omega d/v)\right)\\&+2i\left[\frac{1}{3}(\omega d/v)^3+2\sum_{m=1}^{\infty}\left(\frac{cos(mkd)}{m^3}sin(m\omega d/v)\right.\right.\\&\left.\left.-\omega d/v\frac{cos(mkd)}{m^2}cos(m\omega d/v)\right)\right],\\F_{x(y)}(k,\omega)=&-2\sum_{m=1}^{\infty}\left(\frac{cos(mkd)}{m^3}cos(m\omega d/v)+\omega d/v\frac{cos(mkd)}{m^2}sin(m\omega d/v)\right.\\&\left.-(\omega d/v)^2\frac{cos(mkd)}{m}cos(m\omega d/v)\right)\\&-i\left[-\frac{2}{3}(\omega d/v)^3+2\sum_{m=1}^{\infty}\left(\frac{cos(mkd)}{m^3}sin(m\omega d/v)\right.\right.\\&\left.\left.+\omega d/v\frac{cos(mkd)}{m^2}cos(m\omega d/v)-(\omega d/v)^2\frac{cos(mkd)}{m}sin(m\omega d/v)\right)\right].\end{aligned}\tag{11}$$

Similar to metallic nanochains, $ImF_\alpha(k,\omega) \equiv 0$ (for $\alpha = z, x(y)$), which indicates perfect quenching of the radiation losses at any sphere in the chain (meaning that to each sphere, the amount of energy that comes in from the other spheres is the same as the energy outflow due to Lorentz friction). We can easily verify this property, as the related infinite sums in Eq. (11) can be found analytically.[49] Eq. (10) is highly nonlinear with respect to the complex ω and can be solved both perturbatively in an analytical manner[22] or numerically even beyond the perturbation approach.[21] The solutions determined for $Re\omega$ and $Im\omega$ (i.e., for the self-frequency and damping of the plasmon-polaritons, respectively) can be applied to the axon model as an effective chain of electrolyte spheres with ion concentrations adjusted to the actual neuron parameters.

The resonance frequency ($Re\omega(k)$), its k-derivative, which is the group velocity, and the attenuation rate ($Im\omega(k)$) of the dipole plasmon-polariton modes numbered by the wave vector k, derived by the solution of Eq. (10)[21,22] for electrolyte chains with exemplary concentration, chain size and ion parameters, are plotted in Fig. 4.

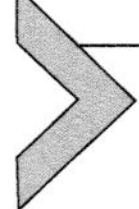

4. Plasmon-polariton model of saltatory conduction: Fitting the plasmon-polariton kinetics to the axon parameters

The bulk plasmon frequency is $\omega_p = \sqrt{\frac{q^2 n 4\pi}{m}}$ (we assumed for the model that the ion charge is $q = 1.6 \times 10^{-19}$ C and the ion mass is $m = 10^4 m_e$, where $m_e = 9.1 \times 10^{-31}$ kg is the mass of an electron) and for a concentration of $n = 2.1 \times 10^{14}$ $1/\mathrm{m}^3$, we obtain the Mie-type frequency for ion dipole oscillations, $\omega_1 \simeq 0.1 \frac{\omega_p}{\sqrt{3\varepsilon_1}} \simeq 4 \times 10^6$ $1/\mathrm{s}$, where the relative permittivity of water is $\varepsilon_1 \simeq 80$ for frequencies in the MHz range[50] (though for higher frequencies, beginning at approximately 10 GHz, this value decreases to approximately 1.7, corresponding to the optical refractive index of water, $\eta \simeq \sqrt{\varepsilon_1} = 1.33$). For a thin axon cord and thus, in the discrete model, for a strongly prolate inner electrolyte cord segment, we reduce the longitudinal Mie-type frequency by a factor of 0.1[51,52] compared with the isotropic spherical case. The axon consists of a cord with a small diameter of $2r$, and this thin cord is wrapped with a myelin sheath of a length of $2a$ per segment; however, for the effective model, we consider fictitious electrolyte spheres of radius a. Thus, the auxiliary concentration n of ions in the fictitious spheres corresponds to an ion concentration in the cord of $n' = \frac{n 4/3 \pi a^3}{2a\pi r^2}$, which yields a typical concentration of ions in a nerve cell of $n' \sim 10$ mM

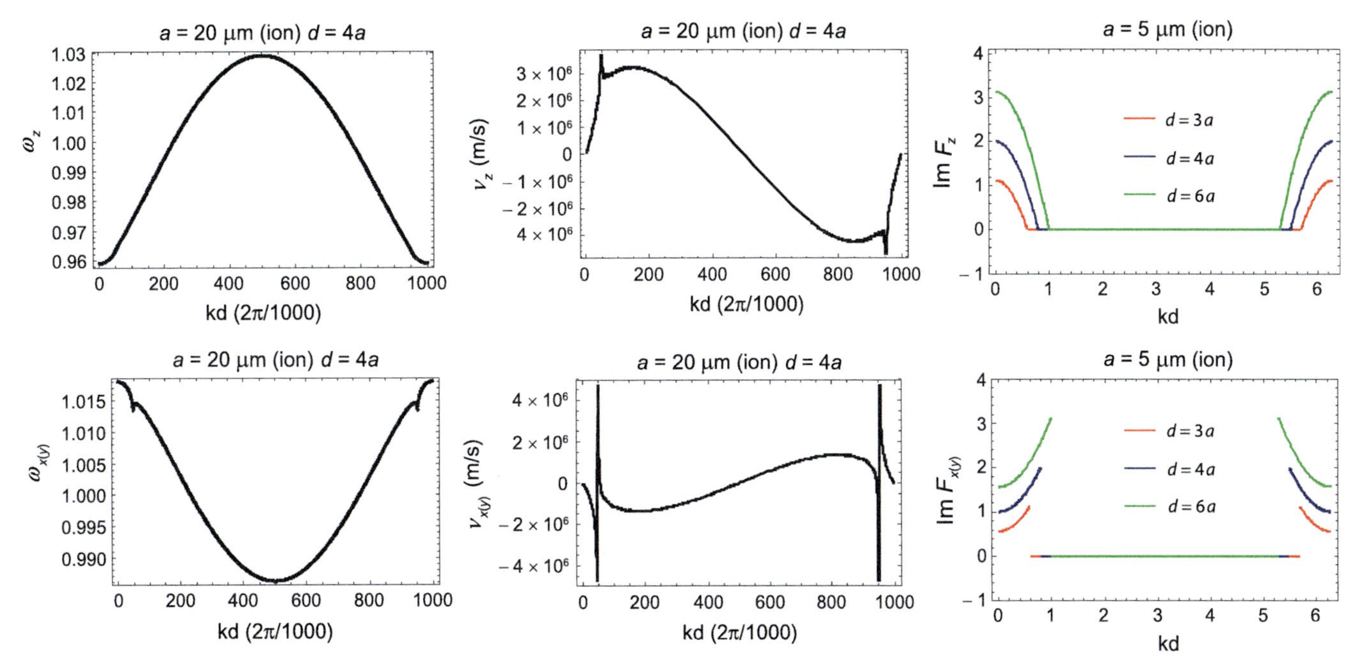

Fig. 4 Exact solution for the self-frequencies of the longitudinally and transversely polarized modes of the plasmon-polaritons in an electrolyte chain (ω in units of ω_1) obtained by solving of Eq. (10) in the region $kd \in [0, 2\pi)$ (*left*) and the corresponding group velocities for both types of polarization, $v_g = \frac{d\omega}{dk}$ (*central*); damping of plasmon-polaritons, i.e., the functions $ImF_z(k; \omega = \omega_1)$ and $ImF_{x(y)}(k; \omega = \omega_1)$ for chains of electrolyte spheres of radius a with separations of $d = 3a$, $4a$, and $6a$, the shift of the singularities toward the band edges with decreasing d/a is noticeable (*right*).

(i.e., $\sim 6 \times 10^{24}$ $1/m^3$). This is because all the ions participating in the dipole oscillation correspond in the sphere model to a much smaller volume in the real system that of the thin cord portion (the insulating myelin sheath consists of a lipid substance without any ions). The insulating, relatively thick myelin coverage creates the periodically broken channel (corrugated conductor–insulator structure) required for plasmon-polariton formation and its propagation (Fig. 5). To reduce the coupling with the surrounding intercellular electrolyte and protect against any leakage of plasmon-polaritons, the myelin sheath must be sufficiently thick, much thicker than what is required merely for electrical insulation. The regulatory role of the myelin sheath thickness is detailed in Appendix A.

For the resulting plasmon frequency in each segment, $\omega_1 \simeq 4 \times 10^6$ 1/s, one can determine the plasmon-polariton self-frequencies in a chain of segments of length $2a = 100$ μm (for a Schwann cell length of $2a$) and for small chain separations of $d/a = 2.01$, 2.1, and 2.2 (corresponding to Ranvier node lengths of 0.5, 5, and 10μm, respectively) within the approach presented above (via the solution of Eq. 10). The derivative of the obtained self-frequencies with respect to the wave vector k determines the group velocities of the plasmon-polariton modes. The results are presented in Fig. 6. We observe that for the electrolyte system parameters listed above, the group velocity of the plasmon-polaritons easily reaches 100 m/s. The longitudinal mode is polarized suitable to the prolate geometry of segments, assuming that the initial postsynaptic AP or that from the synapse hillock predominantly excites longitudinal ion oscillations.

From Fig. 6 we see that it is possible to arrange the wave packet (by selection of appropriate subset of k as shadowed in the right panel of the figure) which can propagate with the appropriately high velocity. This subset of k determines also the frequency of plasmon-polariton as visualized by shadowing in left panel in the figure. The mechanism which selects the appropriate k range is the HH mechanism at nodes of Ranvier and the thickness of the myelin sheath. The frequency of the plasmon polariton must be adjusted to the characteristic time of the opening of Na^+ channels at Ranvier node. MHz frequency allows for synchronization with μs time of gate triggering. Such a mode of plasmon-polariton is stabilized in contrary to modes with larger or lower frequencies. For larger frequency the k region right-shifts, which causes lowering of the velocity of plasmon-polariton—cf. Fig. 6, whereas for lower frequency the k region left-shifts, which also causes lowering of the velocity on the left side of maximum.

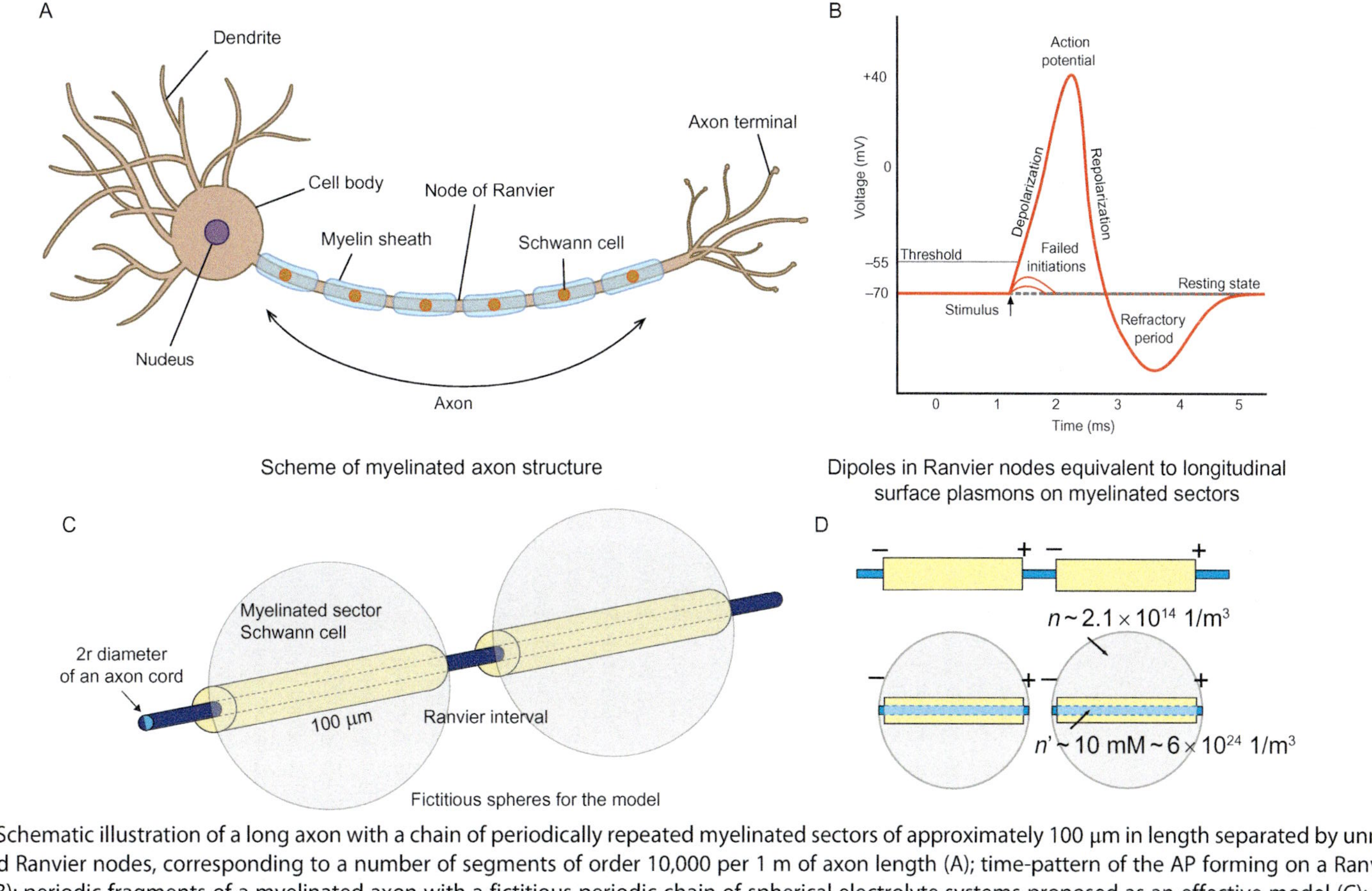

Fig. 5 Schematic illustration of a long axon with a chain of periodically repeated myelinated sectors of approximately 100 μm in length separated by unmyelinated Ranvier nodes, corresponding to a number of segments of order 10,000 per 1 m of axon length (A); time-pattern of the AP forming on a Ranvier node (B); periodic fragments of a myelinated axon with a fictitious periodic chain of spherical electrolyte systems proposed as an effective model (C); the equivalence of polarized Ranvier nodes with longitudinal surface plasmons on myelinated sectors (effective concentration of ions n in the auxiliary sphere corresponds to the actual ion concentration n') (D). *Reproduced from Jacak, W. Propagation of Collective Surface Plasmons in Linear Periodic Ionic Structures: Plasmon Polariton Mechanism of Saltatory Conduction in Axons. J. Phys. Chem. C 2015, 119 (18), 10015 with permission*

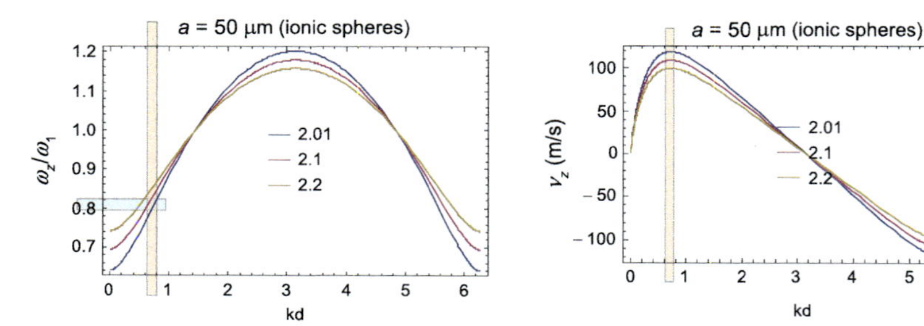

Fig. 6 Solutions for the self-frequencies and group velocities of the longitudinal mode of a plasmon-polariton in the model electrolyte chain; ω is presented in units of ω_1, here $\omega_1 = 4 \times 10^6$ 1/s, for a chain of spheres with radius $a = 50\mu$m and Ranvier separation d, $d/a = 2.01, 2.1$, or 2.2, for an equivalent ion concentration in the inner electrolyte cord of the axon of $n' \sim 10$ mM. Region of k, shadowed in the figure-right panel, corresponds to an optimal wave packet with velocity 100 m/s, the corresponding region of frequency is shadowed in the left panel. Any shift in frequency causes lowering of the group velocity.

The regulatory role pays here the thickness of the myelin layer (as detailed in Appendix A). The dipole oscillating in a myelinated segment of the axon cord induces in the outer intercellular electrolyte the opposite dipole. Both dipoles coupled across the myelin layer of thickness ξ create the pair of coupled oscillators with beating frequency determined by ξ. This beating frequency can be precisely tuned by ξ to required value at which the frequency of plasmon-polariton is synchronized with gating-time of Na^+ channels, thus synchronized with HH cycle at nodes of Ranvier. Such selected mode of plasmon-polariton is continuously supplied with energy form the AP formation at Ranvier nodes. This mode is thus not damped and propagates with the optimal velocity. Other frequency modes are damped as their Ohmic losses are not covered by the AP formation out of synchronization. A thinner myelin sheath causes an increase of the beating frequency of the dipole oscillator pair, which rises the frequency of plasmon-polariton mode but lowers its velocity, as visible from Fig. 6. Too thick myelin is also inconvenient—it causes lowering of beating frequency, thus lower frequency of plasmon-polariton mode and with also reduced velocity (the left side with respect to maximum velocity in Fig. 6). Such an behavior agrees with observations of the saltatory conduction velocity lowering at demyelination syndromes like Multiple Sclerosis. The details of the regulatory role of the myelin thickness are presented in Appendix A.

Note that for $\omega_1 = 4 \times 10^6$ 1/s and $a = 50$ μm, the light-cone interference conditions $kd - \omega_1 d/c = 0$ and $kd + \omega_1 d/c = 2\pi$ are fulfilled for extremely small values of kd and $2\pi - kd$, respectively, (of the order of

10^{-6} for $d/a \in [2, 2.5]$) and thus are negligible with regard to the plasmon-polariton kinetics (though are important for metals[21,23]). The related singularities on the light-cone induced by the far- and medium-field contributions to the dipole interaction are pushed to the borders of the k domain and thus are unimportant for the considered ion system. Hence, the quenching of the radiative losses (i.e., the perfect balance the Lorentz friction in each segment by the radiation income from the other segments in the chain) for the plasmon-polariton modes in the axon model occurs practically throughout the entire $kd \in [0, 2\pi)$ region. Additionally, the aforementioned singularities[21] are characteristic of infinite chains and therefore cannot fully develop because the nerve model electrolyte chains are of a finite length, whereas other effects, such as quenching of irradiation losses, occur for finite chains due to very fast convergence of sums in Eq. (11) with denominators m^2 and m^3 (practically, a chain consisting of only 10 segments exhibits almost the same properties as an infinite chain).

Although the electrolyte system chain model for a myelinated axon appears to be a crude approximation of the real axon structure, it can serve for the comparison of the energy and time scales of plasmon-polariton propagation implied by the model with the observed kinetic parameters of nerve signals. In the model, the propagation of a plasmon-polariton through the axon chain, excited by an initial AP on the first Ranvier node (after the synapse or, for the reverse signal direction, in the neuron cell hillock), sequentially ignites the consecutive Ranvier node blocks of Na^+ and K^+ ion gates. The resulting firing of the AP traverses the axon with an observed velocity of approximately 100 m/s, consistent with the velocity actually observed in myelinated axons (and not possible for ion diffusive current upon the cable model[27,29]). The plasmon-polariton ignition of consecutive Ranvier nodes releases the creation of the same AP pattern aided by the external energy supply at each Ranvier node block by ATPase. Because of the nonlinearity of the HH ion channel block mechanism, the signal growth saturates at a constant level, and the overall timing of each AP spike has the stable shape of the local polarization/depolarization scheme. The permanent supply of energy associated with creation of the AP spikes at sequentially firing nodes of Ranvier contributes to the plasmon-polariton assuring that its amplitude is beyond the activation threshold. The external energy supply (through the conventional ATP/ADP cell mechanism) assisting HH cycle at each node of Ranvier residually compensates the Ohmic losses of the plasmon-polariton mode propagating along the axon and ensures the undamped propagation over an unlimited range. Although the entire signal cycle of the AP on a single Ranvier node block requires several milliseconds (or even longer when one includes the time required to restore steady-state conditions, which, on the other hand, conveniently blocks the reversing of the signal), subsequent nodes are ignited more rapidly, corresponding to the velocity of the plasmon-polariton wave packet triggering the ignition of consecutive Ranvier nodes (in a period of one millisecond ca. 1000 nodes of Ranvier are ignited). Thus, we deal with the firing of the axon, which propagates with the velocity of the ion plasmon-polariton wave packet of ca 100 m/s. The direction of the velocity of the plasmon-polariton wave packet is adjusted to the semi-infinite geometry of the chain (in fact the chain is finite and is excited at one of its ends). The firing of the AP triggered by the plasmon-polariton traverses along the axon in only one direction because the nodes that have already fired have had their Na^+, K^+ gates discharged and require a relatively long time to restore their original status (their require a time of the order of even one second and sufficient energy supply to bring the ion concentrations to their steady values via cross-membrane active ion pumps against the concentration gradient).

The plasmon-polariton scheme described above for the ignition of AP spike formation in the ordered chain of Ranvier nodes along an axon is thus consistent with the saltatory conduction observed in myelinated axons. The observations that firing of the AP can simultaneously move in two opposite directions if a certain central node of Ranvier of a passive axon is ignited, as well as the observation of the maintenance of the firing traverse despite small breaks in the axon cord or a few damaged Ranvier nodes also agree with the collective wave-type plasmon-polariton model of saltatory conduction in contrast to the lack of satisfactory explanations in models based on the cable theory. The maintenance of the plasmon-polariton kinetics despite discontinuities in the axon cord agrees well with the discrete chain model.

In Fig. 7, the group velocity of the plasmon-polariton traversing a firing myelinated axon is plotted for various diameters of the axon internal cord, with a length of 100μm for each myelinated segment wrapped by Schwann cells and Ranvier intervals of 0.5, 5, and 10 μm. The dependence of the group velocity on the length of the Ranvier interval is weak (i.e., negligible at the scale considered, which is consistent with the equivalence of the discrete model for the continuous system if one considers wave-type plasmon-polariton propagation), but the increase in the velocity with increasing internal cord thickness is significant, similarly to linear increase in real axons with increasing diameters.

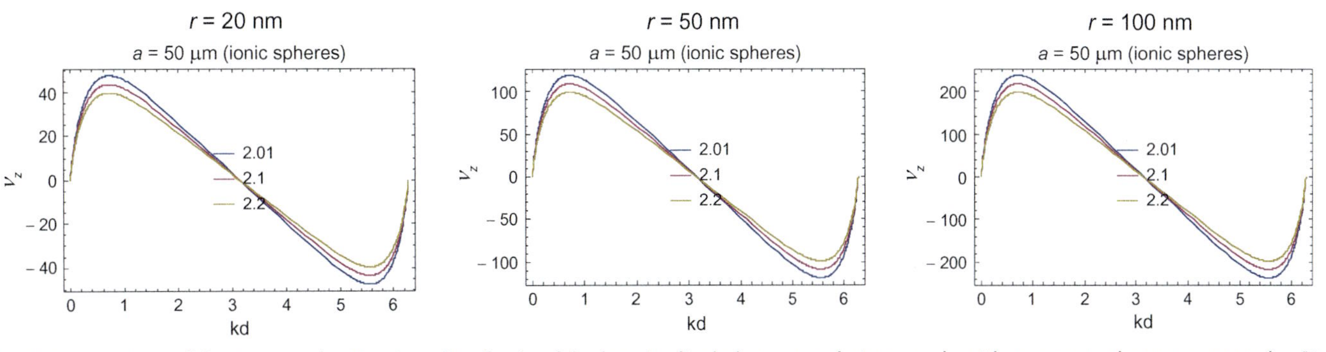

Fig. 7 Comparison of the group velocities, in units of m/s, of the longitudinal plasmon-polariton mode with respect to the wave vector $k \in [0, 2\pi/d)$ within the axon model for a Schwann cell myelinated sectors with a length of 100 μm, Ranvier separations of 0.5, 5, and 10 μm (represented by $d/a = 2.01$, 2.1, and 2.2 in the figure, respectively) and for the axon cord radii of $r = 20$, 50, and 100 nm. *Reproduced from Jacak, W. Propagation of Collective Surface Plasmons in Linear Periodic Ionic Structures: Plasmon Polariton Mechanism of Saltatory Conduction in Axons. J. Phys. Chem. C 2015, 119 (18), 10015 with permission.*

To comment on the appropriateness of the chain model for axons, let us note that even though the thin core of the axon is a continuous ion conducting fiber, the surface electromagnetic field can be closely pinned to the linear conductor similarly as to the Goubau line (well known from microwave technology)[53,54] and if periodically wrapped by dielectric shells, plasmon-polaritons propagate similarly as in a chain. For plasmon-polariton kinetics, the continuity or discontinuity of the conducting fiber is unimportant because we deal here with traversing wave packet of the synchronized plasma oscillations and not of a net current, similarly to Goubau microwave lines, which also have discontinuous segments impossible to be crossed by any current. The Goubau lines maintain their transmittance via discrete disconnected elements. The segments of the thin axon cord wrapped with the myelin shells and with ion concentration typical for neurons, $n' \simeq 10$ mM $\simeq 6 \times 10^{24}$ $1/\mathrm{m}^3$, inside the cell can be equivalently modeled by spheres with a diameter equal to the segment length and the ion concentration of $n \simeq 2 \times 10^{14}$ $1/\mathrm{m}^3$ (for the assumed axon cord diameter of 100 nm), conserving the number of ions in the segment. Such a model is justified by the same structure of the dynamics equation as that for dipole plasmon fluctuations in a chain of spheres, i.e., Eq. (6) and the modification of this equation for prolate spheroid or elongated cylindrical rod chains. This modification resolves itself to the substitution of the isotropic ω_1 frequency in Eq. (6) with frequencies that are different for each polarization, $\omega_{\alpha 1}$, i.e., $\left[\frac{\partial^2}{\partial t^2} + \frac{2}{\tau_{\alpha 0}}\frac{\partial}{\partial t} + \omega_{\alpha 1}^2\right] D_\alpha(ld,t) = \mathcal{A} \sum_{m=-\infty,\ m\neq l}^{m=\infty} E_\alpha\left(md, t - \frac{|l-m|d}{v}\right) + \mathcal{A}E_{L\alpha}(ld,t) + \mathcal{A}E_\alpha(ld,t)$, where $\mathcal{A} = V\frac{nq^2}{m}$ is a shape independent factor proportional to the number of ions at concentration n in the volume of the spheroid with semiaxes a, b, c, $V = \frac{4\pi}{3}abc = \frac{4\pi}{3}a^3$ (the latter for a sphere). Taking into account that the plasmon frequency in a bulk electrolyte with ion concentration n equals to, $\omega_p = \sqrt{\frac{nq^2 4\pi}{m}}$, one can rewrite $\mathcal{A}$ as follows, $\mathcal{A} = \frac{abc\omega_p^2}{3} = \varepsilon a^3 \omega_1^2$ (the latter for a sphere, for which $\omega_1 = \frac{\omega_p}{\sqrt{3\varepsilon}}$). The Ohmic losses can be included via the anisotropic term, $\frac{1}{\tau_{\alpha 0}} = \frac{\mathrm{v}}{2\lambda_B} + \frac{C\mathrm{v}}{2a^\alpha}$, where a^α is the dimension (semiaxis) of the spheroid in the direction α (equal to a, b, c for a spheroid). The first isotropic term in the expression for $\frac{1}{\tau_{\alpha 0}}$ approximates ion scattering losses such as those occurring in the bulk electrolyte (thus is isotropic), whereas the second term describes the losses due to scattering of ions on the anisotropic boundary of the spheroid. This term can be neglected for longitudinal polarization because the neuron cord is continuous along the z direction. The dipole coupling is independent of the shape of the chain elements. The mutual independence of dipole oscillations with distinct polarizations described above follows from the linearity of the dynamics equation (versus the dipole) regardless of the metal or electrolyte conducting elements.

Because the dynamics equation is not affected by the anisotropy, the solutions of the equation for each polarization have the same form as that for the spherical case with the exception of the modification of the related frequency of the dipole oscillations in each direction and the small correction of the orientation dependent contribution of the scattering ratio (this part is related to the boundary scattering of carriers and is not important for longitudinal polarization when the axon cord is continuous). Thus, we can independently renormalize the equation for dipole oscillations for each polarization direction, introducing the resonance oscillation frequency for each direction $\omega_{\alpha 1}$ in a phenomenological manner (these frequencies can be estimated numerically, whereas for a sphere, $\omega_1 = \frac{\omega_p}{\sqrt{3\varepsilon}}$; in general, the longer the semiaxis, the lower the related dipole oscillation frequency is).

The periodic structure of a myelinated axon does not form a chain of electrolyte spheres but rather is a thin electrolyte cord with periodically distributed myelinated sectors separated by very short unmyelinated intervals of Ranvier nodes. The periodic corrugated structure of the dielectric isolation allows, however, for collective plasmon wave-type oscillations, $\sim e^{iqz}$, with q governed by the periodicity. The wave-type propagation has the form of synchronic dipole oscillations of myelinated sectors. These dipole oscillations are equivalent to periodic polarization of Ranvier nodes. The model allows for quantitative estimation of the relevant propagation characteristics and verification of whether plasmon-polariton dynamics fits the observed features of the saltatory conduction in myelinated axons.

The relatively low polarization of Ranvier nodes induced by plasmon-polaritons initiates opening of the Na^+ and K^+ across-membrane ion channels at nodes of Ranvier, which results in a characteristically large AP signal formation due to the transfer of ions through the open gates caused by the difference in ion concentrations on opposite sides of the membrane. The entire HH cycle at a single node of Ranvier requires several milliseconds, but the initial increase in polarization needed to the rapid opening of the Na^+ channel occurs on the microsecond timescale and must be synchronized with the plasmon-polariton frequency. Each Ranvier node activity triggered in series by plasmon-polariton supports, on the other hand, the plasmon-polariton wave packet amplitude to maintain the constant its value despite Ohmic losses (Figs. 8 and 9).

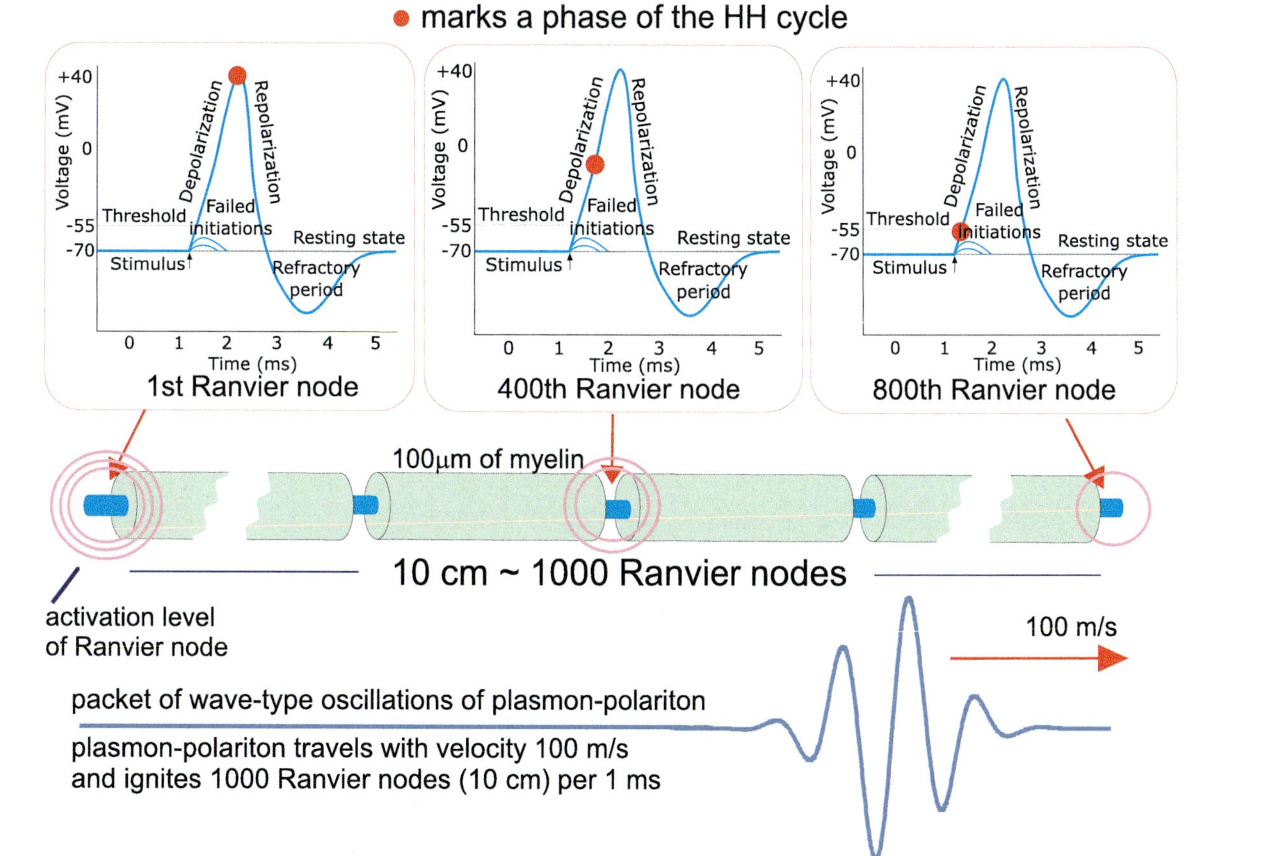

Fig. 8 Schematic presentation of the firing of 10 cm long myelinated axon. The wave packet of plasmon-polariton oscillations travels along the axon with the velocity of 100 m/s and in time of 1 ms ignites 1000 Ranvier nodes (as the myelinated segments have a length of 100 μm each). The Hodgkin–Huxley (HH) cycle of the AP spike creation takes ca. 1 ms, thus consecutive Ranvier nodes are in various phases of HH cycle, as indicated by *red points*. Hence, the whole 10 cm long axon is in firing within the time period of 1 ms.

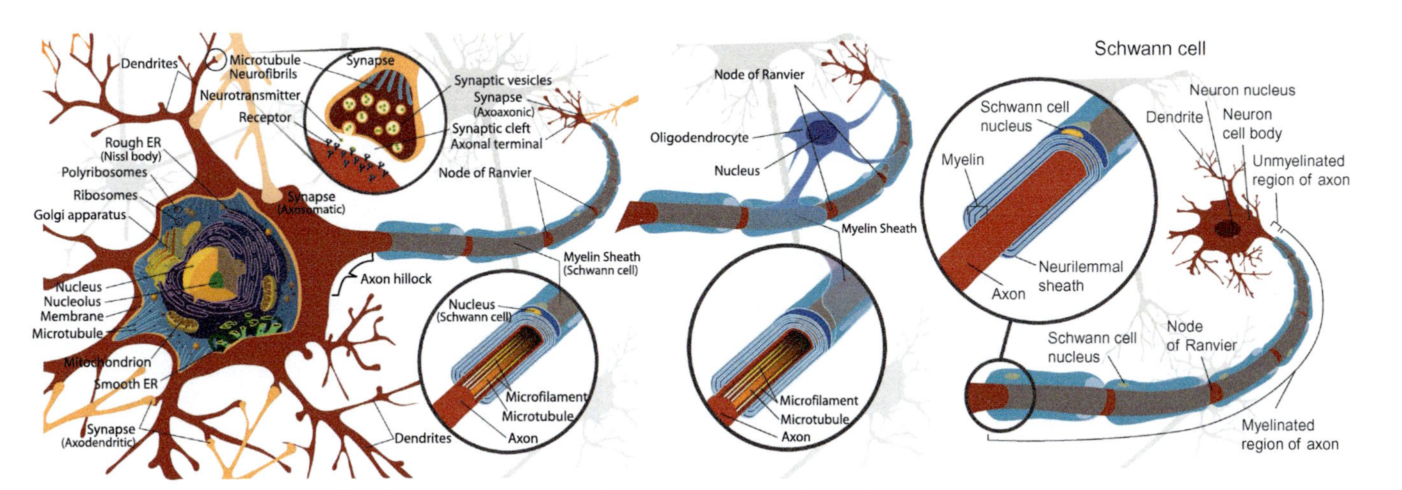

Fig. 9 Imagine of the neuron cell with myelinated axon indicated. In the central neural system the white matter axons are myelinated by oligodendrocytes whereas in the peripheral system by Schwann cells. *Reproduced from Wikipedia Public Domain.*

Plasmon-polaritons do not interact with external electromagnetic waves or, equivalently, with photons (even at adjusted frequency), which is a consequence of the large difference between the group velocity of plasmon-polaritons and the velocity of photons ($c/\sqrt{\varepsilon}$). The resulting large discrepancy between the wavelengths of photons and plasmon-polaritons of the same energy prohibits mutual transformation of these two types of excitations because of momentum–energy conservation constraints. Therefore, plasmon-polariton signaling by means of collective wave-type dipole plasmon oscillations along a chain, i.e., plasmon-polaritons, can be neither detected nor perturbed by external electromagnetic radiation. This also fits well with neuron signaling properties in the PNS and in the white myelinated matter in the CNS. The temperature influences the mean velocity of ions, $\mathsf{v} = \sqrt{\frac{3kT}{m}}$, thereby enhancing the Ohmic losses with increasing temperature (cf. Eq. 7), which in turn strengthens plasmon-polariton damping. Hence, at higher temperatures, higher external energy supplementation is required to maintain the same long-range propagation of plasmon-polaritons with a constant amplitude. This property is also consistent with experimental observations.

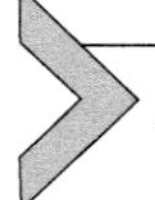

5. Soft plasmonics toward recognition of the different role of gray and white matter in information processing

The electrical activity of the cortex in a brain is referred to the transduction of electric signals through axons and dendrites connected together by electrochemical synapses into a highly entangled web. Not all trajectories for signal travel are accessible (blocked by passive synapses or by their absence) and the activation of some synapses makes selected paths available for communication for longer or shorter time-periods depending on the intensity and repetition of the synapse stimulation. Via the activation of selected synapses, the nervous system creates an entangled pattern resembling a braided web of trajectories. The classification of the resulting entanglement patterns may be done in topological terms of so-called pure braid groups[55] (cf. Appendix B). Elements of these groups allow to distinguish complicated filament muddles in a mathematically rigorous topological-homotopy way. Two braids are homotopic if one can be transformed into the other by continuous deformations without cutting. For N threads, the number of various topologically inequivalent braid patterns is infinite though countable,[55] in practice limited only by size restrictions such as finite filament diameter and length. The variety of braid patterns highly exceed ordinary addressing, for which N filaments gives only $N!$ opportunities, cf. Figs. 10–12.

If one activates electrically such an entangled network by pushing an AC charge current along the threads, one notices that the unique pattern of braiding with a specific number of loops might serve as a multiloop e-m active circuit, such as a distributed coil, cf. Fig. 13. The AC electric current in this braid-coil could generate an individual unique e-m field space-configuration that could fall into resonance with another even distant part of the web representing the same braid group element. Thus, one can associate information messages (elements of the memory) with various elements of the pure braid group imprinted in a nonlocal manner in fragments of the large neuron web of the brain via the activation and temporal consolidation of a certain subset of synapses. This message could be temporarily stored in the neuron-braid as long as the selected engaged synapses are still active (or are ready to be active). This information can be next invoked by the e-m resonance with a similar braid within the brain temporarily created in another brain structure, e.g., in the hippocampus as the result of an influx of neuron signals encoding the information form surroundings via the senses. The new message tentatively creates a temporary braid in the hippocampus which can excite the same braid recorded in the past in the cortex via e-m resonance, which in this way allows for the identification of a new message in the mind.

To utilize the braids in the proposed e-m recognition process, AC electric currents are necessary to initiate the e-m resonance. The role of such AC electricity may play the brainwaves that are constantly observed via EEG in the gray matter of the cerebral cortex. For transduction of neuron signals in nonmyelinated axons and in all dendrites in the cortex, the diffusive movement of ions is employed. This electric current-type communication is described very well by the cable theory.[24–28]

Hence, one can expect that the storage and identification of information take place in the cortex, taking an advantage of the huge information capacity of entanglement of dendrites and nonmyelinated axons creating various braid structures numbered by pure braid group elements. The precise homotopy resolution and quick and complex access to the stored messages in entangled braids[55–57] via the e-m resonance, additionally using a few distinct frequencies of brain waves, might be helpful to better understand the functionality of memory and pattern recognition. All these phenomena occur in the gray matter of the cortex, where signal transduction is accomplished by the diffusion of ions equivalent to an ordinary electric AC current

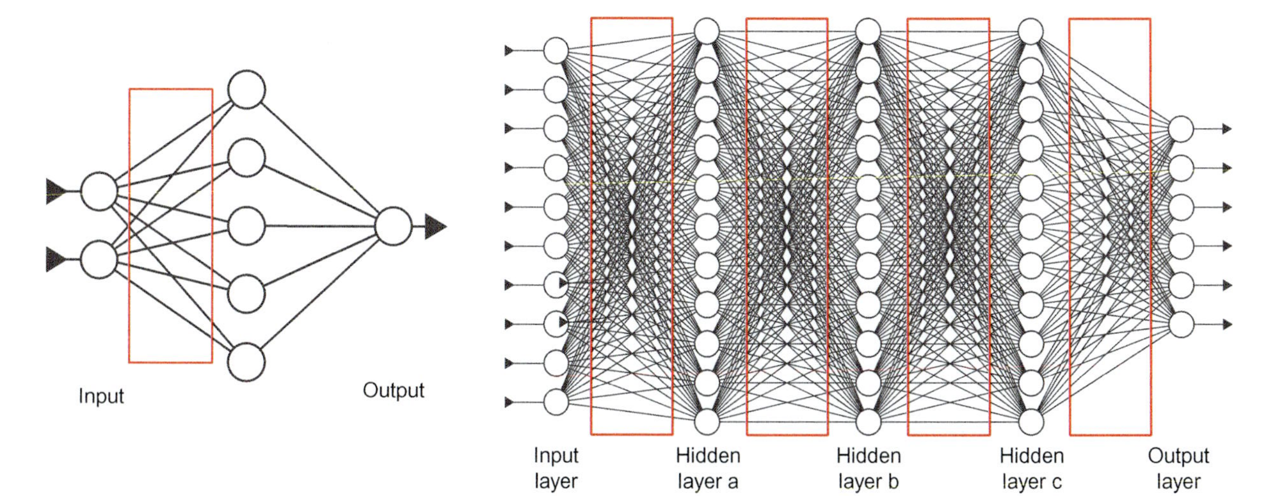

Fig. 10 In conventional neuron web concept, the information has been coded in addressing of connections between registers indicated in the pictures by *dots*. In the braid group concept of the neuron web organization, the information is coded rather in entanglement of paths connected registers (the regions marked by frames); the information resource of homotopy distinct braids of N lines is infinite, whereas the resources of addressing of two N element registers is only $N!$.

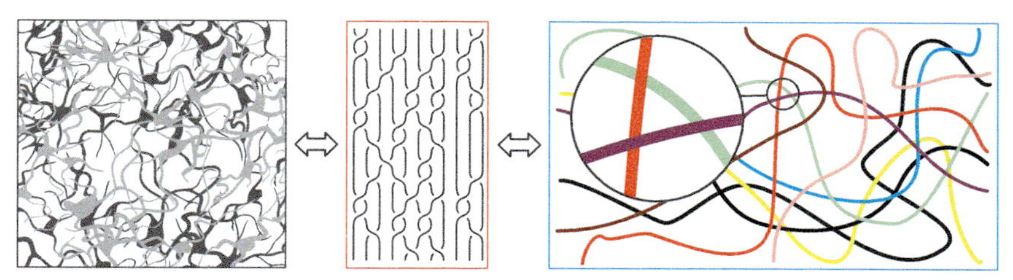

Fig. 11 An example of braidings of neuron paths (dendrites and nonmyelinated axons connect via synapses)—*left panel*; the corresponding element of the pure braid group—*central panel*; the entanglement path pattern cannot be changes by any deformations without cutting—*right panel*.

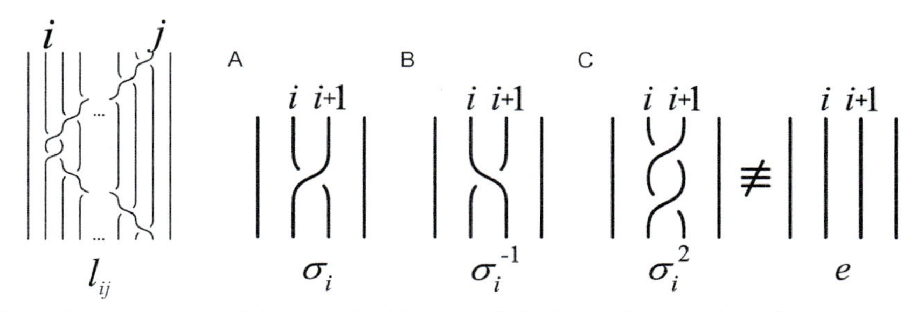

Fig. 12 Generator l_{ij} of the pure braid group (*left*), the exchange of neighboring particles, σ_i (A) and its inverse, σ_i^{-1} (B) and the reason why braids on a plane are complicated, because this inequality holds for 2D manifolds but not for 3D[55] (C).

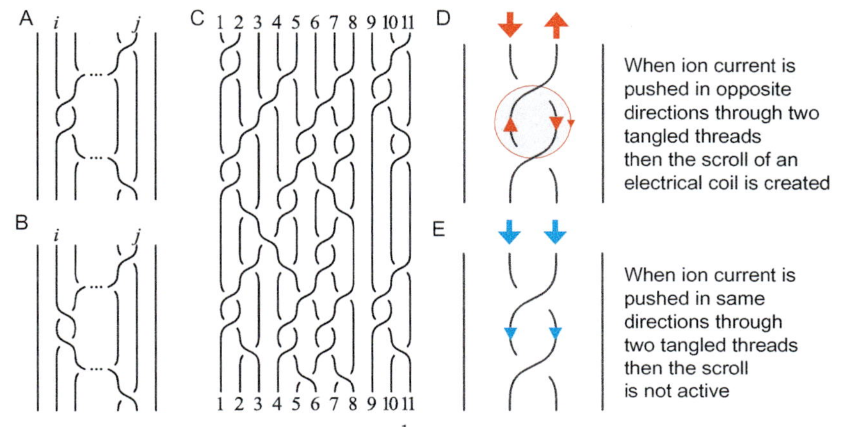

Fig. 13 Generator l_{ij} (A) and its inverse l_{ij}^{-1} (B) of the pure braid group; an example of the $N = 11$ pure braid (C); the scheme of creating of an e-m active (D) (nonactive (E)) scroll of coil by entanglement of threads in a pure braid.

which induces a local e-m field around a braid-coil. Different types of brainwave frequencies might be convenient here to implement simultaneous independent resonance channels for better identification of multiloop coils created by neuron filament braids.

The topological characteristics of entangled filament webs, like neuron webs, may be described in terms of homotopy, reflecting the complexity of braids precisely defined in a mathematical way. The collection of all braids creates the algebraic group being the first homotopy group marked with $\pi_1(\mathcal{A})$ where the space $\mathcal{A}$ is the configuration space of a system of N particles located on the same manifold M, each of which runs its own trajectory on M (for more detail cf. Appendix B).

We propose to identify a fragment of the cortex filaments connected via active synapses predefined by a prior record of a certain information in a web as a specific pattern from the pure braid group. In particular, one can use binary code to record any message in a 3-filament bundle (two of the generators of the corresponding $N = 3$ pure braid group serve to code 0 and 1 bits, and the third generator encodes the end of a word). It is possible to code an N-letter alphabet message in an N-order pure braid group. We demonstrated that the ratio of the information efficiency of an N-order pure braid group to the resources needed to organize a physical web (like a neuron web with N entangled filaments) attains a maximum at $N = 20$–30, which agrees with the number of phones in the majority of languages.[57] This result may suggest that words are coded in a relatively small number of filaments in entangled bunches (Fig. 14).

The braid-like coding of information is topological and nonlocal and is more efficient that the conventional addressing of synapse registers (the register of N synapses offers only $N!$ various addressing connections with another N synapse register, whereas the number of distinct braids of threads linking these registers is infinite, limited only by the physical constraint of the finite size, diameter, and length of the threads).

To identify neuron braids, we propose e-m resonance utilizing the ion currents that flow in the neuron filaments of the gray matter. The different role plays the white matter with myelinated axons corrugated by the periodic myelin sheaths. The saltatory conduction in myelinated axons has the plasmon-polariton character without any e-m signature, thus cannot participate in e-m resonance.

The different electricity of the gray and white matters of neurons causes that the web of neuron filaments in gray matter would serve to encode the information in the entanglement of filaments selected by activation of certain synapses. The e-m signature of a particular braid, when an AC signal is

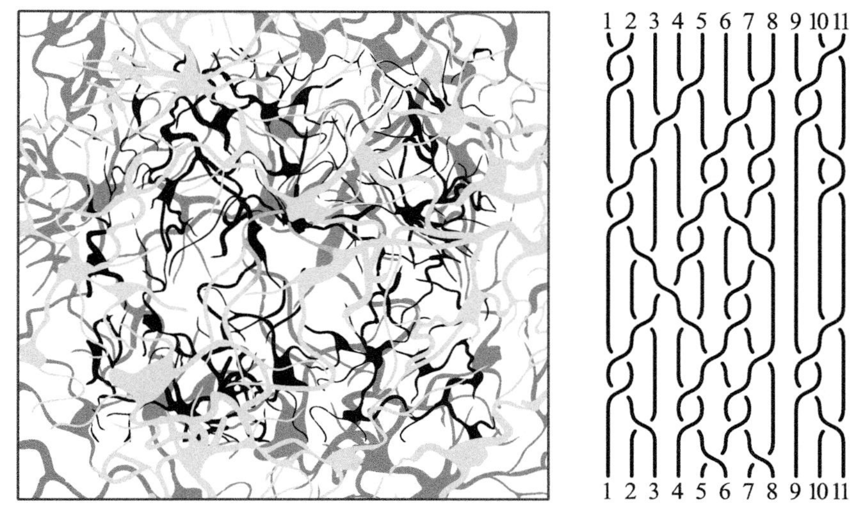

Fig. 14 Various paths in the neuron web are activated when certain synapses are active—the active synapses linking various dendrites and nonmyelinated axons define paths which are highly braided in the brain—these braids can be precisely enumerated by unique elements of the pure braid group.

transduced through the neuron network, can be recognized via the e-m resonance with a similar braid in another part of the cortex. The white matter does not take part in e-m resonance and serves only as the communication channel. Both myelinated and nonmyelinated structures share between them different but complementary roles in the whole nervous system.

To illustrate the information capacity of braids let us consider three filaments linking three steady points with another three fixed point and the information is associated with various types of filament entanglements. The pure braid group describing such entanglements is generated by three generators (cf. Appendix B), which can represent the bits 0 and 1 and the end-mark of a word in the binary code. Similarly, one can consider N filaments linking N points, and the N-element alphabet for the coding of information. Shortening the length of the information code in a N-element alphabet, compared to the length of information coded in binary, will therefore be expressed as $\log_2(N)$. On the other hand, an increase in the number of alphabet characters (the number of lines N) makes the coding structure more complex, which increases the consumption of resources used for building subsequent connections of strands. The function of resource expenditure (energy, materials) necessary to organize a network can be modeled in the power function form, e.g., $g_1 = N, g_2 = N^{\frac{1}{2}}, g_3 = N^{\frac{1}{3}}, g_4 = N^{\frac{1}{4}}$. A function graph $I_i(N) = g_i(N)/(\log_2 N)$—Fig. 15 shows the minimal value, at which

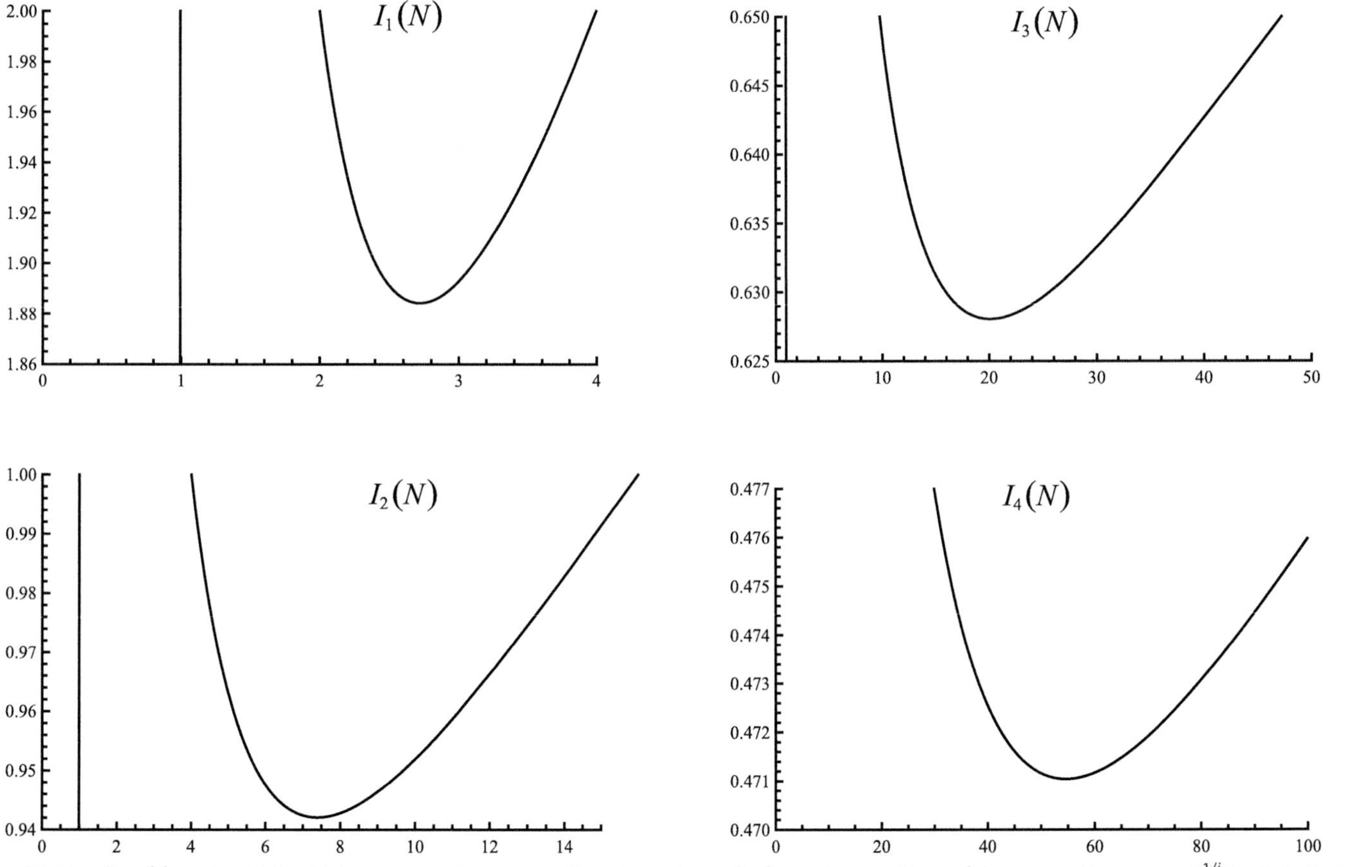

Fig. 15 Graphs of function $I_i(N)$, which expresses the expenditure to code profit, for various scalings of the expenditure, $g_i \sim N^{1/i}$ $i = 1, 2, 3, 4$. *Reproduced from Jacak, J.; Gonczarek, R.; Jacak, L.; Jóźwiak, I. Application of Braid Groups in 2D Hall System Physics. World Scientific: Singapore, 2012 with permission.*

coding in N alphabets (in N strand braids) is optimal in terms of the expenditure-information capacity ratio.

This result seems interesting, as it shows that the information capacity of entanglements of braids reaches maximum at about 20 strands, Fig. 15. A small network, corresponding to an alphabet with about 20 elements is therefore the most information optimal and economical in terms of the expenditures for organizing 3D entangled bundle networks. Surprisingly, there is a coincidence between this number and the number of phonemes (sounds) informatively used in most languages, which may point to a braids, not addressing structure for saving and processing linguistic information in neuronal networks. From the above analysis it follows that optimal are bunches of ca 20 neuron pats to code linguistic terms. This seems to agree with some observations that such a information packages are actually coded in certain small fragments of the neuron web in the brain cortex.

6. Conclusion

The utilization of the radiatively undamped plasmon-polariton propagation in a chain of electrolyte subsystems may explain efficient and long-range saltatory conduction in myelinated axons in the peripheral neural system and in the white matter of the brain and spinal cord. The effective plasmon-polariton model of the triggering of AP firing along an axon myelinated by Schwann cells separated by nodes of Ranvier fits well with the high conduction velocity observed at the saltatory conduction in agreement with the temperature and size dependence (with respect to a diameter of the axon) of the conduction velocity. This coincidence together with the immunity of plasmon-polaritons to external e-m perturbations or detection support the reliability of the new model proposed for the saltatory conduction in myelinated axons, which is very efficient, quick, and energy frugal despite the poor ordinary conductivity of axons.

Worth noting is some quantum aspect of plasmon-polaritons, which are synchronized in a wave form simultaneous plasmon oscillations in all segments of a chain. The plasmon oscillations in a particular segment are, however, of quantum nature (regardless of whether electrons or ions), though seem to be quite ordinary oscillations of charge fluctuations. Plasmons are coherent oscillations of all charges in the subsystem and this coherence can be understood exclusively in quantum terms (e.g., of random phase approximation by Pines[14] and Bohm[15]).

The different electricity of myelinated axons in comparison to dendrites and nonmyelinated axons indicates a different role of the gray and white matter in information processing in a brain. Emphasizing these differences we propose a topological concept of the memory functionality utilizing braids and their e-m signature in the brain cortex to store and to identify information. The e-m passive white matter does not take a part in such e-m activity and plays only the communication role.

Acknowledgment

This study was supported by NCN project P.2018/31/B/ST3/03764.

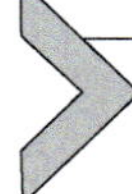

Appendix A. The role of the thickness of the myelin sheath

It is known that the thickness of the myelin layer is an essential factor deciding on the proper functioning of myelinated axons. This thickness is greater than that needed to only isolation because the role of myelin is more specific not related with electrical isolation. In the case of plasmon dipole oscillating in the myelinated segment, as the part of the plasmon-polariton, the ion oscillations in the axon cord excite the oppositely directed also oscillating dipole of ions in the outer electrolyte, in the cave surrounding the myelinated segment, cf. Fig. A.1. Coupling of both dipoles across the myelin

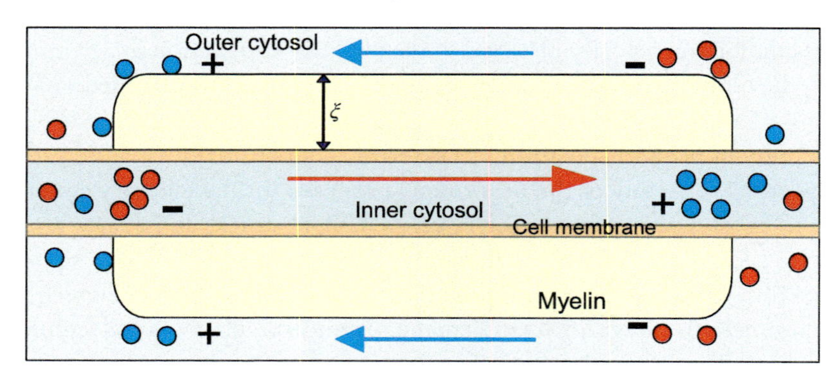

Fig. A.1 Cartoon of a single myelinated segment—polarization of longitudinal dipole type of the inner cytoplasm inside the axon cell (*red arrow*) induces local opposite polarization of the outer cytoplasm (*blue arrows*). Both dipoles oscillate as the pair of coupled oscillators across the insulating myelin layer. The coupling is weak for sufficiently thick myelin sheath and causes slow beats due to oscillator coupling. At the nodes of Ranvier the coupling across much thinner cell membrane is strong and causes quick beating out of a resonance with ion gate timing, thus damped. Slow beating oscillations of coupled dipoles are ranged only to the myelinated sector.

sheath of thickness, ξ, is of the near-field coupling form. By $d_1(t)$ let us denote the longitudinal dipole in the axon rod in this myelinated segment, and by d_2 the dipole in the outer cytosol cave adjacent to the myelinated segment and induced by d_1 dipole. Oppositely directed d_2 dipole is activated by d_1 and vice verse. Equation for the dynamics of this subsystem is as follows,

$$\begin{aligned}\frac{d^2 d_1}{dt^2} + \frac{2}{\tau_1}\frac{dd_1}{dt} + \omega_1^2 d_1 = \mu_1 E_1,\\ \frac{d^2 d_2}{dt^2} + \frac{2}{\tau_2}\frac{dd_2}{dt} + \omega_2^2 d_2 = \mu_1 E_2,\end{aligned} \tag{A.1}$$

where $\frac{1}{\tau_i}$ is the damping rate of the surface plasmon i-th dipole, ω_i its self-frequency, μ_i is the longitudinal polarizability of the rod for $i = 1$ and of the cave in surrounding cytosol for $i = 2$ (the polarizability for a sphere equals to $\frac{a^3 4\pi n q^2}{3M}$, a—the sphere radius, n—concentration of ions, M—ion mass, q—ion charge). The electrical fields induced by dipoles in the near-field zone are, $E_{1(2)}(t) = -\frac{1}{\varepsilon\xi^3} d_{2(1)}(t - \frac{\xi}{v})$, where ξ is the thickness of the myelin sheath, $v = \frac{c}{\sqrt{\varepsilon}}$, and these fields describe mutual interaction of dipoles d_1 and d_2, i.e., the electrical induction caused by opposite dipoles in the near-field coupling approximation. Simplifying (for illustration) by the assumption, $\mu_1 = \mu_2 = \mu$, $\omega_1 = \omega_2 = \omega_0$, and neglecting here the damping and retardation, Eq. (A.1) attains the shape,

$$\begin{aligned}\frac{d^2 d_1}{dt^2} + \omega_0^2 d_1 = -\frac{\mu}{\varepsilon\xi^3} d_2,\\ \frac{d^2 d_2}{dt^2} + \omega_0^2 d_2 = -\frac{\mu}{\varepsilon\xi^3} d_1,\end{aligned} \tag{A.2}$$

This is the equation for two coupled harmonic oscillators. It has the self-frequencies, for assumed solution, $d_1(t) = Ae^{i\Omega t + \phi}$ and $d_2(t) = Be^{i\Omega t + \psi}$, given by,

$$det\begin{bmatrix} -\Omega^2 + \omega_1^2 & , \frac{\mu}{\varepsilon\xi^3} \\ \frac{\mu}{\varepsilon\xi^3} & , -\Omega^2 + \omega_1^2 \end{bmatrix} = 0, \tag{A.3}$$

i.e., $\Omega_1^2 = \omega_0^2 + \frac{\mu}{\varepsilon\xi^3}$, $\Omega_2^2 = \omega_0^2 - \frac{\mu}{\varepsilon\xi^3}$.

For the initial condition suitable for excitation of the considered segment of the axon, i.e., $d_1(0) = D$, $d_2(0) = 0$, $\frac{dd_{1(2)}}{dt}(0) = 0$, one gets the beating with low frequency, $\Omega_1 - \Omega_2 \simeq \frac{\mu}{\varepsilon\xi^3\omega_0}$. For sufficiently large ξ we get thus slow oscillations required for the time scale of opening of the Na^+ ion channels at nodes of Ranvier to trigger the HH cycle. The period of the igniting signal (with an amplitude beyond the threshold for Na^+ ion channel opening) cannot be lower than the characteristic time of the activation of these ion channels.

The frequency for longitudinal plasmon is reduced in strongly elongated axon rod segment with the aspect ratio $\sim 10^{-3}$ and additionally reduced by an appropriate increase of ξ due to the beating effect and finally achieves the value of $\sim 10^6$ 1/s, resulting in the saltatory conduction velocity ~ 100 m/s. Thus we see, that the myelin layer thickness controls the velocity of the saltatory conduction and simultaneously accommodates the frequency of the igniting signal oscillation (of plasmon-polariton wave packet) to the time scale of triggering of the ion channels at nodes of Ranvier (being of order of a microsecond). Only this frequency is selected and the corresponding plasmon-polariton wave packet is strengthened by synchronized HH cycles in contrary to other nonsynchronized frequency modes of the plasmon-polariton. Nonsynchronized modes are quickly damped due to Ohmic losses.

At the Ranvier node the inner and outer cytosol are separated by the thin bare cell membrane, thus dipole coupling across the thinner barrier is stronger and the corresponding beating quicker. The quick component of this beating is also quenched as not synchronized with the time scale of electrically gated Na^+ channels, in contrary to slow component of the trans-myelin beating. Via synchronization with HH cycle ignition time scale, this selected mode of plasmon-polariton is continuously supplemented in energy by HH cycles and simultaneously is able to ignite the HH cycle on consecutive Ranvier nodes along the chain of myelinated segments on arbitrary large distances. This explains why only the myelinated segments oscillate and the frequency of related plasmon-polariton wave packet is precisely tuned by the myelin thickness, ξ. The oscillation of longitudinal surface plasmon on the myelinated segment of axon rod is equivalent with opposite dipole oscillation (polarization) of the node of Ranvier (but it is not its own self-oscillation). This is schematically shown in Fig. 5D.

The described mechanism of selection of the low self-frequency of plasmon oscillations of the myelinated fragment allows for accommodation of the frequency of plasmon-polariton (via the equation defining dynamics of plasmon-polariton in the chain of myelinated segments, i.e., Eq. (10)). Not all frequency modes of plasmon-polariton are persistent. Ohmic damping causes their attenuation on a short distance unless the energy is supplied by AP formation at consecutive nodes of Ranvier. This requires, however, a perfect coincidence of the plasmon-polariton mode frequency with timing of opening of Na^+ ion channels across the axon membrane at nodes of Ranvier. Too quick oscillations are not able to trigger the opening of the gates which eliminates corresponding modes of plasmon-polariton. Too low oscillations reduce velocity of corresponding nodes and are retarded with respect to the optimal ones. This simple mechanism selects the optimal and persistent wave packet of plasmon-polariton with the velocity observed in the saltatory conduction. This wave packet is supplied with energy by AP formation at consecutive nodes of Ranvier. One can state that the thickness of the myelin sheath is a control factor for synchronization of plasmon polariton with the AP formation mechanism needed for overcoming of the Ohmic attenuation of the plasmon-polariton ignition signal over arbitrary large distances. Reducing of the myelin sheath thickness increases frequency of the plasmon-polariton which causes, however, de-synchronization with opening time of Na^+ channels, which perturbs plasmon-polariton kinetics, just like at Multiple Sclerosis. Appendix

Appendix B. Braid groups—Preliminaries

Mathematical formalism especially suitable to describe entangled web of neurons in the brain cortex (and in the gray matter of cerebellum or spinal cord) is the method of pure braid groups. These groups collect classes of homotopic trajectory loops in the configuration space of N particles (selected points). Loops are homotopic if one loop can be continuously transformed by a deformation without cutting into another one. Loops in configuration space of N particles can be represented by strands between ordered in line N points and the same ordered points as shown in geometrical presentation in e.g., Figs. 12 and 13.[a] [55] The configuration space of N particles depends on the mathematical space M (called as manifold) on which all N particles are placed (e.g., a plane, 3D space, sphere, and so on),

[a] The geometric presentation consists in depicting of some initial points position of all N particles on above and the same positions of these points in the bottom, linked by N arbitrarily entangled strands representing trajectories of particles.

$$F_N(M) = M^N - \Delta, \tag{B.1}$$

where $M^N = M \times M \times \cdots \times M$ is N-th fold normal product of the M manifold and Δ is the subset of M^N of its diagonal points, when at least two particle positions coincide. Subtracting of the diagonal subset from M^N assures that the number of particles (and of strands in graphical presentation) is conserved. Loops can be added one to another one creating in this way new loops. This operation satisfies the condition required for an algebraic group structure. The braid groups are very well known in algebraic topology and are the first homotopy groups $\pi_1(F_N(M))$[58]. First homotopy group $\pi_1(\mathcal{A})$, can be defined for any space $\mathcal{A}$ and it displays the topology of $\mathcal{A}$. If $\mathcal{A}$ is simply connected space,[b] then its first homotopy group is trivial, $\pi_1(\mathcal{A}) = e$, where e is the neutral element of the group. First homotopy groups of multiply connected spaces are nontrivial. The first homotopy group $\pi_1(\mathcal{A})$ is often called also as the fundamental group of the space $\mathcal{A}$.[58]

If as the space $\mathcal{A}$ one takes a configuration space, $F_N(M)$, then we arrive with the $\pi_1(F_N(M))$ which is the pure braid group of N particles. It is interesting that the $\pi_1(F_N(M))$ essentially depends on the manifold M. For three dimensional manifolds $\pi_1(F_N(M)) = e$, i.e., all braids in 3D space can be disentangled without any cutting of strands. However, for two-dimensional M, like a plane, the braid group $\pi_1(F_N(M))$ is nontrivial infinite group. It means that it exists an infinite number of distinct braids, i.e., of distinct entanglements of N strands, which cannot be transformed one into another by any continuous deformations without cutting. Such nontrivial topological property means that arbitrary entanglement of strands linking N fixed points representing in graphical presentation N points on a plane numbered (in an arbitrary but fixed manner) by indices $1, \ldots, N$ cannot be disentangled if presents a nontrivial braid. The geometrical presentation of braids for particles on a plane can be 1-1 mapped onto specific and unique entanglement patterns of strands in 3D between N fixed steady initial and final points.[55]

This property of pure braids for N particle on the plane perfectly fits to three dimensional web of strands linking N fixed start points with another N fixed points, cf. Fig. B.1. These fixed start points can be numbered by index $i = 1, \ldots, N$ and the same for the fixed end points. Then the family of all graphical presentations of elements of the braid group, $\pi_1(F_N(M)) = \pi_1(M^N - \Delta)$, for $M = R^2$ (a plane), is 1-1 equivalent (isomorphic) to a

[b] The space $\mathcal{A}$ is simply connected if any loop in this space can be continuously deformed (shrunk) to a point; for example the plane R^2 is simply connected, but the plane with a hole is not simply connected, because a loop encircling a hole is not contractible to a point.

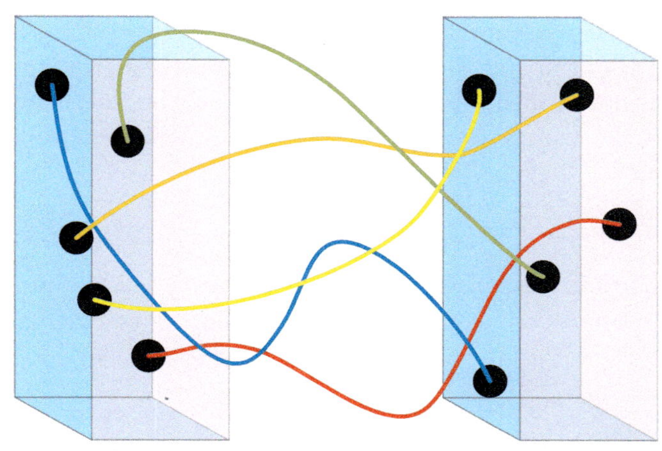

Fig. B.1 Fragment of the neuron web; selected but fixed start points are linked by neuron filaments with the same number of fixed final points; important is the braid representing entanglement of strands in the bundle.

collection of arbitrarily entangled bundles of N strands linking two arbitrary but fixed 3D distribution of start points with also arbitrary but fixed 3D distribution of end points. Such entangled 3D webs can be associated with fragments of the cortex web of neurons. Various types of entanglement of linking strands may be regarded as information carriers, which can be assigned by the corresponding braids from the pure braid group, $\pi_1(F_N(M))$, with $M = R^2$. The information capacity of such a bundle is infinite. Note that the conventional addressing of N start points to other N end points offers only the information capacity equal to $N!$, which is negligible small in comparison to numbers of braids even if the number of entanglements is finite for each strand due to constraints imposed by the geometry and size of real neuron filaments with nonzero thickness and length.

We postulate thus that such fragments of the cortex can serve as the information carriers for a memory storage by the physical realization of particular braids utilizing nonmyelinated axons, dendrites and synapses to organize strands linking N initial and N final points.

Each such bundle is unique and can be represented by an appropriate element of the pure braid group, $\pi_1(F_N(M))$, with $M = R^2$. This group has a countable structure of multi-cyclic group with generators.[55] The generators correspond to the simplest entanglement of two trajectories of a particle pair, i-th and j-th ones, whereas the rest of particle trajectories remain untangled. The generators l_{ij} can be expressed by σ_i (exchanges of neighboring particles i-th and $(i+1)$-th ones, not conserving their positions,[55] σ_i are thus not elements of the pure braid group but are convenient to mathematically express the generators and next any element from the pure braid group),

$$l_{ij} = \sigma_{j-1} \cdot \sigma_{j-2} \ldots \sigma_{i+1} \cdot \sigma_i^2 \cdot \sigma_{i+1}^{-1} \ldots \sigma_{j-2}^{-1} \cdot \sigma_{j-1}^{-1}, \quad 1 \geq i \geq j \geq N-1. \tag{B.2}$$

The generators l_{ij}, presented geometrically in Figs. 12 and 13, must fulfill the following relations[55,57] in order to generate the pure group, $\pi_1(F_N(M))$, with $M = R^2$.

$$l_{rs}^{-1} \cdot l_{ij} \cdot l_{rs} = \begin{cases} l_{ij}, & i < r < s < j \\ l_{ij}, & r < s < i < j \\ l_{rj} \cdot l_{ij} \cdot l_{rj}^{-1}, & r < i = s < j \\ l_{rj} \cdot l_{sj} \cdot l_{ij} \cdot l_{sj}^{-1} \cdot l_{rj}^{-1}, & i = r < s < j \\ l_{rj} \cdot l_{sj} \cdot l_{rj}^{-1} \cdot l_{sj}^{-1} \cdot l_{ij} \cdot l_{sj} \cdot l_{rj} \cdot l_{sj}^{-1} \cdot l_{rj}^{-1}, & r < i < s < j. \end{cases} \tag{B.3}$$

Possibility to represent of any element of our pure braid group by the group-sum of generators is convenient to distinguish between various entanglements in an arbitrary bundle of neuron filaments.

Such an coding of information in braids of pure braid group realized by activation of some synapses in the cortical web of neurons is the nonlocal and topological effect.

The advantage of the coding of information in nonlocal entanglements of N neuron strands (created by connected axons and dendrites via synapses) would be a simplified model of memory storage system in the cortex. Via all paths in the neuron web are continuously pushing brainwaves, being the AC electrical ion current consisting of several frequency components (between 0.5 and 100 Hz). The selected bundle of N neuronal path entangled as the corresponding braid acts as a multiloop coil of electrical conductors with AC current. Dendrites and nonmyelinated axons operate according to the cable theory with ordinary electrical diffusive currents of ions. Such currents have an electromagnetic signature with low frequency corresponding to frequency of brainwaves. If in another brain region (e.g., hippocampus) arises a new braid, let say, due to signal income from senses, then via the electromagnetic resonance the same braids already stored in the cortex is instantly recognized as fallen into an electromagnetic resonance and thus featured (strengthened) electrically between other segments, i.e., other braids not fitting to the new braid excitation. This would be a rough scheme for the memory mechanism.

References

1. Hodgkin, A.; Huxley, A. A Quantitative Description of Membrane Current and its Application to Conduction and Excitation in Nerve. *J. Physiol. (London)* **1952**, *117*, 500.
2. Rall, W.; Burke, R.; Holmes, W.; Jack, J.; Redman, S.; Segev, I. Matching Dendritic Neuron Models to Experimental Data. *Physiol. Rev.* **1992**, *72*, S159.
3. Rall, W. Cable Theory for Dendritic Neurons. In: *Methods in Neuronal Modeling: From Synapses to Networks*; Koch, C., Segev, I., Eds.; MIT Press, Cambidge, MA, 1989.
4. Rall, W. Core Conductor Theory and Cable Properties of Neurons. In: *Handbook of Physiology, Section 1, The Nervous System*; Kandel, E. Ed.; American Physiol. Society: Bethesda, 1977.
5. Jack, J.; Noble, D.; Tsien, R. *Electric Current Flow in Excitable Cells*. Clarendon Press, Oxford, 1983.
6. Cooley, J.; Dodge, F. Digital Computer Solutions for Excitation and Propagation of the Nerve Impulse. *Biophys J.* **1966**, *6*, 583.
7. FitzHugh, R. Dimensional Analysis of Nerve Models. *J. Theor. Biol.* **1973**, *40*, 517.
8. Barnes, W. L.; Dereux, A.; Ebbesen, T. W. Surface plasmon subwavelength optics. *Nature* **2003**, *424*, 824.
9. Brongersma, M. L.; Hartman, J. W.; Atwater, H. A. Electromagnetic Energy Transfer and Switching in Nanoparticle Chain Arrays Below the Diffraction Limit. *Phys. Rev. B* **2000**, *62*, R16356.
10. Maier, S. A. *Plasmonics: Fundamentals and Applications*. Springer VL: Berlin, 2007.
11. de Abajo, F. J. G. Optical Excitations in Electron Microscopy. *Rev. Mod. Phys.* **2010**, *82*, 209.
12. Pitarke, J. M.; Silkin, V. M.; Chulkov, E. V.; Echenique, P. M. Theory of Surface Plasmons and Surface-Plasmon Polaritons. *Rep. Prog. Phys.* **2007**, *70*, 1.
13. Berini, P. Long-Range Surface Plasmon Polaritons. *Adv. Opt. Photonics* **2009**, *1*, 484.
14. Pines, D. *Elementary Excitations in Solids*. ABP Perseus Books: Massachusetts, 1999.
15. Pines, D.; Bohm, D. A Collective Description of Electron Interactions: II. Collective vs Individual Particle Aspects of the Interactions. *Phys. Rev.* **1952**, *85*, 338.
16. Jacak, W. A. Plasmons in Finite Spherical Electrolyte Systems: RPA Effective Jellium Model for Ionic Plasma Excitations. *Plasmonics* **2016**, *11*, 637.
17. Jacak, W. Propagation of Collective Surface Plasmons in Linear Periodic Ionic Structures: Plasmon Polariton Mechanism of Saltatory Conduction in Axons. *J. Phys. Chem. C* **2015**, *119*(18), 10015.
18. Maier, S. A.; Atwater, H. A. Plasmonics: Localization and Guiding of Electromagnetic Energy in Metal/Dielectric Structures. *J. Appl. Phys.* **2005**, *98*, 011101.
19. Huidobro, P. A.; Nesterov, M. L.; Martin-Moreno, L.; Garcia-Vidal, F. J. Transformation Optics for Plasmonics. *Nano Lett.* **2010**, *10*, 1985.
20. Citrin, D. Coherent Excitation Transport in Metal-Nanoparticle Chains. *Nano. Lett.* **2004**, *4*, 1561.
21. Jacak, W. Exact Solution for Velocity of Plasmon-Polariton in Metallic Nano-Chain. *Opt. Express* **2014**, *22*, 18958.
22. Jacak, W. A. On Plasmon Polariton Propagation Along Metallic Nano-Chain. *Plasmonics* **2013**, *8*, 1317.
23. Jacak, W.; Krasnyj, J.; Chepok, A. Plasmon-Polariton Properties in Metallic Nanosphere Chains. *Materials* **2015**, *8*, 3910.

24. Ermentrout, G. B.; Terman, D. H. *Mathematical Foundations of Neuroscience. Interdisciplinary Applied Mathematics.*; Vol. 35. Springer: New York, Dordrecht, Heidelberg, London, 2010.
25. Dayan, P.; Abbott, L. F. *Theoretical Neuroscience: Computational and Mathematical Modeling of Neural Systems. Computational Neuroscience.* MIT: Cambridge, MA, 2001.
26. Izhikevich, E. M. *Dynamical Systems in Neuroscience: The Geometry of Excitability and Bursting.* Computational Neuroscience MIT, MIT: Cambridge, MA, 2007.
27. Debanne, D.; Campanac, E.; Białowąs, A.; Carlier, E.; Alcaraz, G. Axon physiology. *Physiol. Rev.* **2011**, *91*, 555.
28. Thomson, W. On the Theory of the Electric Telegraph. *Proc. R. Soc. London* **1854**, 7, 382.
29. Brzychczy, S.; Poznański, R. *Mathematical Neuroscience.* Academic Press: San Diego, 2011.
30. Scurfield, A.; Latimer, D. C. A Computational Study of the Impact of Inhomogeneous Internodal Lengths on Conduction Velocity in Myelinated Neurons. *PLoS One* **2018**, *13, e0191106.*
31. Fribance, S.; Wang, J.; Roppolo, J. R.; de Groat, W. C.; Tai, C. Axonal model for temperature stimulation. *J. Comput. Neurosci.* **2016**, *41*, 185.
32. Richardson, A. G.; McIntyre, C. C.; Grill, W. M. Modelling the Effects of Electric Fields on Nerve Fibres: Influence of the Myelin Sheath. *Med. Biol. Eng. Comput.* **2000**, *38*, 438.
33. Waxman, S.; Bennett, M. Relative Conduction Velocities of Small Myelinated and Non-Myelinated Fibres in the Central Nervous System. *Nat. New Biol.* **1972**, *238*, 217.
34. Goldman, L.; Albus, J. Computation of Impulse Conduction in Myelinated Fibers; Theoretical Basis of the Velocity-Diameter Relation. *Biophys J.* **1968**, *8*, 596.
35. Moore, J. W.; Joyner, R. W.; Brill, M. H.; Waxman, S. D.; Najar-Joa, M. Simulations of Conduction in Uniform Myelinated Fibers. Relative Sensitivity to Changes in Nodal and Internodal Parameters. *Biophys J.* **1978**, *21*, 147.
36. Song, X.; Wang, H.; Chen, Y.; Lai, Y. C. Emergence of an Optimal Temperature in Action-Potential Propagation Through Myelinated Axons. *Phys Rev. E* **2019**, *100*, 032416.
37. Keener, J.; Sneyd, J. *Mathematical Physiology.* Springer, 2009.
38. Zayats, A. V.; Smolyaninov, I. I.; Maradudin, A. A. Nano-Optics of Surface Plasmon Polaritons. *Phys. Rep.* **2005**, *408*, 131.
39. Citrin, D. S. Plasmon Polaritons in Finite-Length Metal-Nanoparticle Chains: The Role of Chain Length Unravelled. *Nano Lett.* **2005**, *5*, 985.
40. Mie, G. Beiträge zur Optik trüber Medien, speziell kolloidaler Metallösungen. *Ann. Phys.* **1908**, *330*(3), 377–445.
41. Jacak, J.; Krasnyj, J.; Jacak, W.; Gonczarek, R.; Chepok, A.; Jacak, L. Surface and Volume Plasmons in Metallic Nanospheres in Semiclassical RPA-Type Approach; Near-Field Coupling of Surface Plasmons With Semiconductor Substrate. *Phys. Rev. B* **2010**, *82*, 035418.
42. Landau, L. D.; Lifshitz, E. M. *Field Theory.* Nauka: Moscow, 1973.
43. Jackson, J. D. *Classical Electrodynamics.* John Willey and Sons Inc.: New York, 1998.
44. Jacak, W. A. Size-Dependence of the Lorentz Friction for Surface Plasmons in Metallic Nanospheres. *Opt. Express* **2015**, *23*, 4472.
45. Citrin, D. Plasmon-Polariton Transport in Metal-Nanoparticle Chains Embedded in a Gain Medium. *Opt. Lett.* **2006**, *31*, 98.
46. Markel, V. A.; Sarychev, A. K. Propagation of Surface Plasmons in Ordered and Disordered Chains of Metal Nanospheres. *Phys. Rev. B* **2007**, *75*, 085426.
47. Zhao, L. L.; Kelly, K. L.; Schatz, G. C. The Extinction Spectra of Silver Nanoparticle Arrays: Influence of Array Structure on Plasmon Resonance Wavelength and Width. *J. Phys. Chem. B* **2003**, *107*, 7343.
48. Zou, S.; Janel, N.; Schatz, G. C. Silver Nanoparticle Array Structures That Produce Remarkably Narrow Plasmon Lineshapes. *J. Chem. Phys.* **2004**, *120*, 10871.

49. Gradshteyn, I. S.; Ryzhik, I. M. *Table of Integrals Series and Products*. Academic Press, Inc.: Boston, 1994.
50. Meissner, T.; Wentz, F. The Complex Dielectric Constant of Pure and Sea Water from Microwave Satellite Observations. *IEEE THRS* **2004**, *42*, 1836.
51. El-Brolossy, T. A.; Abdallah, T.; Mohamed, M. B.; Abdallah, S.; Easawi, K.; Negm, S.; Talaat, H. Shape and Size Dependence of the Surface Plasmon Resonance of Gold Nanoparticles Studied by Photoacoustic Technique. *Eur. Phys. J. Special Topics* **2008**, *153*, 361.
52. Liz-Marzan, L. M. *Tuning Nanorod Surface Plasmon Resonances. SPIE.org Newsroom* **2007**. https://spie.org/news/0798-tuning-nanorod-surface-plasmon-resonances.
53. Goubau, G. Surface Waves and Their Application to Transmission Lines. *J. Appl. Phys.* **1950**, *21*, 119.
54. Sommerfeld, A. Über die fortpflanzung elektrodynamischer Wellen langs eines Drahts. *Ann. Phys. Chem.* **1899**, *67*, 233.
55. Birman, J. S. *Braids, Links and Mapping Class Groups*. Princeton UP: Princeton, 1974.
56. Mermin, D. The Topological Theory of Defects in Ordered Media. *Rev. Mod. Phys.* **1979**, *51*, 591.
57. Jacak, J.; Gonczarek, R.; Jacak, L.; Jóźwiak, I. *Application of Braid Groups in 2D Hall System Physics*. World Scientific: Singapore, 2012.
58. Spanier, E. *Algebraic Topology*. Springer VL: Berlin, 1966.

CHAPTER FIVE

Nonequilibrium quantum brain dynamics

Akihiro Nishiyama[a], Shigenori Tanaka[a], and Jack A. Tuszynski[b,c,*]

[a]Graduate School of System Informatics, Kobe University, Nada-ku, Kobe, Japan
[b]Department of Oncology, Cross Cancer Institute, Edmonton, AB, Canada
[c]Department of Physics, University of Alberta, Edmonton, AB, Canada
*Corresponding author: e-mail address: jack.tuszynski@gmail.com

Contents

Abstract

We provide an overview of our recent studies on nonequilibrium quantum brain dynamics (QBD). QBD is essentially the application of quantum electrodynamics to a system that is immersed in water molecules containing dynamically coupled electric dipoles associated with essential biomolecules within the structure of the brain: phospholipids and proteins. We adopt a model of memory storage due to the breakdown of rotational symmetry of dipolar modes in QBD. Furthermore, we describe the dynamics of the QBD system using the Schrödinger-like equations for coherent electric dipole fields, the Klein–Gordon equations for coherent electric fields, and the Kadanoff–Baym equations for incoherent dipoles and photons. Nontrivial results are obtained by adopting quantum field theory to describe open quantum systems including networks. Nonequilibrium properties of the model of QBD are described by the use of numerical simulations to obtain time-dependent solutions.

Advances in Quantum Chemistry, Volume 82
ISSN 0065-3276
https://doi.org/10.1016/bs.aiq.2020.08.003

1. Introduction

What is memory? In simplest terms, memory is the ability of a physical system to write, store, and retrieve (read) information. In the context of human brain, it is generally assumed that memory is not localized in specific brain areas but, instead, emerges due to the strength of synaptic connections between interacting neurons, which are involved in electrical activity (spiking). In general, higher cognitive functions including memory, perception, attention, decision-making, learning and consciousness are associated with synchronized oscillations of large neuronal groups, known as neuronal synchrony, and reflecting mainly dendritic-somatic synchrony. This type of large-scale coherent synchronization is believed to cause a cascade of cognitive processes related to perceptual binding and consciousness including the maintenance of working memory, and perceptual stabilization.

In terms of physical properties, memory can be characterized as having the property of diversity, long-termed but imperfect stability, and non-locality, which has been suggested in earlier experiments.[1] Memory storage has been considered distributed and holographic, i.e., encoded at various scales and locations.[2,3] It is worth mentioning in this connection that spatial location uses $1/f$ representations, which is a hallmark of scale-free networks. Moser, Moser, and O'Keefe were awarded the 2014 Nobel Prize in Medicine and Physiology for their discovery of fractal-like "grid cells" mapping spatial location in the brain.[4–8] They found that spatial grids of an animal's immediate and larger scale surroundings are represented in several layers of hexagonal grid patterns in entorhinal cortex, each layer in a different scale. Scale-free architectures such as the ones the brain is suspected of exhibiting are inherently robust and resistant to random disruptions providing functional stability. In human brain, evidence suggests neural network structures, temporal dynamics, and representation of mental states are all "scale-free," with self-similar patterns repeating at multiple spatio-temporal levels and locations.

Perhaps as a result of this scale-free spatio-temporal property of memory storage in human brain, each memory does not disappear when particular domains of neurons are damaged or destroyed (the nonlocal nature). Conventional neuroscience has failed in giving us a concrete picture of what memory is, in spite of many efforts which date back to the end of 19th century. Some neuroscientists might claim that synaptic connections are essential in memory and learning. However, they still do not answer how the limited numbers of connections among neurons induce mass delocalized excitations as has been empirically described but not quantitatively modeled by conventional neuroscience. In addition, it is worth noting that synaptic connections are transient and remodeled on a time scales of a few days. It is, therefore, in our opinion necessary to describe the mass action in the brain dynamics beyond the concepts found in conventional neuroscience.[9] Furthermore, the memory models based on the concept of synaptic connections among neurons are unable to explain how the processes of learning or memory may take place in a single-cell organism. Recent results in cell biology have demonstrated that single-cell organisms can learn without the presence of synapses and hence without synaptic activation taking place.[10] The observed learning mechanism indicates that a single cell cannot be viewed as a one-bit system but a complex information processing unit. This observation also leads to the question Why human beings would not make use of the mechanism of memory formation and information processing developed in single-cell organisms, which do not possess synapses? It stands to reason that evolutionary development would not only retain but refine and improve upon an earlier achievement as has been the case across biological species consistent with Darwinian theory of evolution. This would lead us to conclude that a neuron, as a single cell, should be viewed as a much more complicated and sophisticated information storage/processing unit than a single bit switch than can either be in an on- or off-state. Consequently, microscopic physical properties of a single cell need to be closely examined with an eye on their functional utility in the context of information storage and processing. Based on the very special physical properties such as regular geometry, multifunctionality, electrical conductivity, and mechanical stability we believe that cytoskeletal protein filaments, especially microtubules, might be best suited to play a significant role in memory formation and information processing. Hence, in our opinion a new approach is required to describe memories in biological systems. This new approach should include microscopic-level excitations that appear to be outside the scope of present-day conventional neuroscience.

Quantum field theory of the brain namely quantum brain dynamics (QBD) has been one of the most detailed physics-based formulations intended to describe how memory is formed in the brain.[11,12] The historical timeline of QBD is described as follows. Quantum field theory (QFT) of the brain originated with the monumental work by Ricciardi and Umezawa in 1967.[13] They proposed to use the concept of spontaneous breakdown of symmetry (SBS), equivalent to the emergence of macroscopic order, which is grounded in quantum field theory and applied it to the description of learning and memory in the brain. The concept of SBS can emerge in QFT with infinitely many unitarily inequivalent representations, which can describe the diversity of vacua (ground states), while standard quantum mechanics in contrast to QFT cannot describe the presence of diverse vacua.[14] Unitarily inequivalent representations emerge due to the infinite degrees of freedom in quantum fields. As a result of SBS, the long-range correlations via massless Nambu–Goldstone (NG) bosons in the vacua appear in the system.[15–17] The vacua in SBS (the lowest energy state) are described by the condensation of NG bosons. The long-term memory can be understood to correspond to the lowest energy state, while the short-term memory is described by the intermediate meta-stable excited states that can decay over time. In 1968, Fröhlich suggested that quantum coherence with long-range correlations plays a major role in the biological systems where the phenomenon of Bose–Einstein condensation may occur. A system that supports Bose–Einstein condensation behaves as a single entity (the Fröhlich condensate) if the frequencies of the oscillating dipoles are in a narrow range around the resonance frequencies and the coupling constants for the dipole interactions with the heat bath and the energy pump are sufficiently large in order to generate strong nonlinearities that provide suitable conditions for symmetry breaking.[18,19] This idea, if proven experimentally to take place in biological systems, would be of immense value to our understanding of biological organisms and the way they integrate and synchronize the great multitude of seemingly independent activities. A prime example of this type of behavior is the integration of sensory inputs in the human brain and the ability to process information arriving from the external environment with a subsequent unitary response of the entire organism. The key element in the Fröhlich theory is the existence of a nonlinear interaction between the oscillating dipoles in the biological system. It has been recently found to be described by the distance dependence $1/r$ suggested by Preto et al.,[20] which represents the long-range distance dependence than $1/r^3$ originally suggested by Fröhlich.[21] These interacting dipoles are capable of undergoing Bose–Einstein condensation with the associated breakdown of the rotational symmetry resulting in their collective behavior as a giant dipole.[22] In 1971, Pribram applied the concept of holography to brain dynamics, where the phenomenon of electromagnetic coherence found in laser physics plays a significant role.[23]

In 1978–79, Stuart et al. advanced the study of brain dynamics with the discussion about the stable nonlocal memory storage and recall processes in the Takahashi model.[24,25] This research indicated that the brain is most likely a mixed system composed of the electro-chemical subsystem in terms of classical neurons and a subsystem consisting of quantum degrees of freedom described by two types of quanta, namely corticons and exchange bosons. Memory storage within this framework is described by the vacuum state with macroscopic order emerging as a result of the collective interactions involving corticons and exchange bosons. Macroscopic order by SBS is maintained by NG bosons of the system with long-range spatial correlations. Nonlocality and stability of memory storage can be explained by SBS of the quantum system. The creation and annihilation operation involving a finite number of NG bosons represents the memory recall in which macroscopic order (or the vacuum state) in SBS is not affected. What exactly corticons and exchange bosons in physiological systems are was not specifically elucidated at this stage of theory development. In 1976, Davydov and Kislukha proposed a theory of solitary wave propagation along DNA and protein chains, specifically along their alpha-helical structures, which has been called the Davydov soliton.[26] The Fröhlich condensate and the Davydov soliton are found to represent formally equivalent reciprocal manifestations of static and dynamical properties, respectively, in the nonlinear Schödinger equation of an equivalent quantum Hamiltonian.[27] In 1980s, Del Giudice et al. applied QFT approach to biological systems.[28–31] In 1988, they proposed a specific physical mechanism using laser-like phenomena of QED with water electric dipoles as active degrees of freedom.[30] Electric dipole fields can be expanded in a series decomposition in terms of spherical harmonics, which represent eigenstates of the angular momentum operator for dipoles. Adopting a two-state energy level approximation with the ground state and the first excited states in angular momentum, they showed collective modes of dipoles and photons emerge and suggested permanent electric polarization of dipoles. In 1990s, Jibu and Yasue suggested a concrete picture of corticons and exchange bosons in QFT of the brain, which represent water electric dipole fields and photon fields, respectively.[11],[32–35] Quantum field theory of the brain, or quantum brain dynamics (QBD), is in fact an application of quantum electrodynamics (QED) to a biological system immersed in water molecules creating electric dipole fields. It is envisaged that when physical stimuli arrive in the brain, they trigger the breakdown of rotational symmetry of electric dipoles causing these dipoles to be aligned in the same direction and to oscillate in synchrony. The vacua in systems with SBS of dipoles are maintained by massless NG bosons, i.e., dipolar wave quanta. The stability of the macroscopic quantum system and the nonlocal property of the ordered system are realized by long-range correlations via NG bosons. Massless NG bosons in SBS are absorbed into longitudinal modes of photons in QED, and then the photons acquire mass due to the Higgs mechanism. The massive photons are called evanescent photons. Since photons acquire mass, they can stay confined to the coherent domains, the size of which is on the order of 50 μm. Then, external electromagnetic fields are excluded from these coherent domains due to the Meissner effect, which is a characteristic property of superconducting systems resulting in exponential damping of magnetic fields from the condensate. However, superconductivity is essentially an equilibrium phenomenon in condensed matter physics, hence it appears lacking the energy supply as a basis for biological coherence. Furthermore, Jibu et al. adopted at least two types of quantum mechanisms to account for the information transfer processes in QBD. The first mechanism involves super-radiance phenomena in brain dynamics.[34] (The concept of super-radiance itself in N-body systems of a gas was suggested by Dicke in 1954.[36]) Since the eigenstate of a super-radiant state is given by the superposition of dipoles, it shows a cooperative effect of dipoles and photons. Super-radiance has the property of ultra-efficient absorption and emission of photons with the directionality of wave propagation. The intensity of the coherent electromagnetic pulse has a peak on the order of $\sim N^2$ within narrow time scales $\sim 1/N$ (N: the number of dipoles) after a time delay. In a biophysically based model of biological coherence, we can adopt a system composed of microtubules connecting two coherent domains to achieve information transfer based on the super-radiance effect. We make use of the self-induced transparency in microtubules surrounded by water molecules and filling the lumen of microtubules, where the electromagnetic pulse can propagate without decay as if the system inside the microtubule were electromagnetically transparent. A second mechanism has been constructed to adopt quantum tunneling of photons.[35] When the distance between two coherent domains surrounded by an incoherent domain is smaller than the inverse of the mass of evanescent photons, so that the quantum tunneling effects of photons can occur. The quantum tunneling is essentially equivalent to the Josephson effect in two superconducting domains separated by a normal domain. Del Giudice et al. also studied the effect in a biological system.[31] In 1995, Vitiello studied a capacity of memory in QBD[37] and demonstrated that a huge memory capacity is possible by regarding the brain as a dissipative open system, where the mathematical formalism used is that of thermo-field-dynamics with the doubling of the degrees of freedom. Each memory evolves in a classical deterministic trajectory similarly to the emergence of chaos in the dissipative model of the brain.[38] In the limit of the infinite volume of the brain, the overlap among distinct memory states is zero at any time. However, since the volume of the brain is finite, the overlap becomes nonzero, and this might explain the association of different memories in QBD.[39] In 2003, Zheng and Pollack discovered the presence of the so-called exclusion zone (EZ) of water in the vicinity of hydrophilic surfaces by careful and extensive experimentation with numerous materials.[40] The properties of EZ water correspond to those described earlier in theoretical papers on coherent water in QED.[41]

The aim of our work is to show nonequilibrium properties in memory formation as described by models based on quantum brain dynamics. A suitable approach to this problem is to describe nonequilibrium QBD using Schrödinger-like equations for coherent electric dipole fields, the Klein–Gordon equations for coherent electric fields, and the Kadanoff–Baym equations for incoherent particles. These equations are adopted to describe physical phenomena in quantum many-body systems. We can derive aspects of super-radiance and equilibration simultaneously from a previously introduced single Lagrangian representation.[42] Note that the neural wave equations proposed by Pribram in 1991 can be derived from Lagrangian density equations with nonrelativistic charged bosons. The difference between the Schrödinger-like equations for water and neural wave equations for ionic bioplasma is in the involvement of different degrees of freedom (water versus ions). When we represent both the dipolar dynamics of water molecules and ions with photons, we must add extra terms for charged bosons to the system's Lagrangian. The theory of charged bosons is given by the nonrelativistic version of our previous work for relativistic charged bosons.[43] We can also extend the dynamics of quantum fields to network systems in order to describe our thinking processes.[44–46] We believe that with these modern more sophisticated approaches we can appropriately describe nonequilibrium properties within QBD for memory formation.

Our approach in nonequilibrium QBD can be applied to not only memory storage and retrieval mechanisms but also to a more general concept of consciousness in cognitive processes. Memories appear as qualia, the subjective experiences. In case memories are encoded as quanta or quantum fields as shown in QBD, some form of correspondence between qualia and quanta might emerge. An important but open question in this regard is whether reductionism of qualia to quanta is the correct view of the issues or perhaps a global approach based on panpsychism is more appropriate. Whatever the answer to this question, we believe there may soon occur a major paradigm shift in the concept of consciousness once proper integration of advanced quantum field theory concept into the field of consciousness is achieved. It is our hope that this chapter represents a small step in this direction.

This chapter is organized as follows. In Section 2, we provide background material regarding physiology and neuroscience. In Section 3, we introduce our model of nonequilibrium QBD. In Section 4, we describe the framework of memory formation in our models. In Section 5, we describe the perspectives in the description of consciousness in QBD. In Section 6, we summarize this chapter.

2. Brief background in physiology and neuroscience

The issue of the molecular nature of brain activities has been of great interest to not only neurophysiologists but also information scientists and physicists. While much is known about the architecture of the brain, brain cells and subcellular structures and their individual activities in terms of the action potential, ion channels and ionic currents, etc., much less is known about such issues as where memories are stored, which molecular mechanisms are involved in information processing and cognitive functions. Speculations about molecular mechanisms of cognition abound and range from mundane to exotic. In particular, much has been speculated recently about the possibility of some cognitive functions requiring the operation at a level of quantum physics. Experimental determination or even corroboration of such ideas as the Penrose–Hameroff Orch OR theory still appears to be vastly outside the scope of experimental methods and instrumentation available to researchers but this can change soon with the development of more precise tools such as PET, fMRI, and other imaging techniques.

Nonetheless some questions can be addressed using general principle of physics and the known information about brain physiology at a level accessible to modern methods. For example the metabolic rates of human brains have been studied extensively and their energetic demands are well known. It is also known how much of the metabolic energy is required to carry out various molecular-level processes such as protein production, motor protein motion, transcription and translation, etc. It is clear that human brain is a very efficient but energy-supply dependent organ that must be physically represented using nonequilibrium physics approaches. In this chapter we will provide a basis for this methodology within quantum field theory. Finally, most of the brain's functional roles involve motor control, cognition and information storage and processing. Physics has until recently treated information with some neglect focusing instead on entropy and its physically measurable manifestation, heat. Some 70 years ago, one of the most influential physicists of all time, Erwin Schrödinger, wrote a book entitled "What is Life" in which he outlined challenging problems plaguing the interpretation of biological processes using physical reasoning. Entropy reduction was a major focal point of his discussion. While it is now known that entropy reduction that biological cells generate is at the cost of metabolic energy consumption and heat dissipation into the environment, a related problem of information storage and processing and the energy cost of such processes is still worth exploring.

Memory is in general associated with synaptic plasticity but changes in synaptic plasticity occur on a much shorter time scale. Moreover, synaptic proteins are transient, lasting between hours to days. On the other hand, long-term memories can last years or even lifetimes while short-term memories occur on the order of seconds to minutes. Hence, there is a disconnect on the temporal scale. So far, the most well-established and generally accepted experimental model for memory involves long-term potentiation (LTP) in which synaptic activation generates calcium influx, which then triggers activation of the hexagonal holoenzyme calcium-calmodulin kinase II (CaMKII). Each activated CaMKII enzyme molecule extends two sets of six kinase domains and each set is able to encode six bits of synaptic information. This biological information is encoded by phosphorylation in an appropriate hexagonal lattice enabling subsequent functional signaling. Craddock et al.[47] showed by computational modeling how CaMKII enzyme can encode up to 6 bits of memory in hexagonal lattices in neuronal (dendritic) microtubules, which represent major components of the cellular cytoskeleton inside neurons (in addition to actin filaments and intermediate filaments). This model provides a mechanistic molecular-level explanation how the human brain can encode, store and recall memory through phosphorylation states in hexagonal lattices of the microtubule surface, specifically involving serine residues in the C-termini of tubulin monomers. These states can further regulate intra-cellular processes such as motor protein traffic and MAP (microtubule associated protein) connections between neighboring microtubules. It would be sensible to expect that efficient and relatively fast processes such as phosphorylation events inside neurons could play the role of mediating consciousness that includes memory storage, recall and event erasure. Assuming that tubulin phosphorylation states act as binary bits switching on a millisecond time scale, with approximately 10 million tubulin dimers per neuron, this would allow for up to

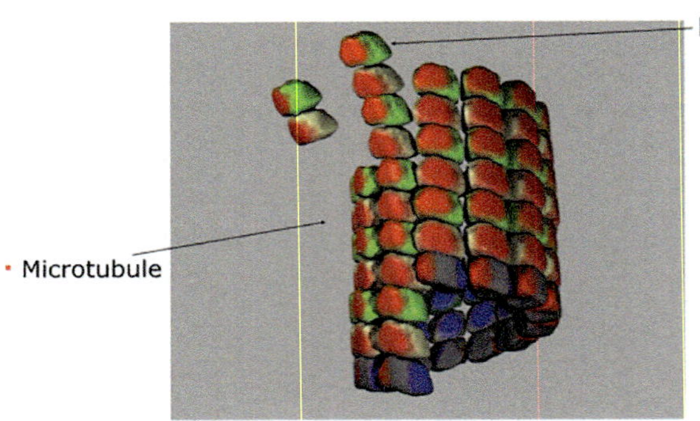

Fig. 1 Microtubule.

10 billion operations per second per neuron, and roughly 10^{21} operations per second for the entire brain giving it an enormous amount of operating computational capacity. Therefore, a mechanism like this would be physiologically advantageous and consistent with our knowledge of molecular biology (Fig. 1).

We need the presence of small but nonzero electric fields to trigger the spontaneous breakdown of symmetry of water molecules' electric dipole fields. The electric fields due to the presence of electrostatic charges and dipole moments in the structure of microtubules are well documented in the literature on this subject.[48,49] These electric fields provide initial conditions for the dynamics of the electric dipoles and photons emerging in the surrounding medium composed of water molecules (potentially EZ water) and ions.

3. Model in nonequilibrium QBD

In this section, we propose a quantitative model to describe nonequilibrium QBD, namely Quantum Electrodynamics with water molecules' electric dipole fields involved in nonlinear interactions. Our approach represents an extension of the theory proposed by Del Giudice et al.[30] We describe two types of degrees of freedom, which are electric dipoles and photons. These are called corticons and exchange bosons, respectively in Refs. 24, 25.

Our Lagrangian represents the sum of a term representing electromagnetic fields, a term of the water electric dipole field $\Psi(x, \theta, \varphi)$ and its complex conjugate $\Psi^*(x, \theta, \varphi)$ with the position x (in time and space) and (θ, φ)

in the polar coordinates and a term given by the inner product of the electric field E_i and the density of dipole moments μ_i with $i = 1, 2, 3$ in 3 + 1 dimensions (3 spatial dimensions and 1 temporal dimension). Electric dipole fields $\Psi(x, \theta, \varphi)$ and $\Psi^*(x, \theta, \varphi)$ provide the information regarding the orientation of electric dipoles given by $u_i = (\sin\theta\cos\varphi, \sin\theta\sin\varphi, \cos\theta)$ where the label represents spatial coordinates $i = 1, 2, 3$. The density of dipole moments is written by $2ed_e \int d\cos\theta d\varphi \Psi^* u_i \Psi$ with $2ed_e = 1.9$ Debye with the elementary charge e and the distance of polarization $d_e = 0.2$ Å. Electric dipole fields $\Psi(x, \theta, \varphi)$ and $\Psi^*(x, \theta, \varphi)$ are expanded into spherical harmonics $Y_{lm}(\theta, \varphi)$ with the azimuthal quantum number l and the magnetic quantum number m which is adopted to describe an atom composed of electrons and a nucleus as $s, p, d \ldots$ orbitals. As shown in Fig. 2, we adopt a two-energy level approximation for angular momentum just taking an s and a p orbital, namely the ground state s (with $l = 0$ and $m = 0$) and the first excited state p (with $l = 1$ and $m = 0, \pm 1$) expressed by the subscripts $\alpha = 0, \pm 1$. The energy difference is $\frac{l(l+1)}{2I}\big|_{l=1} - \frac{l(l+1)}{2I}\big|_{l=0} = \frac{1}{I} = 4$ meV with the moment of inertia denoted as I. We use $\Psi(x, \theta, \varphi) = \psi_s(x) Y_{00}(\theta, \varphi) + \sum_{\alpha=0,\pm 1} \psi_\alpha(x) Y_{1\alpha}(\theta, \varphi)$ and its complex conjugate. Then, the density of dipole moments $\mu_{i=3}$ in the 3-direction is $\frac{2ed_e i}{\sqrt{3}}(\psi_0 \psi_s^* - \psi_s \psi_0^*)$, for example.

Our model describes the dynamics for both coherent fields and quantum fluctuations (incoherent particles) in QBD. Coherent fields can be used to describe order parameters, namely coherent electric fields E_i and a coherent electric dipole moment $\overline{\mu}_i$ given by the coherent electric dipole field $\overline{\psi}_s$, $\overline{\psi}_\alpha$, $\overline{\psi}_s^*$, and $\overline{\psi}_\alpha^*$, in the macroscopic quantum system. Coherent electric dipole

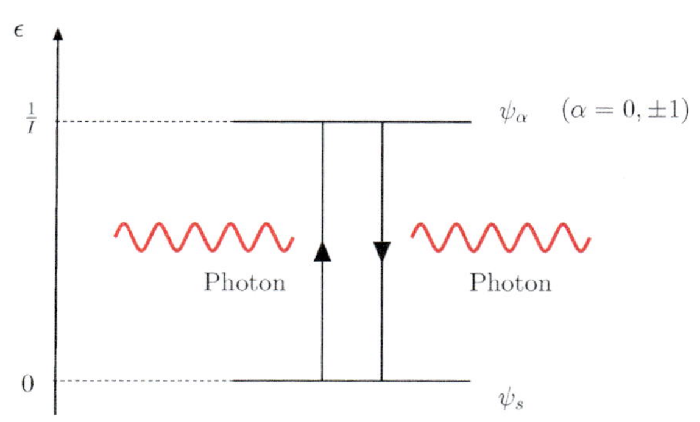

Fig. 2 Energy levels for electric dipole fields. The energy difference is $\frac{1}{I} = 4$meV.

fields $\overline{\psi}_s$ and $\overline{\psi}_\alpha$ obey the Schrödinger-like equations coupled with electric fields E_i with $i = 1, 2, 3$ given by,

$$\left(i\frac{\partial}{\partial x^0} + \frac{\nabla_i^2}{2m}\right)\overline{\psi}_s + \frac{2ed_e i}{\sqrt{6}} \sum_{\alpha=0,\pm1} \left[-\alpha(E_1 + i\alpha E_2) + \sqrt{2}(1-|\alpha|)E_3\right]\overline{\psi}_\alpha = 0, \tag{1}$$

$$\left(i\frac{\partial}{\partial x^0} + \frac{\nabla_i^2}{2m} - \frac{1}{I}\right)\overline{\psi}_\alpha + \frac{2ed_e i}{\sqrt{6}} \left[\alpha(E_1 - i\alpha E_2) - \sqrt{2}(1-|\alpha|)E_3\right]\overline{\psi}_s = 0. \tag{2}$$

Here, $|\overline{\psi}_s|^2$ and $|\overline{\psi}_\alpha|^2$ represent the number density of coherent dipoles in (s) and (α), respectively. The sum $|\overline{\psi}_s|^2 + \sum_\alpha |\overline{\psi}_\alpha|^2$ corresponds to the number density N/V of water molecules in liquid state. Coherent electric fields obey the Klein–Gordon equations as,

$$\left[\left(\tfrac{\partial}{\partial x^0}\right)^2 - \left(\tfrac{\partial}{\partial x^1}\right)^2\right]E_2 = \frac{\overline{\mu}_2}{I^2} + \frac{8(ed_e)^2}{3I}\left(2|\overline{\psi}_1|^2 - |\overline{\psi}_s|^2\right)E_2$$
$$+ \text{(coupled terms with } E_3) + \text{(quantum fluctuations)}, \tag{3}$$

$$\left[\left(\tfrac{\partial}{\partial x^0}\right)^2 - \left(\tfrac{\partial}{\partial x^1}\right)^2\right]E_3 = \frac{\overline{\mu}_3}{I^2} + \frac{8(ed_e)^2}{3I}\left(|\overline{\psi}_0|^2 - |\overline{\psi}_s|^2\right)E_3$$
$$+ \text{(coupled terms with } E_2) + \text{(quantum fluctuations)}, \tag{4}$$

with time x_0, space x_i with $i = 1, 2, 3$, $\overline{\mu}_2 = \frac{4ed_e}{\sqrt{6}}\left(\overline{\psi}_1\overline{\psi}_s^* + \overline{\psi}_s\overline{\psi}_1^*\right)$, and $\overline{\mu}_3 = \frac{2ed_e i}{\sqrt{3}}\left(\overline{\psi}_0\overline{\psi}_s^* - \overline{\psi}_s\overline{\psi}_0^*\right)$ in $E_1 = 0$ and $\overline{\psi}_1 = \overline{\psi}_{-1}$. They are derived as the Euler–Lagrange equations for our system's Lagrangian. On the other hand, the creation-annihilation operations of incoherent photons given by quantum fluctuations can describe the dynamics of memory recall. Note that the dynamics for the incoherent particles (photons and dipoles) was neglected in the preceding work.[30] Incoherent particles are described by the Kadanoff–Baym (KB) equations.[50–52] In QBD, the KB equations for incoherent dipoles are expressed as,

$$i\Delta_0^{-1} - i\Delta^{-1} - i\Sigma = \mathbf{0}, \tag{5}$$

where the 4×4 matrix $i\Delta_0^{-1}(x, y)$ is written by,

$$i\Delta_0^{-1}(x,y) \equiv \begin{bmatrix} i\dfrac{\partial}{\partial x^0} + \dfrac{\nabla_i^2}{2m} & \dfrac{2ed_e i}{\sqrt{6}}\left[-\beta(E_1 + i\beta E_2) + \sqrt{2}(1-|\beta|)E_3\right] \\ \dfrac{2ed_e i}{\sqrt{6}}\left[\alpha(E_1 - i\alpha E_2) - \sqrt{2}(1-|\alpha|)E_3\right] & \left(i\dfrac{\partial}{\partial x^0} + \dfrac{\nabla_i^2}{2m} - \dfrac{1}{I}\right)\delta_{\alpha\beta} \end{bmatrix} \times \delta_C^4(x-y), \tag{6}$$

in closed-time path (CTP) $\mathcal{C}$[53,54] and the 4×4 matrix $\mathbf{\Delta}(x, y)$ is,

$$\Delta(x,y) = \begin{bmatrix} \Delta_{ss}(x,y) & \Delta_{s\beta}(x,y) \\ \Delta_{\alpha s}(x,y) & \Delta_{\alpha\beta}(x,y) \end{bmatrix}, \tag{7}$$

with Green's functions $\Delta_{s\beta}(x,y) = \langle T_{\mathcal{C}} \delta\psi_s(x) \delta\psi_\beta^*(y) \rangle$ with fluctuations of dipole fields $\delta\psi_s = \psi_s - \overline{\psi}_s$, the 4×4 matrix of the self-energy $\mathbf{\Sigma}(x, y)$, time-ordered product $T_{\mathcal{C}}$ in CTP, and $\alpha, \beta = 0, \pm 1$. Similarly the KB equations for incoherent photons are expressed by,

$$iD_0^{-1} - iD^{-1} - i\Pi = 0, \tag{8}$$

with $iD_{0,ij}^{-1}(x,y) \equiv -\delta_{ij}\partial_x^2 \delta_{\mathcal{C}}^{d+1}(x-y)$, and Green's functions $D_{ij}(x,y) = \langle T_{\mathcal{C}} a_i(x) a_j(y) \rangle$ with fluctuations of photon field a_i ($i = 1, 2, 3$) in the gauge fixing $a^0 = 0$, and the self-energy for photons $\Pi_{ij}(x, y)$. We use the KB equations to describe incoherent photons and dipoles in both 3 + 1 dimensions and 2 + 1 dimensions as shown earlier in Ref. 42.

We adopt the super-radiance mechanism to describe information transfer among the interacting subsystems in the brain. Super-radiance represents a cooperative effect of atoms or electric dipoles. It describes ultra-efficient absorption and emission of photons, and the propagation of the electric fields occurs along one direction. We shall consider one atom or one dipole in the excited state within a two-energy level approximation. An atom or a dipole spontaneously decays on the time scale τ. Next, we consider the system of N atoms or dipoles occupying their excited states. We might then consider N atoms or dipoles in their excited states emitting an electromagnetic field with N times larger intensity than that for one atom or one dipole since we can reasonably assume that the decay time should not depend on N. However, this assumption turns out to be wrong. Dicke showed that the decay time of

N atoms depends on the number of atoms N.[36] Super-radiance phenomena mean that N atoms or dipoles correlate with one another in sufficiently high-density systems, and decay on the time scale $\tau_R \sim \tau/N$. The eigenstate in super-radiance is given by the superposition of N excited atoms or dipoles. Since N atoms or dipoles occupying their excited states cooperatively decay on time scales $\tau_R \sim \tau/N$, the intensity of a resultant electromagnetic pulse will scale with the number of atoms or dipoles as N^2. Furthermore, the electromagnetic pulse in the super-radiant case does not decay exponentially, but has a peak at the delay time τ_0 which is dependent on the initial conditions of the associated electric fields. The pulse as a function of time x_0 is written in the form by $\frac{1}{\tau_R} \times \frac{1}{\cosh\left(\frac{x_0 - \tau_0}{\tau_R}\right)}$. It is possible to derive the property of super-radiance for water molecules' electric dipoles from our Lagrangian.[42] Importantly, the corresponding time scale represented by τ_R is 10^{-13} s for water electric dipoles.

It is also possible to describe entropy production for incoherent particles during equilibration. Elementary processes in QBD are depicted in Fig. 3. Coherent electric fields change the angular momenta of incoherent dipoles from 0 to $\frac{1}{I}$, and vice versa, as shown in Fig. 3A. In Fig. 3B, incoherent dipoles in (α) lose energy by emitting photons and change to dipoles in (s). Dipoles in (s) also gain energy by absorbing photons and change to dipoles in (α) in time-reversed processes in Fig. 3B. These processes contribute to the entropy production of the system. Entropy ceases to increase when the distribution functions of incoherent dipoles and photons approach

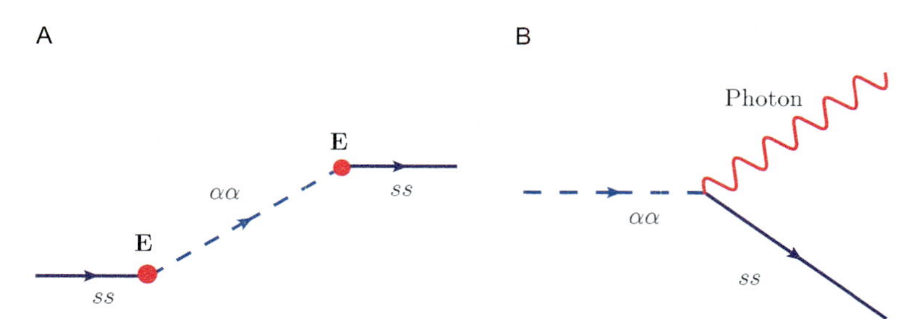

Fig. 3 Dynamical processes for incoherent dipoles and photons. The *solid* (ss) and *dotted lines* ($\alpha\alpha$) represent the propagation of incoherent dipoles in (s) and (α), respectively, and the wavy lines represent the propagation of incoherent photons. (A) Incoherent dipoles in the ground state (s) changes to those in the first excited states (α) by coherent electric fields **E**. The dipoles in (α) change to those in (s). The vertex represents coherent electric fields **E**. (B) Incoherent dipoles in (α) emit photons and change to dipoles in (s).

the Bose–Einstein distribution limit. As a consequence, the temperature of the dipole system and that of the photon system converge to the same value. Similarly, the chemical potential of the dipole system in (s) and that in (α) also converge to the same value, while the chemical potential of the photon system becomes zero since the number of photons changes in the time course of the system's evolution.

Finally, we comment on the tachyonic instability and the Higgs mechanism in the Klein–Gordon equation (4). We can show the tachyonic instability in inverted population where the $Z_0 \equiv |\overline{\psi}_0|^2 - |\overline{\psi}_s|^2 > 0$ for the electric field $E_{i=3}$. Then the electric field increases exponentially $\sim \exp(\Omega x_0)$ with $1/\Omega_{\text{max}} \sim 10^{-14}$ s in $\Omega < \Omega_{\text{max}}$. Here, Ω_{max} is 13×4 meV which corresponds to the value computed in the preceding work.[30] The electric field will continue to increase until the inverted population changes to the normal population $Z_0 \leq 0$. In the normal population $Z_0 \leq 0$, the Higgs mechanism takes place. Then, electric fields oscillate with the frequency $2\pi/\Omega_{\text{Higgs}} \sim 10^{-13}$ s with $\Omega_{\text{Higgs}} \sim \Omega_{\text{max}}$. Moreover, the Meissner effect appears and external electric fields are excluded in the coherent regions. The maximum penetration length $1/\Omega_{\text{Higgs}}$ is om the order of μm.

4. Memory in nonequilibrium QBD

In this section, we describe how our approach leads to memory formation in the nonequilibrium representation of quantum brain dynamics (QBD), which can be characterized by the ability to write, store, and recall information.

4.1 Memory encoding

We consider a physical system composed of water molecules' electric dipoles and photons. Without external stimuli, the directions of these dipole moments are random. The system has rotational symmetry for electric dipoles due to a lack of special direction in space. When external stimuli are present, a phase transition occurs in the system, namely electric dipoles become aligned along the same direction and the rotational symmetry of the dipolar system is spontaneously broken. First, we need a small but nonzero electric field E_i ($i = 1, 2, 3$ in $3 + 1$ dimensions) to trigger the breakdown of this rotational symmetry. As explained above, we might adopt the system of microtubules as the appropriate intra-neuronal source of nonzero electric fields. If nonzero electric fields are present at the initial time, the density of dipole moments $\overline{\mu}_i$ in the direction of these electric fields increases gradually from 0. Next, when dipole moments are nonzero, electric fields E_i tend to be amplified in the direction of the dipole moments. Then the dipole moments tend to be amplified. Finally, all the dipoles will be aligned in the same direction. In these processes, the contribution of quantum fluctuations is important. The quantum fluctuations (or incoherent particles) play the role of energy supply for the electric fields E_i and dipole moments. In the presence of quantum fluctuations in the Klein–Gordon equations for electric fields E_i, the nontrivial dynamical processes have a tendency to evolve toward alignment of all dipoles along the same direction, which can only occur when the critical density of incoherent particles is exceeded. These incoherent particles behave as energy sources. Here, we need to introduce dynamics for not only coherent fields but also incoherent particles (quantum fluctuations). We adopt the Kadanoff–Baym equations to describe the dynamics of quantum fluctuations.

4.2 Memory storage

Memory storage is described by the vacua with macroscopic order in the presence of the spontaneous breakdown of symmetry (SBS), where the dipoles are aligned in the same direction and the rotational symmetry is spontaneously broken. This order is maintained by long-range correlations with massless dipole wave quanta, the so-called Nambu–Goldstone (NG) bosons, which emerge in the process of symmetry breaking. The NG bosons are condensed in the vacuum states as a result of the breakdown of symmetry.

The main criticism with respect to the quantum decoherence effects is commonly raised in regard to memory storage issues. Maximum time scales of long-term memory in QBD must be on the order of human lifetime. We can perform numerical simulations in nonequilibrium QBD and investigate whether quantum decoherence occurs or not in the presence of thermal effects at the physiological temperature. The corresponding dynamics is described by coherent fields in Schrödinger-like equations and the Klein–Gordon equations and for incoherent particles (quantum fluctuations) in the Kadanoff–Baym equations. To describe the entire system more realistically, we can investigate the open system dynamics described by the central region "C" connected with the left "L" and the right "R" reservoirs as elaborated on in detail in a recent publication.[44–46] We set two reservoirs as upstream and downstream in regard to the energy flow. The energy flow appears from the upstream "L" to "C" and from "C" to the downstream reservoir "R." Here, we should check for possible error correction mechanisms in memory storage in open systems described by numerical simulations.

4.3 Memory recall in cognitive processes

In this section, we provide a scenario applicable to cognitive processes involving memory recall, which is based on nonequilibrium quantum brain dynamics (QBD). Coherent states with the condensation of infinite number of particles represent memory storage in this representation. The dynamics of creation-annihilation of a finite number of photons represents the memory recall process. The dynamics of a finite number of incoherent photons and dipoles can be described by the Kadanoff–Baym (KB) equations within our theory. It is possible to investigate the dynamics of coherent fields, the expectation values of quantum fields in coherent states, and incoherent particles simultaneously by solving the corresponding Schrödinger-like equations, the Klein–Gordon equations and the Kadanoff–Baym equations. Since memory recall is described by the dynamics of a finite number of creation-annihilation processes for photons, we are ready to describe our cognitive processes involving memory recall within the nonequilibrium QBD by investigating the dynamics described by the Kadanoff–Baym equations. We can adopt two types of quantum effects, namely super-radiance and quantum tunneling, to describe information transfer processes in the brain.

First, we adopt ultra-efficient emission and absorption of photons in super-radiance with the presence of correlations among dipoles suggested in an earlier version of QBD.[34] The time scale of emission and absorption is on the order of 10^{-13} s in microtubules with the characteristic length 1 μm. This time scale estimate is much shorter than the time scale of short-range interactions. Thus, the information transfer with super-radiance is protected against the destructive effects of thermal fluctuations.

Second, we also adopt quantum tunneling effects of photons as information transfer.[35] We set the physical systems involving quantum tunneling effects in the brain. It is useful to introduce networks of interacting systems as schematically illustrated in Fig. 4 as quantum systems involved in tunneling processes.[45] The network systems represent the central system "C" and the surrounding systems labeled by 1, 2, ...N_{res} affected by the initial change in "C." Importantly, decoherence, entropy production and chemical

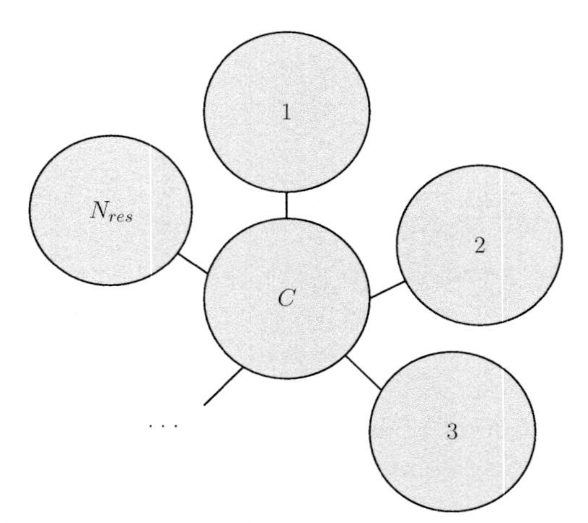

Fig. 4 Network of systems connected by quantum tunneling.

equilibration processes occur in the dynamics of the Klein–Gordon equations for coherent fields and are described by the Kadanoff–Baym equations for incoherent particles (quantum fluctuations) in ϕ^4 model, (not QBD). We can also describe open systems (or networks of systems) within the framework of quantum electrodynamics (QED) for charged bosons.[46] Due to the energy-momentum conservation, the energy ω and the momentum $\mathbf{p}$ of incoherent photons must not change during quantum tunneling processes between two systems, that is $\omega_1 = \omega_2$ and $\mathbf{p}_1 = \mathbf{p}_2$ with subscripts for system 1 and 2. We can consider the simple dispersion relations $\omega_1 = \sqrt{\mathbf{p}_1^2 + M_1^2}$ and $\omega_2 = \sqrt{\mathbf{p}_2^2 + M_2^2}$ with effective mass M_1 and M_2 for system 1 and 2, in the quasi-particle approximation with zero spectral width for incoherent photons in the Kadanoff–Baym equations. The effective mass of photons in QED is dependent on the chemical potential of charged bosons in the Bose–Einstein distribution, or equivalently on the scalar potential.[46] The difference in the effective mass might be used to characterize and quantify the strength of memory recall in the brain, since quantum tunneling for $M_1 \neq M_2$ is not allowed in the case of zero spectral width of photons due to the energy-momentum conservation. In the case of QBD, namely QED with molecules' electric dipoles, a similar discussion regarding the dispersion relations of photons might be possible for quantum tunneling. In Ref. 42, we find that dispersion relations for photons are given by the solution of $(\omega^0)^2 - \mathbf{p}^2 -$ (self-energy) $= 0$ with self-energy $\propto (\omega^0)^2$ by taking the poles of spectral functions of photons. The self-energy, which is given by the product of $(\omega^0)^2$ and the distribution function for incoherent dipoles in the ground state and the first excited state, determines the velocity of photons. These distribution functions are dependent on the intensity of coherent electric fields. Hence, the difference between coherent electric fields can determine the difference between dispersion relations for photons. Due to the energy-momentum conservation in quantum tunneling, this difference between dispersion relations for photons might determine which memory recall process occurs via quantum tunneling. Since incoherent photons involved in quantum tunneling thermalize by collisions with dipoles or transfer to another system with quantum tunneling in QBD, this may significantly affect our results and require a modified approach to the problem.

The dynamics of cognitive process involving memory recall can be investigated in networks of systems when we adopt super-radiance and quantum tunneling. It is described by extending our theory in an isolated system[42] to a theory applicable to networked systems instead.

5. Consciousness in QBD

In this section, we briefly describe the perspective based on the use of the quantum brain dynamics (QBD) in an attempt to offer an insight into the nature of consciousness within the realm of physical mechanisms, especially based on quantum field theory. Memories appear as subjective experiences for humans, i.e., they can be described as qualia. When we remember somebody or something, memories appear in the form of informational representation involving all five senses. For example, a walk in the park on a beautiful summer day may bring memories of bright colors, warm sun rays falling on our skin, smell of flowers, taste of cherries in our mouths and sounds of children at play. In investigating memories, we note that memories have aspects of not only objective and quantifiable information but also subjective qualia, which are dependent on our state of mind at the time of the event that is being recalled. The same scenery may invoke happy memories to some people and sad ones to others, depending on the individual context. If memories with subjective experiences are ordered patterns of quantum excitations of quantum fields, we might arrive at the reductionism of qualia in the subjective world to quanta in objective physical systems. Or, we might arrive at the concept that elementary particles involve qualia as viewed within the philosophical framework of panpsychism.

6. Summary and perspective

In this chapter we have reviewed the body of scientific literature on quantum approaches to human brain functioning spanning more than 50 years and also proposed a novel nonequilibrium quantum field approach to describe memory formation processes within quantum drain dynamics (QBD). We suggested a quantitative framework that is based on the non-equilibrium QBD using Schrödinger-like equations for coherent electric dipole fields, the Klein–Gordon equations for coherent electric fields and the Kadanoff–Baym equations for quantum fluctuations (incoherent particles). Memory storage in QBD involves generation of macroscopic order in quantum many-body systems. Our approach can describe not only the emergence of ordered states representing memory storage but also our cognitive processes with memory recall represented by the action of creation-annihilation operations involving incoherent photons, which are described by the Kadanoff–Baym equations for quantum network systems. The outline of these new and powerful nonequilibrium quantum field theory methods is in our opinion a very promising direction toward a better understanding of the fundamental mechanisms behind the functioning of the human brain.

References

1. Lashley, K. S. *Brain Mechanisms and Intelligence: A Quantitative Study of Injuries to the Brain.* University of Chicago Press, 1929.
2. Lashley, K. S. *In Search of the Engram. In Society for Experimental Biology, Physiological Mechanisms in Animal Behavior. Society's Symposium IV.* Academic Press, 1950;454–482.
3. Pribram, K. H.; Yasue, K.; Jibu, M. *Brain and Perception: Holonomy and Structure in Figural Processing.* Psychology Press, 1991.
4. O'Keefe, J.; Dostrovsky, J. The Hippocampus as a Spatial Map: Preliminary Evidence From Unit Activity in the Freely-Moving Rat. *Brain Res.* **1971**, *34*, 171–175.
5. O'Keefe, J. Place Units in the Hippocampus of the Freely Moving Rat. *Exp. Neurol.* **1976**, *51*(1), 78–109.
6. Fyhn, M.; Molden, S.; Witter, M. P.; Moser, E. I.; Moser, M.-B. Spatial Representation in the Entorhinal Cortex. *Science* **2004**, *305*(5688), 1258–1264.
7. Hafting, T.; Fyhn, M.; Molden, S.; Moser, M.-B.; Moser, E. I. Microstructure of a Spatial Map in the Entorhinal Cortex. *Nature* **2005**, *436*(7052), 801–806.
8. Sargolini, F.; Fyhn, M.; Hafting, T.; McNaughton, B. L.; Witter, M. P.; Moser, M.-B.; Moser, E. I. Conjunctive Representation of Position, Direction, and Velocity in Entorhinal Cortex. *Science* **2006**, *312*(5774), 758–762.
9. Freeman, W. J. *Mass Action in the Nervous System.* Vol. 2004. Citeseer, 1975; Vol. 2004.
10. Boisseau, R. P.; Vogel, D.; Dussutour, A. Habituation in non-neural organisms: evidence from slime moulds. *Proc. R. Soc. B Biol. Sci.* 2016, *283*(1829), 20160446.
11. Jibu, M.; Yasue, K. *Quantum Brain Dynamics and Consciousness.* John Benjamins, 1995.

12. Vitiello, G. *My Double Unveiled: The Dissipative Quantum Model of Brain*. Vol. 32. John Benjamins Publishing, 2001; Vol. 32.
13. Ricciardi, L. M.; Umezawa, H. Brain and Physics of Many-Body Problems. *Kybernetik* **1967**, *4*(2), 44–48.
14. Umezawa, H. *Advanced Field Theory: Micro, Macro, and Thermal Physics*. AIP, 1995.
15. Nambu, Y.; Jona Lasinio, G. Dynamical Model of Elementary Particles Based on an Analogy with Superconductivity. I. *Phys. Rev.* **1961**, *112*, 345.
16. Goldstone, J. Field Theories With Superconductor Solutions. *Il Nuovo Cimento (1955-1965)* 1961, *19*(1), 154–164.
17. Goldstone, J.; Salam, A.; Weinberg, S. Broken Symmetries. *Phys. Rev.* **1962**, *127*(3), 965.
18. Fröhlich, H. Bose Condensation of Strongly Excited Longitudinal Electric Modes. *Phys. Lett. A* **1968**, *26*(9), 402–403.
19. Fröhlich, H. Long-range Coherence and Energy Storage in Biological Systems. *Int. J. Quantum Chem.* **1968**, *2*(5), 641–649.
20. Preto, J.; Pettini, M.; Tuszynski, J. A. Possible Role of Electrodynamic Interactions in Long-Distance Biomolecular Recognition. *Phys. Rev. E* **2015**, *91*(5), 052710.
21. Fröhlich, H. Selective Long Range Dispersion Forces Between Large Systems. *Phys. Lett. A* **1972**, *39*(2), 153–154.
22. Paul, R. Production of Coherent States in Biological Systems. *Phys. Lett. A* **1983**, *96*(5), 263–268.
23. Pribram, K. H. *Languages of the Brain: Experimental Paradoxes and Principles in Neuropsychology*. Prentice-Hall, 1971.
24. Stuart, C. I. J. M.; Takahashi, Y.; Umezawa, H. On the Stability and Non-Local Properties of Memory. *J. Theoretical Biol.* **1978**, *71*(4), 605–618.
25. Stuart, C. I. J. M.; Takahashi, Y.; Umezawa, H. Mixed-System Brain Dynamics: Neural Memory as a Macroscopic Ordered State. *Found. Phys.* **1979**, *9*(3–4), 301–327.
26. Davydov, A. S.; Kislukha, N. I. Solitons in One-Dimensional Molecular Chains. *Phys. Status Solidi (b)* 1976, *75*(2), 735–742.
27. Tuszyński, J. A.; Paul, R.; Chatterjee, R.; Sreenivasan, S. R. Relationship Between Fröhlich and Davydov Models of Biological Order. *Phys. Rev. A* **1984**, *30*(5), 2666.
28. Del Giudice, E.; Doglia, S.; Milani, M.; Vitiello, G. Spontaneous Symmetry Breakdown and Boson Condensation in Biology. *Phys. Lett. A* **1983**, *95*(9), 508–510.
29. Del Giudice, E.; Doglia, S.; Milani, M.; Vitiello, G. A Quantum Field Theoretical Approach to the Collective Behaviour of Biological Systems. *Nuclear Phys. B* **1985**, *251*, 375–400.
30. Del Giudice, E.; Preparata, G.; Vitiello, G. Water as a Free Electric Dipole Laser. *Phys. Rev. Lett.* **1988**, *61*(9), 1085.
31. Del Giudice, E.; Smith, C. W.; Vitiello, G. Magnetic Flux Quantization and Josephson Systems. *Phys. Scripta* **1989**, *40*, 786–791.
32. Jibu, M.; Yasue, K. A Physical Picture of Umezawa's Quantum Brain Dynamics. *Cybernet. Syst. Res.* **1992**, *92*, 797–804.
33. Jibu, M.; Yasue, K. Intracellular Quantum Signal Transfer in Umezawa's Quantum Brain Dynamics. *Cybernet. Syst.* **1993**, *24*(1), 1–7.
34. Jibu, M.; Hagan, S.; Hameroff, S. R.; Pribram, K. H.; Yasue, K. Quantum Optical Coherence in Cytoskeletal Microtubules: Implications for Brain Function. *Biosystems* **1994**, *32*(3), 195–209.
35. Jibu, M.; Yasue, K. What is Mind? Quantum Field Theory of Evanescent Photons in Brain as Quantum Theory of Consciousness. *INF* **1997**, *21*(3), 471–490.
36. Dicke, R. H. Coherence in Spontaneous Radiation Processes. *Phys. Rev.* **1954**, *93*(1), 99.
37. Vitiello, G. Dissipation and Memory Capacity in the Quantum Brain Model. *Int. J. Modern Phys. B* **1995**, *9*(08), 973–989.

38. Vitiello, G. Classical Chaotic Trajectories in Quantum Field Theory. *Int. J. Modern Phys. B* **2004**, *18*(4–5), 785–792.
39. Sabbadini, S. A.; Vitiello, G. Entanglement and Phase-Mediated Correlations in Quantum Field Theory. Application to Brain-Mind States. *Appl. Sci.* **2019**, *9*(15), 3203.
40. Zheng, J.-M.; Pollack, G. H. Long-Range Forces Extending From Polymer-Gel Surfaces. *Phys. Rev. E* **2003**, *68*(3), 031408.
41. Del Giudice, E.; Voeikov, V.; Tedeschi, A.; Vitiello, G. The Origin and the Special Role of Coherent Water in Living Systems. In: *In The Fields of the Cell*; Fels, D., Cifra, M., Scholkmann, F., Eds.; Research Signpost, 2014. ISBN: 978-81-308-0544-3, pp 95–111.
42. Nishiyama, A.; Tanaka, S.; Tuszynski, J. A. Non-Equilibrium Quantum Brain Dynamics: Super-Radiance and Equilibration in 2+1 Dimensions. *Entropy* **2019**, *21*(11), 1066. https://doi.org/10.3390/e21111066.
43. Nishiyama, A.; Tuszynski, J. A. Nonequilibrium Quantum Electrodynamics: Entropy Production During Equilibration. *Int. J. Modern Phys. B* **2018**, *32*(24), 1850265.
44. Nishiyama, A.; Tuszynski, J. A. Non-equilibrium $\phi4$ Theory in Open Systems as a Toy Model of Quantum Field Theory of the Brain. *Ann. Phys.* **2018**, *398*, 214–237.
45. Nishiyama, A.; Tuszynski, J. A. Non-Equilibrium $\phi4$ Theory for Networks: Towards Memory Formations With Quantum Brain Dynamics. *J. Phys. Commun.* **2019**, *3*(5), 055020.
46. Nishiyama, A.; Tanaka, S.; Tuszynski, J. A. Non-Equilibrium Quantum Electrodynamics in Open Systems as a Realizable Representation of Quantum Field Theory of the Brain. *Entropy* **2020**, *22*(1), 43.
47. Craddock, T. J. A.; Tuszynski, J. A.; Hameroff, S. Cytoskeletal Signaling: Is Memory Encoded in Microtubule Lattices by CaMKII Phosphorylation?*PLoS Comput. Biol.* **2012**, *8*(3), e1002421.
48. Tuszynski, J. A.; Carpenter, E. J.; Huzil, J. T.; Malinski, W.; Luchko, T.; Luduena, R. F. The Evolution of the Structure of Tubulin and Its Potential Consequences for the Role and Function of Microtubules in Cells and Embryos. *Int. J. Dev. Biol.* **2006**, *50*(2/3), 341.
49. Tuszyński, J. A.; Brown, J. A.; Crawford, E.; Carpenter, E. J.; Nip, M. L. A.; Dixon, J. M.; Satarić, M. V. Molecular Dynamics Simulations of Tubulin Structure and Calculations of Electrostatic Properties of Microtubules. *Math. Comput. Model.* **2005**, *41*(10), 1055–1070.
50. Baym, G.; Kadanoff, L. P. Conservation Laws and Correlation Functions. *Phys. Rev.* **1961**, *124*(2), 287.
51. Kadanoff, L. P.; Baym, G. *Quantum Statistical Mechanics: Green's Function Methods in Equilibrium Problems*. WA Benjamin, 1962.
52. Baym, G. Self-Consistent Approximations in Many-Body Systems. *Phys. Rev.* **1962**, *127*(4), 1391.
53. Schwinger, J. Brownian Motion of a Quantum Oscillator. *J. Math. Phys.* **1961**, *2*(3), 407–432.
54. Keldysh, L. V. Diagram Technique for Nonequilibrium Processes. *Sov. Phys. JETP* **1965**, *20*(4), 1018–1026.

CHAPTER SIX

Quantum protein folding☆

Liaofu Luo[a,*] and Jun Lv[b,*]

[a]Faculty of Physical Science and Technology, Inner Mongolia University, Hohhot, China
[b]Center for Physics Experiment, College of Science, Inner Mongolia University of Technology, Hohhot, China
*Corresponding authors: e-mail address: lolfcm@imu.edu.cn; lujun@imut.edu.cn

Contents

Abstract

To establish a fundamental framework for understanding biomolecules from the first principle of quantum mechanics, the manuscript reviews the work of authors' group on protein and RNA folding to demonstrate the existence of a common quantum mechanism in the conformational transition of biomolecules. Based on the general equation of the conformation-transitional rate several theoretical results are deduced and compared with experimental data through bioinformatics methods. The main results we obtained are: The temperature dependence and the denaturant concentration dependence of the protein folding rate are deduced and compared with experimental data. The quantitative relation between protein folding rate and torsional mode number (or chain length) is deduced and the obtained formula can be applied to RNA folding as well. The quantum transition theory of two-state protein is successfully generalized to multistate protein folding. Then, how to make direct experimental tests on the quantum property of the conformational transition of biomolecule is discussed,

☆The present chapter was written based on the work appeared in *Quantitative Biology* 2017, 5 (2), 143–158, https://doi.org/10.1007/s40484-016-0087-9, see Ref. 73.

Advances in Quantum Chemistry, Volume 82
ISSN 0065-3276
https://doi.org/10.1016/bs.aiq.2020.01.001

which includes the study of protein photo-folding and the observation of the fluctuation of the fluorescence intensity emitted from the protein folding/unfolding event. The above results show that the quantum mechanics provides a unifying and logically simple theoretical starting point in studying the conformational change of biological macromolecules. The far-reaching results in practical application of the theory are expected.

1. Introduction

There are huge numbers of variables in a biological system. What are the fundamental variables in the life processes at the molecular level? Since the classical works of B. Pullman and A. Pullman on nucleic acids,[1] it is generally accepted that the mobile π electrons play an important role in the biological activities of macromolecules. However, the quantum biochemistry cannot treat a large class problems relating to the conformational variation of biological macromolecules such as protein folding, RNA folding, signal transduction and gene expression regulation, etc. In fact, for a macromolecule consisting of n atoms there are $3n$ coordinates if each atom is looked as a point. Apart from 6 translational and rotational degrees of freedom there are $3n$-6 coordinates describing molecular shape. It has been proved that the bond lengths, bond angles and torsion (dihedral) angles form a complete set to describe the molecular shape. The molecular shape is the main variables responsible for conformational change. However, to our knowledge, there is no successful approach to the quantum motion of molecular shape.

For a complex system consisting of many dynamical variables the separation of slow/fast variables is the first key step in the investigation. In his synergetics Haken proposed that the long-living systems slave the short-living ones, or briefly, the slow variables slave the fast ones. He indicated that the fast variables can be adiabatically eliminated in classical statistical mechanics.[2] However, what is the slow variable for a molecular biological system? The typical chemical bond energy is several electron volts (e.g., 3.80 ev for C—H bond, 3.03 ev for C—N bond, 6.30 ev for C=O dissociation). The CG hydrogen bond energy is 0.2 ev and the TA hydrogen bond energy is 0.05 ev in nucleic acids. The energy related to the variation of bond length and bond angle is in the range of 0.4–0.03 ev. While the torsion vibration energy is 0.03–0.003 ev, the lowest in all forms of biological energies. In terms of frequency, the stretching and bending frequency is 10^{14}–10^{13} Hz while that for torsion is 7.5×10^{12}–7.5×10^{11} Hz. Interestingly, the torsion energy is even lower than the average thermal energy per atom at room temperature (0.04 ev in 25°C); the torsion angles are easily changed even at physiological temperature. Therefore, the torsion motion can be looked as the slow variable and others including mobile π electron, chemical binding, stretching and bending, etc. are fast variables. Moreover, different from stretching and bending the torsion potential generally has several minima that correspond to several stable conformations. Therefore, it is reasonable to assume that the torsions are slaving slow variables for the biomolecule system, the molecular conformations can be defined by torsion states and the conformational change is essentially a quantum transition between them.[3–5]

Although the knowledge of protein structure is in a phase of remarkably growth the dynamical mechanism of protein folding is still unclear. The molecular dynamics (MD) provides a computer method to simulate the mechanism only at the coarse grained level since it is mainly based on the classical mechanics. Simultaneously, due to the large computational cost the limited results currently obtained by atomic-level MD simulations are still inadequate for giving an answer on the basic folding mechanism. A noted problem is why the protein folding rate always exhibits the curious non-Arrhenius temperature dependence (the logarithm folding rate is not a decreasing linear function of $1/T$).[6] The non-Arrhenius peculiarity was conventionally interpreted by the nonlinear temperature dependence of the configurational diffusion constant on rough energy landscapes,[7] by the temperature dependence of hydrophobic interaction[8,9] or by introducing the number of denatured conformation depending on temperature to interpret the difference between folding and unfolding.[6] All these explanations are inconclusive. Recent experimental data indicated very different and unusual temperature dependencies of the folding rates existing in the system of λ_{6-85} mutants[10] and in some de novo designed ultrafast folding protein.[11] These unusual Arrhenius plots of ultrafast folders provide an additional kinetic signature for protein folding, indicating the necessity of searching for a new folding mechanism. Another problem is related to the Levinthal's paradox. In a recent work Garbuzynskiy et al. reported that the measured protein folding rates fall within a narrow triangle (called Golden triangle).[12] Why protein folding rates are confined in such a narrow kinetic region? To explain these longstanding problems it seems that a novel physical model on the folding mechanism is required. Based on the idea that the molecular conformation is defined by torsion state and the folding/unfolding is essentially a quantum transition between them, through adiabatically elimination of fast variables we are able to obtain a set of fundamental equations to describe the rate of the conformational transition of macromolecule.[3,4] Our experience shows that the adiabatically elimination is an effective tool to deal with the multiscale problems such as protein folding. By use of these equations we have successfully explained the non-Arrhenius temperature dependence of the folding rate for each protein. Moreover, the statistical investigation of 65 two-state protein folding rates (which were studied by Garbuzynskiy et al.) shows these fundamental equations are consistent with experimental data.[5,13]

Apart from protein folding, conformation transition occurs in many other molecular biological processes. Both RNA and protein are biological macromolecules. Common themes of RNA and protein folding were indicated.[14,15] It is expected that both obey the same dynamical laws and have a unifying folding mechanism. In a recent work Hyeon et al. reported that RNA folding rates are determined by chain length.[16] However, Hyeon's relation of RNA folding rates vs chain length is obtained empirically and its generalization to protein is problematic. Based on quantum folding theory[4,5] we can make comparative studies on protein and RNA folding and deduce a theoretical formula on RNA folding rate. Our results show that the quantum theory serves a logic foundation for understanding this universality between protein and RNA folding.

In the article we shall sketch the deduction of the general rate equation from quantum transition theory at first. Then, main results in developing quantum folding theory and applying it to the protein and RNA folding problems will be given. It includes (1) the temperature dependence of the folding rate; (2) the denaturant concentration dependence of the protein folding rate; (3) the torsion mode number dependence of the protein folding rate; (4) the chain length dependence of the RNA folding rate; (5) the folding rate of multistate protein. In Sections 8.1–8.3 several discussions are devoted: the protein photo-folding, the quantum coherence and the direct experimental test on the quantum property of the conformational change of macromolecules, and some remarks on the application of the quantum folding theory.

2. General formula for conformational transition rate

Suppose the quantum state of a macromolecule is described by a wave function $\Psi(\theta, x, t)$, where $\{\theta\}$ the torsion angles of the molecule and $\{x\}$ the set of fast variables including the stretching-bending coordinates and the frontier electrons of the molecule, etc. For stationary state the time factor can be split out, namely $\Psi(\theta, x, t) = M(\theta, x)\exp\left(\frac{-iEt}{\hbar}\right)$ (E—the total energy of molecule) where $M(\theta, x)$ satisfies the partial differential equation, written in operator form

$$\left(H_{tor} + H_{fv}\right)M(\theta, x) = EM(\theta, x) \tag{1}$$

H_{tor} and H_{fv} are slow-variable and fast-variable Hamiltonian operators, respectively. Both they are deduced from the classical Hamiltonian (energy as a function of coordinate and conjugate momentum) with the substitution of the jth momentum by the operator $\frac{\partial}{\partial x_j}$ · H_{tor} can be written explicitly as

$$H_{tor} = -\sum \frac{\hbar^2}{2I_j}\frac{\partial^2}{\partial\theta_j^2} + U_{tor}(\theta) \tag{1a}$$

where I_j the jth torsion moment($j = 1, \ldots, N$, N—number of the torsion modes), U_{tor}—torsion potential averaged over fast variables. Similarly, H_{fv} is a function of conjugate pair variables x and $\frac{\partial}{\partial x}$ depending on θ as parameter,

$$\begin{aligned} H_{fv} = -\sum\left(\frac{\hbar^2}{2I_j^{bn}}\frac{\partial^2}{\left(\partial x_j^{bn}\right)^2} + \frac{\hbar^2}{2I_j^{st}}\frac{\partial^2}{\left(\partial x_j^{st}\right)^2} + \frac{\hbar^2}{2m_e}\frac{\partial^2}{\left(\partial x_j^{el}\right)^2}\right) \\ + U_{st,bn}(x, \theta) + U_{mf}(x, \theta) \end{aligned} \tag{1b}$$

where x_j^{bn}, x_j^{st}, x_j^{el} are fast variable coordinates describing bending, stretching and electron position, respectively, I_j^{bn}—the jth bending moment, I_j^{st}—the jth stretching inertia, m_e—electron mass, $U_{st,bn}$—the stretching-bending potential, and U_{mf}—the mean field of electrons. If more fast variables are considered, the corresponding new terms should be added in Eq. (1b). Inserting Eqs. (1a) and (1b) into Eq. (1) we obtain the explicit form of the partial differential equation.

$$\begin{aligned} -\Bigg[\sum\left(\frac{\hbar^2}{2I_j}\frac{\partial^2}{\partial\theta_j^2} + \frac{\hbar^2}{2I_j^{bn}}\frac{\partial^2}{\left(\partial x_j^{bn}\right)^2} + \frac{\hbar^2}{2I_j^{st}}\frac{\partial^2}{\left(\partial x_j^{st}\right)^2} + \frac{\hbar^2}{2m_e}\frac{\partial^2}{\left(\partial x_j^{el}\right)^2}\right) \\ -U_{tor}(\theta) - U_{st,bn}(x, \theta) - U_{mf}(x, \theta)\Bigg]M(\theta, x) = EM(\theta, x) \end{aligned} \tag{2}$$

In adiabatic approximation the wave function is expressed as

$$M(\theta, x) = \psi(\theta)\varphi(x, \theta) \tag{3}$$

and the two factors satisfy partial differential equations

$$H_{fv}\varphi_\alpha(x, \theta) = \varepsilon_\alpha(\theta)\varphi_\alpha(x, \theta) \tag{4}$$

$$(H_{tor} + \varepsilon_\alpha(\theta))\psi_{kn\alpha}(\theta) = E_{kn\alpha}\psi_{kn\alpha}(\theta) \tag{5}$$

respectively, where α denotes the quantum number of fast-variable wave function φ, and (k, n) refers to the conformational (indicating which minimum the wave function is localized around) and the vibrational state of torsion wave function ψ.

Because $M(\theta, x)$ is not a rigorous eigenstate of Hamiltonian $H_{tor} + H_{fv}$, there exists a transition between adiabatic states that results from the off-diagonal elements

$$\langle k'n'\alpha'|H'|kn\alpha\rangle = \int \psi^+_{k'n'\alpha'}(\theta) \sum_j - \frac{\hbar^2}{2I_j} \left\{ \int \varphi^+_{\alpha'} \left(\frac{\partial^2 \varphi_\alpha}{\partial \theta_j^2} + 2 \frac{\partial \varphi_\alpha}{\partial \theta_j} \frac{\partial}{\partial \theta_j} \right) dx \right\} \psi_{kn\alpha}(\theta) d\theta \tag{6}$$

Through tedious calculation we obtain the rate of conformational transition[3,4]

$$W = \frac{2\pi}{\hbar^2 \bar{\omega}'} I'_V I'_E$$

$$I'_V = \frac{\hbar}{\sqrt{2\pi}\delta\theta} \exp\left(\frac{\Delta G}{2k_B T}\right) \exp\left(-\frac{(\Delta G)^2}{2\zeta k_B T}\right) (k_B T)^{\frac{1}{2}} \left(\sum_j^N I_j\right)^{\frac{1}{2}}$$

$$I'_E = \sum_j^M \left| a^{(j)}_{\alpha'\alpha} \right|^2 = M\bar{a}^2 \tag{7}$$

in which $\zeta = \bar{\omega}^2 (\delta\theta)^2 \sum_j^N I_j$ as $z = \frac{k_B T}{\hbar^2} (\delta\theta)^2 I_0 \gg 1$ and the condition $z \gg 1$ is generally satisfied in protein folding problem. I'_V is slow-variable factor and I'_E fast-variable factor of the transitional rate, N is the number of torsion modes participating coherently in a quantum transition, I_j denotes the inertial moment of the atomic group of the jth torsion mode (I_0 denotes its average), $\bar{\omega}$ and $\bar{\omega}'$ are, respectively, the initial and final frequency parameters ω_j and ω'_j of torsion potential averaged over N torsion modes, $\delta\theta$ is the averaged angular shift between initial and final torsion potential, ΔG is the free energy decrease per molecule between initial and final states, M is the number of torsion angles correlated to fast variables, $\bar{a}^2$ is the square of the matrix element of the fast-variable Hamiltonian operator, or, more accurately, its change with torsion angle, averaged over M modes,

$$\bar{a}^2 = \frac{1}{M}\sum_j^M |a_{\alpha'\alpha}^{(j)}|^2$$

$$a_{\alpha'\alpha}^{(j)} = \frac{i\hbar}{I_j^{1/2}} \frac{h_{\alpha'\alpha}^{(j)}}{\varepsilon_\alpha^{(0)} - \varepsilon_{\alpha'}^{(0)}} \quad (\alpha' \neq \alpha)$$

$$h_{\alpha'\alpha}^{(j)} = \left(\frac{\partial H_{fv}}{\partial \theta_j}\right)_0 \tag{8}$$

$\varepsilon_\alpha^{(0)}$, $\varepsilon_{\alpha'}^{(0)}$ are the eigenvalues of H_{fv}. Eqs. (7) and (8) are basic equations for conformational transition.

Note: The adiabatic approximation is the only important approximation in the above deduction. The approximation is applicable because the energy gap between the eigenvalues $\varepsilon_\alpha^{(0)}$ of fast-variable Hamiltonian is generally larger than the torsion energy.[17] The torsion potential U_{tor} is a function of a set of torsion angles $\theta = \{\theta_j\}$. Its form is dependent of solvent environment of the molecule. Suppose the interaction between water (or other solvent, ion and denaturant) molecules (their coordinates denoted by r) and macromolecule is V (r, θ, x), its average over r in a given set of experimental conditions (including chemical denaturants, solvent conditions, etc.)[18] can be expressed by $\langle V(r,\theta,x)\rangle_{av} = V_1(\theta) + V_2(x,\theta)$ where $V_1(\theta)$ is x-independent part of the average interaction. Then we define $U_{tor}(\theta) = U_{tor,vac}(\theta) + V_1(\theta)$ and $H_{fv} = H_{fv,vac} + V_2(x, \theta)$ in the basic equations (1) and (2) with $U_{tor,vac}$ denoting the torsion potential in vacuum and $H_{fv,vac}$ the fast-variable Hamiltonian in vacuum. Therefore, although the influence of solvent is a difficult problem in molecular dynamics approach, it has been taken into account automatically in the present theory by redefining the torsion potential U_{tor} and fast-variable Hamiltonian H_{fv} in the basic equations.

The torsion mode number N describes the coherence degree of multi-torsion transition in the folding. For protein we assume that N is obtained by numeration of all main-chain and side-chain dihedral angles on the polypeptide chain except those residues on its tail which does not belong to any contact. Each residue in such contact fragment contributes two main-chain dihedral angles and, for nonalanine and -glycine, it contributes 1–4 additional side-chain dihedral angles. For nucleic acid the torsion number can be estimated by chain length. Following IUB/IUPAC there are seven torsion angles for each nucleotide, namely

$$\begin{aligned}&\alpha(\mathrm{O3'-P-O5'-C5'}),\\&\beta(\mathrm{P-O5'-C5'-C4'}),\\&\gamma(\mathrm{O5'-C5'-C4'-C3'}),\\&\delta(\mathrm{C5'-C4'-C3'-O3'}),\\&\varepsilon(\mathrm{C4'-C3'-O3'-P}),\\&\varsigma(\mathrm{C3'-O3'-P-O5'})\end{aligned}$$

and χ(O4′–C1′–N1–C2) (for Pyrimidine) or χ(O4′–C1′–N9–C2) (for Purine), of which many have more than one advantageous conformations (potential minima). If each nucleotide has q torsion angles with multi-minimum in potential then the torsion number $N = qL$, where L is chain length.

3. Temperature dependence of protein folding

From the rate Eq. (7) we have deduced a temperature dependence law of protein folding rate $W(T)$

$$\ln W(T) = \frac{S}{T} - RT + \frac{1}{2}\ln T + const. \tag{9}$$

where *const* means temperature-independent term. The law explains the curious non-Arrhenius behavior of the rate-temperature relationships in protein folding/unfolding experiments.[4,5] Several examples are given in Figs. 1–3.

The comprehensive comparisons of the theoretical predictions with experimental rates were made in Ref. 5 for all two-state proteins whose temperature dependence data were available. The strong curvature of folding rate on Arrhenius plot is due to the R term in Eq. (9) which comes from the square free energy $(\Delta G)^2$ in rate Eq. (7) for theoretical $\ln W$. Since the factor $(\Delta G)^2$ occurs only in the rate equation of quantum theory the good agreement between theory and experiments affords support to the concept of quantum folding. Moreover, in this theory the universal non-Arrhenius characteristics of folding rate are described by only two parameters S and R and these parameters are related to the known folding

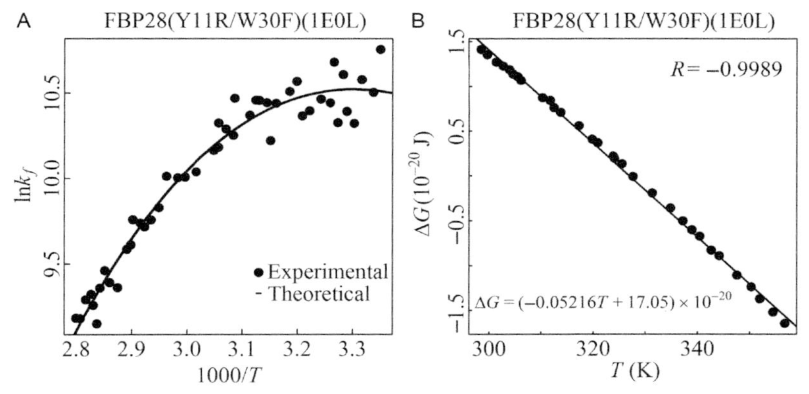

Fig. 1 Model fit to overall folding rate k_f vs temperature T and the temperature dependence of free energy decrease ΔG for protein FBP28 (PDB code: 1E0L). (A) Experimental logarithm folding rates vs temperature are shown by "•,", and *solid lines* are theoretical model fits to the folding rate. (B) Linear fits of free energy decrease ΔG (experimental data shown by "•") with temperature T are checked, where R is the Pearson correlation coefficient. All experimental data are taken from Ref. 19 (k_f in unit s^{-1}, ΔG in unit 10^{-20}J, T in unit Kelvin).

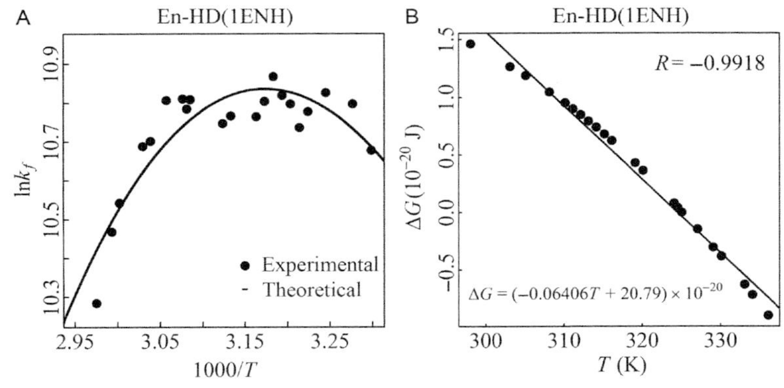

Fig. 2 Model fit to overall folding rate k_f vs temperature T and the temperature dependence of free energy decrease ΔG for protein En-HD (PDB code: 1ENH). Figure caption is put as in Fig. 1 but experimental data taken from Ref. 20.

dynamics. All parameters related to torsion potential defined in this theory (such as torsion frequency $\bar{\omega}$ and $\bar{\omega}'$, averaged angular shift $\delta\theta$ and energy gap ΔE between initial and final torsion potential minima, etc.) can be determined, calculated consistently with each other for all studied proteins.[5] Furthermore, in this theory the folding and unfolding rates are correlated

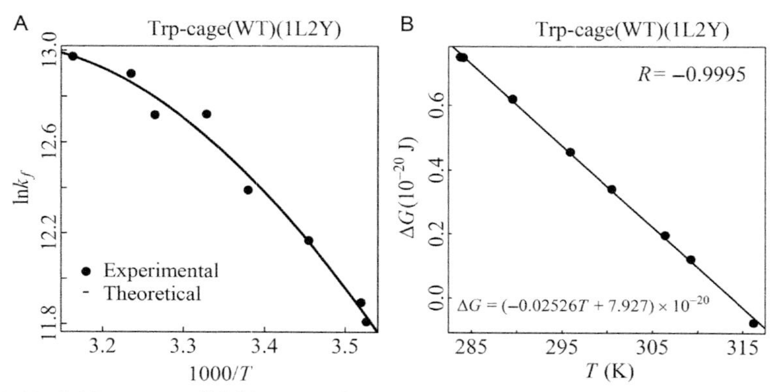

Fig. 3 Model fit to overall folding rate k_f vs temperature T and the temperature dependence of free energy decrease ΔG for protein Trp-cage (PDB code: 1L2Y). Figure caption is put as in Fig. 1 but experimental data taken from Ref. 21.

with each other, needless of introducing any further assumption as in Ref. 6. An interesting relation we obtained is

$$R'/R = (\bar{\omega}/\bar{\omega}')^2 \tag{10}$$

(slope S and curvature R in folding are denoted as S', R' in unfolding). The equation explains why for some proteins the plots of $\ln W$ vs $1/T$ are strongly curved but almost linear for their unfolding by $\bar{\omega} < \bar{\omega}'$.

Assuming the free energy change ΔG in a temperature interval lower than melting temperature T_c has been measured and expressed as

$$\begin{aligned} \Delta G &= \Delta G_0 + \Delta G_1(T - T_c) \\ &= \alpha + \beta T \\ \alpha &= \Delta G_0 - \Delta G_1 T_c, \beta = \Delta G_1 \end{aligned} \tag{11}$$

We know that the above linear relation was tested by experiments for many proteins (three examples shown in Figs. 1B, 2B, and 3B).[10, 11,19–28] By using α and β given from experiments we can rededuce the temperature dependence of folding rate. In fact, by inserting Eq. (11) into Eq. (7) one easily obtains Eq. (9) and in the equation the slope and curvature parameters S and R are given by

$$\begin{aligned} S &= \frac{(\Delta G_0 - \Delta G_1 T_c)}{2k_B}\left(1 - \frac{\Delta G_0 - \Delta G_1 T_c}{\zeta}\right) = \frac{\alpha}{2k_B}\left(1 - \frac{\alpha}{\zeta}\right) \\ R &= \frac{(\Delta G_1)^2}{2\zeta k_B} = \frac{\beta^2}{2\zeta k_B} \end{aligned} \tag{12}$$

Eliminating ζ in Eq. (12) a universal relation between R and S is deduced,

$$R = \left(\frac{\alpha}{2k_B} - S\right)\frac{\beta^2}{\alpha^2} \tag{13}$$

Following the similar deduction the relation between slope and curvature parameters for unfolding rate is obtained,

$$R' = -\left(\frac{\alpha}{2k_B} + S'\right)\frac{\beta^2}{\alpha^2} \tag{14}$$

With the aid of known α and β from the temperature dependence of free energy, constraints on slope and curvature parameters can be obtained by Eqs. (13) and (14) for folding/unfolding rate.

In addition to Eqs. (13) and (14), from the relation between folding rate W and unfolding rate W_u

$$\ln\left(\frac{W}{W_u}\right) = \frac{\Delta G}{k_B T} + \frac{(\Delta G)^2}{2k_B T\varepsilon}\left(\frac{\bar{\omega}^2 - \bar{\omega}'^2}{\bar{\omega}'^2}\right) + \ln\frac{\bar{\omega}}{\bar{\omega}'} \tag{15}$$
$$(\varepsilon = N\bar{\omega}^2(\delta\theta)^2 I_0)$$

(ε is defined as ζ in Eq. (7) when $z \gg 1$) we obtain the relation between R and R'

$$\frac{\Delta G_0}{k_B T_c} + \frac{(\Delta G_0)^2}{2k_B T_c\varepsilon}\left(\frac{R'}{R} - 1\right) + \frac{1}{2}\ln\frac{R'}{R} = 0 \tag{16}$$

Here T_c is defined by $W = W_u$ and as $T = T_c$ one has $\Delta G = \Delta G_0$.

Eq. (13) was proved at precision higher than 90% by using experimental data on S, R, α, and β of 15 two-state proteins.[29] Eqs. (14) and (16) can also be tested by experimental data. It is interesting to note that $\Delta G(T_c) \equiv \Delta G_0 \neq 0$, which has been proved by the linear temperature dependence of free energy.[29] If ΔG_0 were 0 then R' would equal R from Eq. (16). So the inequality between R' and R indicated by experiments would require $\Delta G_0 \neq 0$. Of course, ΔG_0 is a small quantity in the order of $k_B T_c \left|\ln \frac{\bar{\omega}}{\bar{\omega}'}\right|$ that can be seen from Eq. (15). Usually the free energy ΔG at given temperature T was measured through $\Delta G = k_B T \ln(k_f/k_u)$ (the experimental value of W and W_u denoted as k_f and k_u respectively) in literature. However, from Eq. (15), this determination of free energy is not accurate as T near T_c where $\ln(k_f/k_u)$ is a small quantity and the term proportional to $\ln \bar{\omega}/\bar{\omega}'$ cannot be neglected. So, we argue that the present measurement of free energy at temperature T near T_c, namely ΔG_0, is not accurate. We expect Eq. (16) will be experimentally tested by more precise measurement data of ΔG_0 and R' value.

Note: The temperature dependence Eq. (9) for protein folding rate can be generalized to other macromolecular conformational changes as long as the slow variable is torsion in the process. We have demonstrated if only electrons serve as the fast variables the folding rate of a torsion-electron system takes the nearly same form of temperature dependence as Eq. (9) except the unimportant term $\frac{1}{2}\ln T$ changes to $-\frac{1}{2}\ln T$.[3,4] Generally speaking, for fast-variables with energy level spacing $\Delta\varepsilon_\alpha \gg k_B T$ the temperature variation of several tens of degrees would not markedly change the statistical distribution of fast-variable quantum states and therefore, the temperature dependence Eq. (9) still holds. The unusual temperature response of intrinsically disordered protein may be attributed to the change of protein helicity.[30] The thermal sensitivity of secondary structure content induces the shift of the energy level of the protein and causes the change of the temperature dependence of its folding rate.

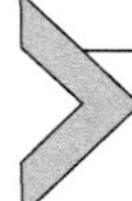

4. Denaturant concentration dependence of protein folding

The denaturant concentration dependence of protein folding rate can be discussed in the same framework. By setting $\Delta G = \Delta G^{(0)} + mc$, $c =$ [*denaturant*], $m = \frac{\partial(\Delta G)}{\partial c}$ at given temperature we obtain

$$\begin{aligned} \ln W &= \ln W^{(0)} + m_f c + m'_f c^2 \\ \ln W_u &= \ln W_u^{(0)} + m_u c + m'_u c^2 \end{aligned} \tag{17}$$

from Eq. (7). Here $W^{(0)}$ and $W_u^{(0)}$ represent the folding and unfolding rates in water, m_f and m_u are the respective slopes of the folding and unfolding arms, m'_f and m'_u describe their curvature

$$\begin{aligned} m_f &= \frac{m}{2k_BT}\left(1 - \frac{2\Delta G^{(0)}}{\zeta}\right); \qquad m_u = -\frac{m}{2k_BT}\left(1 + \frac{2\Delta G^{(0)}}{\zeta'}\right) \\ m'_f &= -\frac{m^2}{2k_BT\zeta}; \qquad m'_u = -\frac{m^2}{2k_BT\zeta'} \\ &(\zeta = NI_0\bar{\omega}^2(\delta\theta)^2, \qquad \zeta' = NI_0\bar{\omega}'^2(\delta\theta)^2) \end{aligned} \tag{18}$$

One may use the four parameters, m_f, m_u, m'_f, and m'_u to fit the experimental data of denaturant dependence. Setting $\frac{\zeta'}{\zeta} = \frac{\bar{\omega}'^2}{\bar{\omega}^2} = 1 + \delta$ ($|\delta|$ is a small quantity) one has

$$
\begin{aligned}
m_f + m_u &= \frac{-m\Delta G^{(0)}}{k_B T\zeta}(2 - \delta) \\
m'_f - m'_u &= \frac{-m^2\delta}{2k_B T\zeta}
\end{aligned}
\tag{19}
$$

If δ is near 0 and neglectable then $m'_f = m'_u = m'$ and $\zeta = \zeta' = 2\Delta G^{(0)} \frac{m_u - m_f}{m_f + m_u}$. In this case only three parameters m_f, m_u, and m' are needed for fitting the experimental data. By using chevron plot (the relation of the relaxation rate constant $\ln k_{obs} = \ln(W + W_u)$ vs denaturant concentration c) to fit the experiments[18] two examples are given in Fig. 4. The classical fits are taken from Ref. 18 where only the linear terms are retained in the chevrons, while the quantum fits with three or four parameters are based on the aforementioned Eqs. (17)–(19). From Fig. 4 one finds the quantum fits are obviously better than the classical linear fits. In the meantime we notice that the polynomial fitting Eq. (17) was reported recently on the frataxin folding in denaturant urea which is in good agreement with experiments.[31] The polynomial fitting is easily understood in our quantum approach.

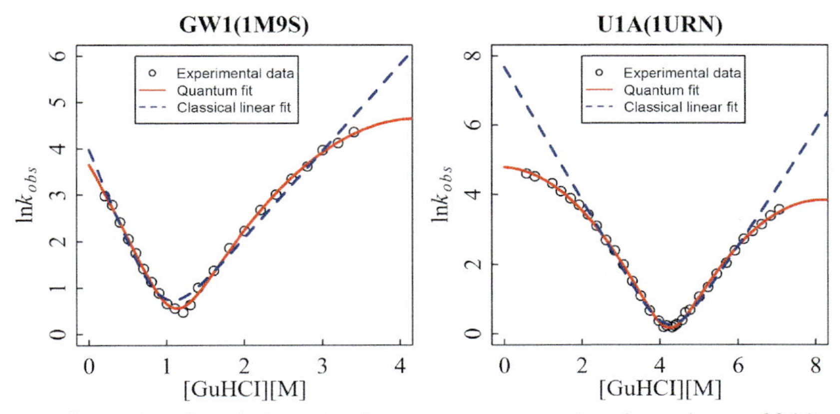

Fig. 4 The statistical analysis on the denaturant concentration dependence of folding/unfolding rates for protein GW1(PDB code: 1M9S) and U1A(PDB code: 1URN). Experimental logarithm relaxation rates $\ln k_{obs}$ are shown by "o." Quantal and classical model fits to the experimental rates are given by *solid* and *dashed lines*, respectively. The data of experimental folding/unfolding rates and classical fits are taken from Ref. 18.

5. *N*-dependence of the protein folding rate

The protein and RNA folding rate is dependent of the number N of the torsion modes participating in the transition. As seen from Eq. (7) the dependence comes mainly from two factors: the fast-variables factor I_E' and the free energy ΔG.

For a large class of conformational change problems such as the common protein and RNA folding, the chemical reactions and electronic transitions are not involved and the fast variables include only bond lengths and bond angles of the macromolecule. In this case an approximate relation of the fast-variable factor I_E' with respect to torsion number N can be deduced. When the kinetic energy in $H_{f\nu}$ is neglected as compared with interaction potential $U_{f\nu}(|\mathbf{r}-\mathbf{r}'|,\theta)$ one has

$$
\begin{aligned}
a_{\alpha'\alpha}^{(j)} &= \frac{i\hbar}{I_j^{1/2}}\frac{1}{\varepsilon_\alpha^{(0)}-\varepsilon_{\alpha'}^{(0)}}\int \varphi_{\alpha'}^{*}(\mathbf{r},\mathbf{r}')\left(\frac{\partial U_{f\nu}}{\partial\theta_j}\right)_{\theta_0}\varphi_\alpha(\mathbf{r},\mathbf{r}')d^3\mathbf{r}d^3\mathbf{r}' \\
&= \frac{i\hbar}{I_j^{1/2}}\frac{1}{V(\varepsilon_\alpha^{(0)}-\varepsilon_{\alpha'}^{(0)})}\int\left(\frac{\partial U_{f\nu}}{\partial\theta_j}\right)_{\theta_0} d^3(\mathbf{r}-\mathbf{r}')
\end{aligned}
\tag{20}
$$

In the above deduction of the second equality the fast-variable wave function $\varphi_\alpha(\mathbf{r},\mathbf{r}')$ has been assumed to be a constant and normalized in the volume V. As energy and volume are dependent of the size of the molecule one may assume energy $\varepsilon_\alpha^{(0)}$ and $U_{f\nu}$ proportional to the interacting-pair number (namely N^2) and V proportional to N. However, because only a small fraction of interacting-pairs are correlated to given $\theta_j (j=1,\ldots,N)$, $\left(\frac{\partial U_{f\nu}}{\partial\theta_j}\right)_{\theta_0}$ does not increase with N. So, one estimates $a_{\alpha'\alpha}^{(j)}\approx N^{-3}$. On the other hand, the integral $\int\left(\frac{\partial U_{f\nu}}{\partial\theta_j}\right)_{\theta_0} d^3(\mathbf{r}-\mathbf{r}')$ may depend on the molecular structure. For example, the high helix content makes the integral increasing. It was indicated that a protein with abundant α helices may have a quite oblong or oblate ellipsoid, instead of spheroid, shape and this protein has higher folding rate.[5,12] Therefore, apart from the factor N^{-3} there is another independent structure-related factor f in $a_{\alpha'\alpha}^{(j)}$. Furthermore, assuming M proportional to N, one obtains

$$
M\bar{a}^2 = cfN^{-5} \tag{21}
$$

where f is a structure-related shape parameter. It means the fast-variable factor I'_E is inversely proportional to N^5. The result is consistent with the direct fit of experimental data of 65 proteins to a power law $M\bar{a}^2 \approx N^{-d}$ that gives d equal about 5.5 to 4.2.[5] Finally from Eqs. (7) and (21) we obtain the N-dependence of folding rate

$$\ln W = \frac{\Delta G}{2k_BT} - \frac{(\Delta G/k_BT)^2}{2\rho N} - 5.5 \ln N + \ln c_0 f \qquad (22)$$

where

$$\rho = I_0 \bar{\omega}^2 (\delta\theta)^2 / (k_B T)$$

is a torsion-energy-related parameter and c_0 is an N-independent constant, proportional to c. The relationship of ln W with N given by Eq. (22) can be tested by the statistical analyses of 65 two-state protein folding rates k_f, which is shown in Fig. 5. The details can be found in Refs. 5 and 29. The theoretical logarithm rate ln W is in good agreement with the experimental ln k_f.

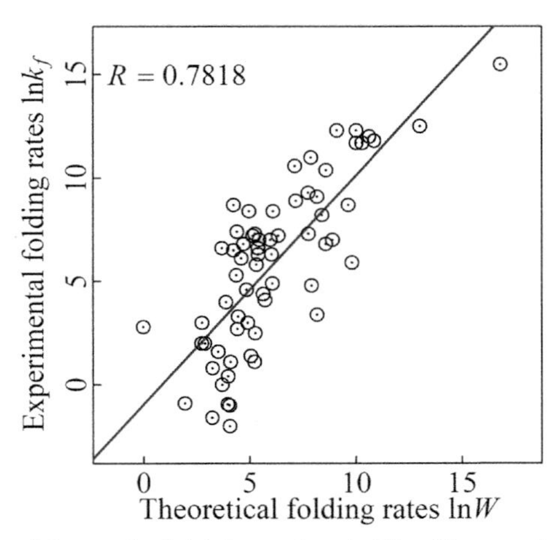

Fig. 5 Comparison of theoretical folding rates ln W with experimental folding rates ln k_f for 65 two-state proteins. Experimental rates for 65 two-state proteins are taken from the database published in Ref. 12 where the folding experiments were carried out at temperature around 25°C and the rates were extrapolated to denaturant-free case. The details of the parameter choice in theoretical calculation can be found in Refs. 5 and 29. The linear regression is given by the *solid line* with correlation $R = 0.7818$.

To find the relation between free energy ΔG and torsion number N we consider the statistical relation of free energy combination $\frac{\Delta G}{2k_B T} - \frac{(\Delta G)^2}{2(k_B T)^2 \rho N}$ that occurs in rate equation (7) or (22). Assuming the free energy differences are measured under a "standard" set of experimental conditions[18] and setting

$$\frac{\Delta G}{2k_B T} - \frac{(\Delta G)^2}{2(k_B T)^2 \rho N} = y, \qquad -\frac{1}{N} = x \tag{23}$$

for the 65-protein dataset[12] we find a good linear relation $y = A + Bx$ existing where A and B are two statistical parameters describing the free energy distribution. The correlation R in the linear regression is near to 0.8 for $\rho =$ 0.065–0.075 and reaches maximum $R = 0.7966$ at $\rho = 0.069$. Thus, by single-ρ-fit we obtain the best-fit statistical relation of free energy for two-state proteins as

$$y = 4.306 + 541.1x; \qquad (\rho = 0.069) \tag{24}$$

(Fig. 6A). As the variation of ρ for different proteins is taken into account the linear regression between free energy combination y and torsion number x will be further improved.

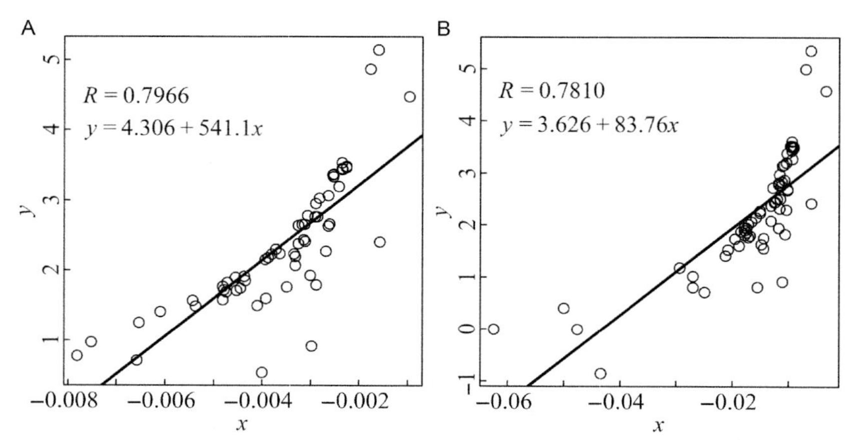

Fig. 6 Statistical relation of free energy with sequence length for two-state proteins. Experimental data are taken from 65-protein set.[5,12] Five proteins in the set denatured by temperature have been omitted in our statistics. In (A) y and x are defined as Eq. (23); in (B) y and x defined as Eq. (25). The linear regression between y and x is plotted by the *solid line* and R means the correlation coefficient.

Because N increases linearly with the length L of polypeptide chain, instead of Eq. (23), by setting

$$\frac{\Delta G}{2k_BT} - \frac{(\Delta G)^2}{2(k_BT)^2\rho_L L} = y, \qquad -\frac{1}{L} = x \tag{25}$$

we obtain the best-fit statistical relation of free energy for two-state proteins (Fig. 6B) as

$$y = 3.626 + 83.76x; \qquad (\rho_L = 0.28) \tag{26}$$

and the correlation coefficient $R = 0.781$(Fig. 6B).

About the relationship of free energy ΔG with torsion number N or chain length L several proposals were proposed in literatures. One statistics was done by assuming the linear relation between ΔG and $\sqrt{N}$, $\Delta G = a\sqrt{N} - b$ $(b \neq 0)$.[5] Another was based on the assumed relation of ΔG *vs* $(Lg + \sigma B_L L^{2/3})$.[12] By the statistics on 65 two-state proteins in the same dataset we demonstrated that the correlation R between free energy and N or L is 0.67 for the former and 0.69 for the latter,[5] both lower than the correlation shown in Fig. 6. We shall use the statistical relation of the free-energy combination, Eq. (23), vs N in the following studies on RNA folding.

In virtue of Eqs. (22)–(24) we obtain an approximate expression for transitional rate $\ln W$ vs N for protein folding

$$\begin{aligned} \ln W &= A - \frac{B}{N} - D\ln N + const \\ D &= 5.5, A = 4.306, B = 541.1 \end{aligned} \tag{27}$$

6. *N*-dependence of the folding rate for RNA molecule

The quantum folding theory of protein is applicable in principle for each step of the conformational transition of RNA molecule. Although recent experiments have revealed multistages in RNA collapse, the final search for the native structure within compact intermediates seems a common step in the folding process. Moreover, the step exhibits strong cooperativity of helix assembly.[14,15] Because the collapse transition prior to the formation of intermediate is a fast process and the time needed for the former is generally shorter than the latter, the calculation of the transition from intermediate to native fold can be directly compared with the experimental data of total rate. By using $N = qL$ (L is the chain length of RNA), instead of Eq. (27), we have

$$\ln W = A - \frac{B'}{L} - D\ln L + c' \qquad (B' = B/q, c' = const - D\ln q) \quad (28)$$

as an alternative expression of folding rate vs chain length. Eq. (28) is deduced from quantum folding theory with some statistical consideration and it predicts the rate W increasing with L, attaining the maximum at $L_{max} = B'/D$, then decreasing with power law L^{-D}.

In a recent work Hyeon and Thirumalai[16] indicated that the chain length determines the folding rates of RNA. They obtained a good empirical relation between folding rate and chain length L in a dataset of 27 RNA sequences. Their best-fit result is

$$\ln W_H = 14.3 - 1.15 \times L^{0.46} \quad (29)$$

Both Eqs. (28) and (29) give us the relation between RNA folding rate and chain length. Comparing the theoretical folding rates $\ln W$ or $\ln W_H$ with the experimental folding rates $\ln k_f$ in 27 RNA dataset the results are shown in Fig. 7. We found that the theoretical equation (28) can fit the experimental data on RNA folding rate equally well as the empirical equation (29). By using the best-fit value of B' and D the correlation between

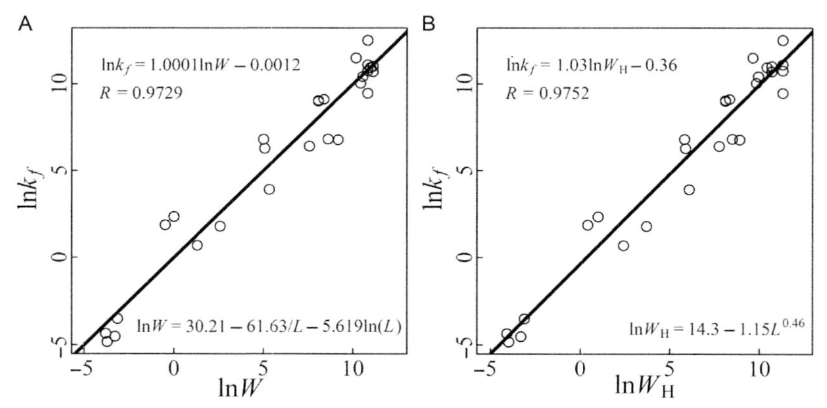

Fig. 7 Comparison of experimental folding rates $\ln k_f$ with theoretical folding rates $\ln W$ (A) or $\ln W_H$ (B) for 27 RNA molecules. Experimental rates are taken from Ref. 16. Theoretical rates are calculated from Eqs. (28) and (29) and shown in the *lower right* of two panels. The regression equation between $\ln k_f$ and $\ln W$ (or $\ln W_H$) is given in the *upper left* of the panel and plotted by the *solid line*. R means the correlation coefficient of the regression analysis.

$\ln W$ (calculated from Eq. (28)) and $\ln k_f$ is $R = 0.9729$ (Fig. 7A), while the correlation between $\ln W_H$ (calculated from Eq. (29)) and $\ln k_f$ is $R = 0.9752$ (Fig. 7B). However, in Fig. 7B the slope of the regression line is 1.03 and the line deviates from origin by -0.36, while in Fig. 7A the slope is 1.0001, very close to 1 and the line deviates from origin only by -0.0012. The reason lies in: although the two equations have the same overall accuracy in fitting experimental data, for large L the errors $Er = \left|\ln W - \ln k_f\right|$ calculated from Eq. (28) are explicitly lower than $Er_H = \left|\ln W_H - \ln k_f\right|$ from Eq. (29) (see Table 1). It means the folding rate lowers down with increasing L as $L^{-D}(D \cong 5.5)$ at large L (a long-tail existing in the W–L curve) rather than a short tail as $\exp\left(-\lambda\sqrt{L}\right)$ assumed in Ref. 16. The long-tail form of folding rate can be used to estimate the rate for long sequence and explain some small-probability events in pluripotency conversion of gene.

There are two independent parameters in RNA folding rate Eq. (28), B' and D, apart from the additive constant. As seen from Fig. 7A we obtain the best-fit D value $D_f = 5.619$ on the 27-RNA dataset, close to $D = 5.5$ predicted from a general theory of quantum folding. Simultaneously we obtain the best-fit B' value $B'_f = 61.63$. The B'_f value derived from RNA folding can be compared with the $B = 541.1$ from protein folding (Eq. 27) if each nucleotide in RNA containing seven torsion angles and the free energy difference between protein and RNA are taken into account.

Table 1 Errors of RNA folding rates in two theoretical models compared with experimental data.

	1	2	3	4	5	6	7	8
L	125	160	205	225	368	377	409	414
$k_f(s^{-1})$	6	2	10.5	6.5	0.03	0.011	0.008	0.013
Er_H	0	0.17	3.27	3.37	1.52	0.71	1.06	1.66
Er	0	0.18	3.15	3.17	0.44	0.42	0.30	0.26

$Er_H = \left|\ln W_H - \ln k_f\right|$, $Er = \left|\ln W - \ln k_f\right|$.

1, hairpin ribozyme; 2, P4–P6 domain (*Tetrahymena* ribozyme); 3, *Azoarcus* ribozyme; 4, *Bacillussubtilis* RNase P RNA catalytic domain; 5, Ca.L-11 ribozyme; 6, *Escherichiacoli* RNase P RNA; 7, *B.subtilis* RNase P RNA and 8, *Tetrahymena* ribozyme. Errors Er_H and Er of RNA folding rates in two theoretical models are listed. Experimental rates k_f's can be found in Ref. 16. The errors of eight RNAs with lengths larger than 120 in 27-RNA dataset were analyzed in the table. The errors of the first RNA hairpin ribozyme were normalized to zero in two models.

7. Multistate protein folding

For a long time the multistate folding mechanism was unclear in theory. Kamagata et al. indicated that the folding rates of nontwo-state proteins show the similar dependence on the native backbone topological parameters as for the two-state proteins.[32] Quantum folding theory provides a unified point on the folding mechanism of nontwo-state and two-state proteins. Zhang and Luo proposed that the multistate folding can be viewed as a joint of several quantum transitions with independent degrees of freedom of torsion angle. So the total collapse rate can be expressed by the formula of two-state folding but with an additional factor indicating the time delay in intermediate state.[33]

Three-state protein is one kind of the multistate protein that was discussed in detail in Ref. 32. Assuming polypeptide chain divided into to two parts and denoting initial state as $|i_1 i_2\rangle$ and final state as $\langle f_1 f_2|$. The transitional matrix element of first-order perturbation $\langle f_1 f_2|(H_1' + H_2')|i_1 i_2\rangle = 0$ where the perturbation Hamiltonian $H_j'(j = 1, 2)$ acts on the jth part only. The matrix element of second-order perturbation is proportional to

$$\sum_{m_1 m_2} \frac{1}{E_{i_1 i_2} - E_{m_1 m_2}} \langle f_1 f_2|(H_1' + H_2')|m_1 m_2\rangle \langle m_1 m_2|(H_1' + H_2')|i_1 i_2\rangle$$
$$= \left(\frac{1}{E_{i_1} - E_{f_1}} + \frac{1}{E_{i_2} - E_{f_2}} \right) \langle f_1|H_1'|i_1\rangle \langle f_2|H_2'|i_2\rangle$$

So the transitional probability W of the three-state protein is proportional to the product of those of the two partial two-state proteins, W_1 from i_1 to f_1 and W_2 from i_2 to f_2, respectively. By using Eq. (22) for two-state protein, namely

$$\ln W_i = \frac{\Delta G_i}{2k_B T} - \frac{(\Delta G_i / k_B T)^2}{2\rho N_i} - 5.5 \ln N_i + \ln c_0 f_i \quad (i = 1, 2) \tag{30}$$

neglecting the difference between structure-related shape parameters f_1 and f_2 in Eq. (30) and denoting N and ΔG of the three-state protein as

$$N = N_1 + N_2, \quad N_1 = rN, \quad N_2 = (1 - r)N$$
$$\Delta G = \Delta G_1 + \Delta G_2, \quad \Delta G_1 = r' \Delta G, \quad \Delta G_2 = (1 - r')\Delta G$$

we obtain

$$\ln W = \frac{\Delta G}{2k_BT} - \frac{(r'\Delta G/k_BT)^2}{2\rho rN} - \frac{((1-r')\Delta G/k_BT)^2}{2\rho(1-r)N} \\ -5.5\ \ln N - 5.5(\ln r + \ln(1-r)) + \ln f + const \tag{31}$$

Eq. (31) is the theoretical folding rate for any three-state protein. As $r = r'$ we have the same dependence of three-state folding rate on T and N as that of two-state protein. The formula can easily be generalized to other m-state proteins with $m > 3$. The statistical comparisons of the theoretical rates $\ln W_f$ of multistate protein folding with experimental rates $\ln k_f$ in a dataset of 38 multistate proteins[34–68] are given in Fig. 8. Since most of the multistate proteins in databases are three-state proteins we use Eq. (31) to calculate $\ln W_f$ directly. It gives the average error MAE $= \langle | \ln k_f - \ln W_f | \rangle = 1.68$ and the correlation between theoretical and experimental rates $R = 0.8098$.

The success of the above calculation shows that a unified quantum folding mechanism does exist for multistate and two-state protein and the folding rate of multistate protein can be obtained in a simple way through calculating the product of the rates of two-state proteins.

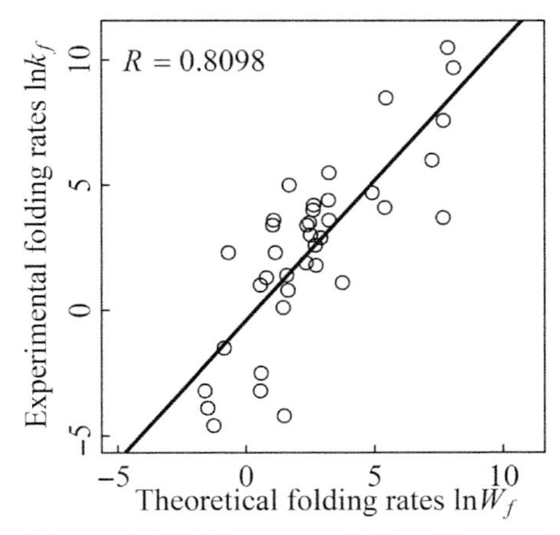

Fig. 8 Comparison of theoretical folding rates $\ln W_f$ with experimental folding rates $\ln k_f$ for 38 multistate proteins. Experimental rates for 38 proteins are taken from Refs. 34–68. In calculating $\ln W_f$ the parameters $r = r' = 0.35$ and $\rho = 0.07$ are assumed. The regression between $\ln k_f$ and $\ln W_f$ is plotted by the *solid line* and R means the correlation coefficient.

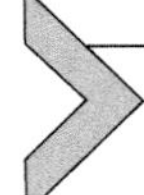

8. Discussions

8.1 Protein photo-folding

Studies in Sections 3–7 show that the quantitative results of many difficult problems in classical theory can easily be deduced from the idea of quantum folding and they can be successfully tested by experimental data. To explore the fundamental physics underlying protein folding more deeply and clarify the quantum nature of the folding mechanism more clearly we have studied the protein photo-folding processes, namely, the photon emission or absorption in protein folding and the inelastic scattering of photon on protein (photon-protein resonance Raman scattering). In these processes the emission or absorption of a photon by atomic electrons is coupled to protein's conformational change. To simplify the discussion we assume only electrons serve as the fast variables of the protein. After the first-principle-calculation based on quantum electrodynamics and quantum folding theory the rates and cross sections of these processes have been deduced.[69] Moreover, these photo-folding processes can be compared with common protein folding without interaction of photons (nonradiative folding). It is demonstrated that there exists a common factor (thermo-averaged overlap integral of vibration wave function) $I_V = \frac{1}{k_B T} I'_V$ (Eq. 7) for protein folding and protein photo-folding. Based on this finding it is predicted that the stimulated photo-folding rates and the resonance fluorescence cross section show the same temperature dependence as protein folding.

Due to the coupling between protein structure and electron motion, the electronic transition will inevitably lead to the structural relaxation or conformational changes of the protein. Therefore the spectrum of protein photo-folding includes information of several kinds of quantum transitions: the electron energy level transition, the transition between vibrational energy levels of the molecule, the transition between rotational energy levels of the molecule and the transition between different molecular conformations. Conformational transitions are somewhat like the rotational transitions, but the rotational transition refers to the whole molecule, while the conformational transition is related only to the dihedral angle rotation of local atomic groups. Because of the coupling with vibration and torsional transition the spectral line of electronic transition is broadened to a band which includes abundant vibration spectrum without and with conformational transition. The width of the spectral band is determined by the torsion vibration frequency, in the order of 10^{13} s^{-1}, one hundredth or thousandth of the electronic transition frequency. Each spectral band includes a large amount of spectral lines and forms an abundant structure. The transition between torsion vibration states in different conformations is a kind of forbidden transition since the overlap integral between initial and final torsional wave functions is very small, about 10^{-5}. So, the width of these extra-narrow spectral lines is five orders smaller than the natural linewidth. This is an important prediction of quantum folding theory.[69]

From the experimental point of view, to observe the extra-narrow spectral line a high-precision and high-resolution spectroscopy is needed. The spectral resolution of femtosecond Raman spectroscopy (FSRS) is $10\ \mathrm{cm}^{-1}$, corresponding to $\Delta\nu = 3 \times 10^{11}\ \mathrm{s}^{-1}$.[70] This resolution is already close to the range of the width that the spectral line of ultranarrow conformational transitions can be searched.

The particular form of the same temperature dependence for protein nonradiative folding and photo-folding and the abundant structure of the photo-folding spectral band consisting of many narrow lines are two main results deduced from protein photo-folding theory. These results are closely related to the fundamental concepts of quantum mechanics. First, they imply the existence of a set of quantum oscillators in the transition process and these oscillators are mainly of torsion vibration type of low frequency. Second, they imply in protein folding that quantum tunneling does exist which means the nonlocality of state and the quantum coherence of conformational–electronic motion. More experimental tests on above two predictions are waited for.

8.2 Quantum coherence and experimental tests on quantum property of macromolecular conformational transition

So far we have discussed the quantum transition between torsion states in macromolecules. A fundamental problem is: due to quantum entanglement with the environment the decoherence possibly makes the quantum picture ceasing to be effective for a macromolecular system. It was estimated that the decoherence time $\tau_D = \tau_R \left(\frac{\hbar}{\Delta x \sqrt{2mk_B T}} \right)^2$ in a simple model where τ_R means the relaxation time, $\frac{\hbar}{\sqrt{2mk_B T}}$ the thermal de Broglie wavelength, Δx the separation of position and m the particle mass. This leads to the decoherence time inversely proportional to particle mass m.[71] For atoms or molecules in water the stochastic collisions with water molecules cause the decoherence of the objects. It was indicated that the decoherence time is lower than the dissipation time by a factor m_w/m (the mass ratio of water molecule to solute).[72] However, the above estimates hold only for the center-of-mass motion of particles. The protein or nucleic acid molecule is a compact aggregate of atoms. The coherence of the motion in internal degrees of freedom may be preserved for a much longer time. How to estimate the coherence of the constituents which are bound in the macromolecule? Set the range of movement of the constituent denoted by d and the thermal de Broglie wavelength of the constituent denoted by λ. For electrons in hydrogen atom, $\lambda = 1$ nm and $d = 0.05$ nm, one has $d/\lambda = 0.05$ and the motion is coherent. For C atoms bounded in a simple organic molecule, $\lambda = 0.007$ nm and $d = 0.15$ nm (d calculated from the C—C or C=C bond length), one has $d/\lambda = 21$. Although the d/λ ratio in this case is much higher than the atom's electrons the quantum coherence of C atoms has been well established by the observed molecular vibrational spectrum. For torsion of atomic groups in protein, $d/\lambda = 0.01$ (inertial moment 10^{-37} g cm^2 and room temperature are taken) and $d = 0.1$ (from the averaged angular shift between two minima of torsion potential), one has $d/\lambda = 10$. For N or C atoms bounded in RNA, $\lambda = 0.007$ nm and $d = 0.1$ nm, one has $d/\lambda = 14$. Since they have d/λ value smaller than the case of bounded C atoms in organic molecule it is reasonable to assume the quantum coherence existing in the latter two cases. However, for C atoms freely moving in a macroscopic scale (say 1 cm) the ratio d/λ is about 10^9, much larger than above cases, and the quantum coherence is definitely destroyed. Therefore, the ratio d/λ can serve as a measure to determine the boundary between quantum and classical motions and by use of the measure one may recognize that the macromolecular conformational motion is basically a quantum event.

What should be cut away by Occam's Razor, classical or quantum, in studying the conformational change of biological macromolecules? What rules, classical or quantum, are obeyed by the macromolecular conformational motion? The final solution of the problem needs more direct experimental evidences.

Whether protein folding is quantum or classical can be directly tested by the observation of the instantaneous nature of the folding event. The instantaneousness is characteristic of the quantum transition. When one says the folding rate 1ms it does not mean the folding continues 1 ms but rather means on average 1000 instantaneous folding events are observed stochastically in 1 s. So, the observation of the instantaneous change of the torsion angle in protein folding provides a clue to solve the puzzling problem.

Qiu et al. used laser temperature-jump spectroscopy to measure the folding rate of the 20-residue Trp-cage protein. They found the fluorescence intensity (FI) increasing rapidly from 11.5 to 14 mV in 4 μs and determined the folding rate 4 μs.[21] Which law, classical or quantum, the folding/unfolding event obeys in the duration of 4 μs? Let us consider the case in terms of the gradual decrease of relevant molecular concentrations. In the beginning, the fluorescence intensity will be weakened accompanying with the lowering of Trp-cage concentration, but the shape of the FI-t (fluorescence intensity vs time) curve remains unchanged. As the concentration decreased to very low, the single-molecule motion can be observed and the fluctuation appears. For quantum folding, the torsion takes only two possible values corresponding to folding and unfolding states, respectively. The Trp fluorescence can be measured only in unfolding state. So, in the duration of 4 μs the fluorescence randomly appears and each occurrence corresponds to one unfolding event. However, if the folding/unfolding obeys the laws of classical physics the torsional angle changes continuously and the Trp fluorescence can be recorded only when the protein reaches the unfolding state. Thus the fluorescence will be measured near the end of 4 μs unfolding process. Two pictures are different from each other. We suggest making the observation of fluorescence fluctuation to test whether the folding obeys the classical or quantum law.

This is an important experiment. If the test has a positive result, then firstly, the boundary between classical and quantum physics will be modified and the applicability of quantum mechanics will be expanded to the internal degrees of freedom of macromolecules; secondly, many strange phenomena of sudden change that is of great importance in molecular biology and molecular genetics will be explained from the idea of quantum transition.

8.3 Some remarks on the application of quantum folding theory

The idea of quantum folding is useful in dealing with practical application problems. In Ref. 73 two examples, glucose transport across membrane and induced pluripotency in stem cell, are given. It shows that the proposed idea and theory can help us to establish a set of reaction equations regarding the problem and determine the rate constants in the biological network. When the theory is employed to the conformational change of DNA molecular chain the cooperativity of nucleotides should be considered. For example, in the discussion on induced pluripotency in stem cell we assumed the torsion transition from differentiate to pluripotent state is of uphill-type (from torsion-ground to torsion excited) while the reverse is of downhill-type (from torsion-excited to torsion-ground). It leads to the small probability of pluripotency conversion as observed in some experiments.[74] However, if the vibrations of torsional coordinate around the potential minimum V_A (torsion-ground) and V_B (torsion-excited, $V_A < V_B$) are considered, then the cooperative transition of DNA chain beneficial to uphill-type process can occur. Suppose the vibration frequency is ω_A or ω_B around V_A or V_B respectively, we proved the stem cell activation is dependent of the relative size of E_A and E_B,

$$E_A - E_B = V_A - V_B - k_B T \ln Y_{A/B}$$
$$k_B T \ln Y_{A/B} = \frac{\hbar}{2}(\omega_B - \omega_A)\text{ctnh}\frac{\hbar\omega_A}{2k_B T}$$

If $E_A - E_B < 0$ the DNA chain is condensed in ground state A, while if $E_A - E_B > 0$ the chain condensed in excited state B. That is, the change of vibration frequency would cause the phase transition of DNA chain.[75] The recently observed high efficiency of the metabolic reprogramming of muscle stem cell[76,77] and some RNA-based reprogramming of human primary fibroblasts[78] can be explained by the frequency-induced phase transition and this gives further supports on the quantum folding theory.

Acknowledgments

Authors are indebted to Drs. Zhao Judong, Zhang Ying, and Zhang Lirong for their numerous discussions and Dr Bao Yulai for his help in literature searching. The work is partly supported by the Inner Mongolia Autonomous Region Natural Science Foundation, Nos. 2015MS0331, 2016MS0306, and 2019LH01004.

References

1. Pullman, B.; Pullman, A. *Quantum Biochemistry*; Wiley Interscience: New York, 1963.
2. Haken, H. *Advanced Synergetics*; Springer: Berlin, 1983.
3. Luo, L. F. Protein Folding as a Quantum Transition Between Conformational States. *Front. Phys. China* **2011**, *6*(1), 133–140. https://doi.org/10.1007/s11467-010-0153-0.
4. Luo, L. F. Quantum Theory on Protein Folding. *Sci. China Phys. Mech. Astron.* **2014**, *57*(3), 458–468. https://doi.org/10.1007/s11433-014-5390-8.
5. Lv, J.; Luo, L. F. Statistical Analyses of Protein Folding Rates From the View of Quantum Transition. *Sci. China Life Sci.* **2014**, *57*(12), 1197–1212. https://doi.org/10.1007/s11427-014-4728-9.
6. Ghosh, K.; Ozkan, S. B.; Dill, K. A. The Ultimate Speed Limit to Protein Folding is Conformational Searching. J. Am. Chem. Soc. (39), 11920–11927, 10.1021/ja066785b.
7. Bryngelson, J. D.; Onuchic, J. N.; Socci, N. D.; Wolynes, P. G. Funnels, Pathways, and the Energy Landscape of Protein Folding: A Synthesis. *Proteins* **1995**, *21*(3), 167–195. https://doi.org/10.1002/prot.340210302.

8. Chan, H. S.; Dill, K. A. Protein Folding in the Landscape Perspective: Chevron Plots and Non-Arrhenius Kinetics. *Proteins* **1998**, *30*(1), 2–33. https://doi.org/10.1002/(SICI)1097-0134(19980101)30:1¡2::AID-PROT2¿3.0.CO;2-R.
9. Akmal, A.; Muñoz, V. The Nature of the Free Energy Barriers to Two-State Folding. *Proteins* **2004**, *57*(1), 142–152. https://doi.org/10.1002/prot.20172.
10. Yang, W. Y.; Gruebele, M. Rate-Temperature Relationships in λ-Repressor Fragment λ_{6-85} Folding. *Biochemistry* **2004**, *43*(41), 13018–13025. https://doi.org/10.1021/bi049113b.
11. Zhu, Y.; Alonso, D. O. V.; Maki, K.; Huang, C. Y.; Lahr, S. J.; Daggett, V.; Roder, H.; DeGrado, W. F.; Gai, F. Ultrafast Folding of α3D: A De Novo Designed Three-Helix Bundle Protein. *Proc. Natl. Acad. Sci. U. S. A.* **2003**, *100*(26), 15486–15491. https://doi.org/10.1073/pnas.2136623100.
12. Garbuzynskiy, S. O.; Ivankov, D. N.; Bogatyreva, N. S.; Finkelstein, A. V. Golden Triangle for Folding Rates of Globular Proteins. *Proc. Natl. Acad. Sci. U. S. A.* **2013**, *110*(1), 147–150. https://doi.org/10.1073/pnas.1210180110.
13. Luo, L. F.; Lu, J. *Temperature Dependence of Protein Folding Deduced From Quantum Transition. arXiv e-prints* **2011**, http://arXiv.org/pdf/1102.3748, arXiv:1102.3748.
14. Thirumalai, D.; Hyeon, C. RNA and Protein Folding: Common Themes and Variations. *Biochemistry* **2005**, *44*(13), 4957–4970. https://doi.org/10.1021/bi047314+.
15. Woodson, S. A. Compact Intermediates in RNA Folding. *Annu. Rev. Biophys.* **2010**, *39*(1), 61–77. https://doi.org/10.1146/annurev.biophys.093008.131334.
16. Hyeon, C.; Thirumalai, D. Chain Length Determines the Folding Rates of RNA. *Biophys. J.* **2012**, *102*(3), L11–L13. https://doi.org/10.1016/j.bpj.2012.01.003.
17. Eyring, H.; Lin, S. H.; Lin, S. M. *Basic Chemical Kinetics*. John Wiley & Sons: New York, 1980.
18. Maxwell, K. L.; Wildes, D.; Zarrine-Afsar, A.; De Los Rios, M. A.; Brown, A. G.; Friel, C. T.; Hedberg, L.; Horng, J. C.; Bona, D.; Miller, E. J.; Valle-Blisle, A.; Main, E. R.; Bemporad, F.; Qiu, L.; Teilum, K.; Vu, N. D.; Edwards, A. M.; Ruczinski, I.; Poulsen, F. M.; Kragelund, B. B.; Michnick, S. W.; Chiti, F.; Bai, Y.; Hagen, S. J.; Serrano, L.; Oliveberg, M.; Raleigh, D. P.; Wittung-Stafshede, P.; Radford, S. E.; Jackson, S. E.; Sosnick, T. R.; Marqusee, S.; Davidson, A. R.; Plaxco, K. W. Protein Folding: Defining a "Standard" Set of Experimental Conditions and a Preliminary Kinetic Data Set of Two-State Proteins. *Protein Sci.* **2005**, *14*(3), 602–616. https://doi.org/10.1110/ps.041205405.
19. Nguyen, H.; Jäger, M.; Moretto, A.; Gruebele, M.; Kelly, J. W. Tuning the Free-Energy Landscape of a WW Domain by Temperature, Mutation, and Truncation. *Proc. Natl. Acad. Sci. U. S. A.* **2003**, *100*(7), 3948–3953. https://doi.org/10.1073/pnas.0538054100.
20. Mayor, U.; Johnson, C. M.; Daggett, V.; Fersht, A. R. Protein Folding and Unfolding in Microseconds to Nanoseconds by Experiment and Simulation. *Proc. Natl. Acad. Sci. U. S. A.* **2000**, *97*(25), 13518–13522. https://doi.org/10.1073/pnas.250473497.
21. Qiu, L.; Pabit, S. A.; Roitberg, A. E.; Hagen, S. J. Smaller and Faster: The 20-Residue Trp-Cage Protein Folds in 4 μs. *J. Am. Chem. Soc.* **2002**, *124*(44), 12952–12953. https://doi.org/10.1021/ja0279141.
22. Dimitriadis, G.; Drysdale, A.; Myers, J. K.; Arora, P.; Radford, S. E.; Oas, T. G.; Smith, D. A. Microsecond Folding Dynamics of the F13W G29A Mutant of the B Domain of Staphylococcal Protein A by Laser-Induced Temperature Jump. *Proc. Natl. Acad. Sci. U. S. A.* **2004**, *101*(11), 3809–3814. https://doi.org/10.1073/pnas.0306433101.
23. Kuhlman, B.; Luisi, D. L.; Evans, P. A.; Raleigh, D. P. Global Analysis of the Effects of Temperature and Denaturant on the Folding and Unfolding Kinetics of the N-Terminal Domain of the Protein L9. *J. Mol. Biol.* **1998**, *284*(5), 1661–1670. https://doi.org/10.1006/jmbi.1998.2246.

24. Manyusa, S.; Whitford, D. Defining Folding and Unfolding Reactions of Apocytochrome b5 Using Equilibrium and Kinetic Fluorescence Measurements. *Biochemistry* **1999**, *38*(29), 9533–9540. https://doi.org/10.1021/bi990550d.
25. Bunagan, M. R.; Yang, X.; Saven, J. G.; Gai, F. Ultrafast Folding of a Computationally Designed Trp-Cage Mutant: Trp2-Cage. *J. Phys. Chem. B* **2006**, *110*(8), 3759–3763. https://doi.org/10.1021/jp055288z.
26. Jäger, M.; Nguyen, H.; Crane, J. C.; Kelly, J. W.; Gruebele, M. The Folding Mechanism of a β-Sheet: The WW Domain. *J. Mol. Biol.* **2001**, *311*(2), 373–393. https://doi.org/10.1006/jmbi.2001.4873.
27. Wang, T.; Zhu, Y.; Gai, F. Folding of a Three-Helix Bundle at the Folding Speed Limit. *J. Phys. Chem. B* **2004**, *108*(12), 3694–3697. https://doi.org/10.1021/jp049652q.
28. Spector, S.; Raleigh, D. P. Submillisecond Folding of the Peripheral Subunit-Binding Domain. *J. Mol. Biol.* **1999**, *293*(4), 763–768. https://doi.org/10.1006/jmbi.1999.3189.
29. Luo, L. F.; Lv, J. *Quantitative Relations in Protein and RNA Folding Deduced From Quantum Theory. bioRxiv* **2016**, https://doi.org/10.1101/021782. https://www.biorxiv.org/content/10.1101/021782v3.full.
30. Uversky, V. N. Unusual Biophysics of Intrinsically Disordered Proteins. *Biochim. Biophys. Acta* **2013**, *1834*(5), 932–951. https://doi.org/10.1016/j.bbapap.2012.12.008.
31. Bonetti, D.; Toto, A.; Giri, R.; Morrone, A.; Sanfelice, D.; Pastore, A.; Temussi, P.; Gianni, S.; Brunori, M. The Kinetics of Folding of Frataxin. *Phys. Chem. Chem. Phys.* **2014**, *16*(14), 6391–6397. https://doi.org/10.1039/c3cp54055c.
32. Kamagata, K.; Arai, M.; Kuwajima, K. Unification of the Folding Mechanisms of Non-Two-State and Two-State Proteins. *J. Mol. Biol.* **2004**, *339*(4), 951–965. https://doi.org/10.1016/j.jmb.2004.04.015.
33. Zhang, Y.; Luo, L. F. The Dynamical Contact Order: Protein Folding Rate Parameters Based on Quantum Conformational Transitions. *Sci. China Life Sci.* **2011**, *54*(4), 386–392. https://doi.org/10.1007/s11427-011-4158-x.
34. Cavagnero, S.; Dyson, H.; Wright, P. E. Effect of H Helix Destabilizing Mutations on the Kinetic and Equilibrium Folding of Apomyoglobin. *J. Mol. Biol.* **1999**, *285*(1), 269–282. https://doi.org/10.1006/jmbi.1998.2273.
35. Golbik, R.; Zahn, R.; Harding, S. E.; Fersht, A. R. Thermodynamic Stability and Folding of GroEL Minichaperones. *J. Mol. Biol.* **1998**, *276*(2), 505–515. https://doi.org/10.1006/jmbi.1997.1538.
36. Banachewicz, W.; Johnson, C. M.; Fersht, A. R. Folding of the Pit1 Homeodomain Near the Speed Limit. *Proc. Natl. Acad. Sci. U. S. A.* **2011**, *108*(2), 569–573. https://doi.org/10.1073/pnas.1017832108.
37. Marianayagam, N. J.; Khan, F.; Male, L.; Jackson, S. E. Fast Folding of a Four-Helical Bundle Protein. *J. Am. Chem. Soc.* **2002**, *124*(33), 9744–9750. https://doi.org/10.1021/ja016480r.
38. Lw, C.; Weininger, U.; Zeeb, M.; Zhang, W.; Laue, E. D.; Schmid, F. X.; Balbach, J. Folding Mechanism of an Ankyrin Repeat Protein: Scaffold and Active Site Formation of Human CDK Inhibitor p19INK4d. *J. Mol. Biol.* **2007**, *373*(1), 219–231. https://doi.org/10.1016/j.jmb.2007.07.063.
39. Calosci, N.; Chi, C. N.; Richter, B.; Camilloni, C.; Engström, Å.; Eklund, L.; Travaglini-Allocatelli, C.; Gianni, S.; Vendruscolo, M.; Jemth, P. Comparison of Successive Transition States for Folding Reveals Alternative Early Folding Pathways of Two Homologous Proteins. *Proc. Natl. Acad. Sci. U. S. A.* **2008**, *105*(49), 19241–19246. https://doi.org/10.1073/pnas.0804774105.
40. Schreiber, G.; Fersht, A. R. The Refolding of Cis- and Trans-Peptidylprolyl Isomers of Barstar. *Biochemistry* **1993**, *32*(41), 11195–11203. https://doi.org/10.1021/bi00092a032.
41. Burns, L. L.; Dalessio, P. M.; Ropson, I. J. Folding Mechanism of Three Structurally Similar β-Sheet Proteins. *Proteins* **1998**, *33*(1), 107–118. https://doi.org/10.1002/(SICI)1097-0134(19981001)33:1¡107::AID-PROT10¿3.0.CO;2-P.

42. Dalessio, P. M.; Ropson, I. J. β-Sheet Proteins With Nearly Identical Structures Have Different Folding Intermediates. *Biochemistry* **2000**, *39*(5), 860–871. https://doi.org/10.1021/bi991937j.
43. Gianni, S.; Guydosh, N. R.; Khan, F.; Caldas, T. D.; Mayor, U.; White, G. W. N.; DeMarco, M. L.; Daggett, V.; Fersht, A. R. Unifying Features in Protein-Folding Mechanisms. *Proc. Natl. Acad. Sci. U. S. A.* **2003**, *100*(23), 13286–13291. https://doi.org/10.1073/pnas.1835776100.
44. Gianni, S.; Calosci, N.; Aelen, J. M.; Vuister, G. W.; Brunori, M.; Travaglini-Allocatelli, C. Kinetic Folding Mechanism of PDZ2 From PTP-BL. *Protein Eng. Des. Sel.* **2005**, *18*(8), 389–395. https://doi.org/10.1093/protein/gzi047.
45. Calloni, G.; Taddei, N.; Plaxco, K. W.; Ramponi, G.; Stefani, M.; Chiti, F. Comparison of the Folding Processes of Distantly Related Proteins. Importance of Hydrophobic Content in Folding. *J. Mol. Biol.* **2003**, *330*(3), 577–591. https://doi.org/10.1016/S0022-2836(03)00627-2.
46. Liu, C.; Gaspar, J. A.; Wong, H. J.; Meiering, E. M. Conserved and Nonconserved Features of the Folding Pathway of Hisactophilin, a β-Trefoil Protein. *Protein Sci.* **2002**, *11*(3), 669–679. https://doi.org/10.1110/ps.31702.
47. Parker, M. J.; Dempsey, C. E.; Lorch, M.; Clarke, A. R. Acquisition of Native β-Strand Topology During the Rapid Collapse Phase of Protein Folding. *Biochemistry* **1997**, *36*(43), 13396–13405. https://doi.org/10.1021/bi971294c.
48. Forsyth, W. R.; Matthews, C. R. Folding Mechanism of Indole-3-Glycerol Phosphate Synthase From Sulfolobus Solfataricus: A Test of the Conservation of Folding Mechanisms Hypothesis in $(\beta\alpha)_8$ Barrels. *J. Mol. Biol.* **2002**, *320*(5), 1119–1133. https://doi.org/10.1016/S0022-2836(02)00557-0.
49. Maki, K.; Cheng, H.; Dolgikh, D. A.; Shastry, M.; Roder, H. Early Events During Folding of Wild-Type Staphylococcal Nuclease and a Single-Tryptophan Variant Studied by Ultrarapid Mixing. *J. Mol. Biol.* **2004**, *338*(2), 383–400. https://doi.org/10.1016/j.jmb.2004.02.044.
50. Parker, M. J.; Spencer, J.; Jackson, G. S.; Burston, S. G.; Hosszu, L. L. P.; Craven, C. J.; Waltho, J. P.; Clarke, A. R. Domain Behavior During the Folding of a Thermostable Phosphoglycerate Kinase. *Biochemistry* **1996**, *35*(49), 15740–15752. https://doi.org/10.1021/bi961330s.
51. Parker, M. J.; Spencer, J.; Clarke, A. R. An Integrated Kinetic Analysis of Intermediates and Transition States in Protein Folding Reactions. *J. Mol. Biol.* **1995**, *253*(5), 771–786. https://doi.org/10.1006/jmbi.1995.0590.
52. Ogasahara, K.; Yutani, K. Unfolding-Refolding Kinetics of the Tryptophan Synthase α Subunit by CD and Fluorescence Measurements. *J. Mol. Biol.* **1994**, *236*(4), 1227–1240. https://doi.org/10.1016/0022-2836(94)90023-X.
53. Jennings, P. A.; Finn, B. E.; Jones, B. E.; Matthews, C. R. A Reexamination of the Folding Mechanism of Dihydrofolate Reductase From Escherichia Coli: Verification and Refinement of a Four-Channel Model. *Biochemistry* **1993**, *32*(14), 3783–3789. https://doi.org/10.1021/bi00065a034.
54. Matouschek, A.; Kellis, J. T.; Serrano, L.; Bycroft, M.; Fersht, A. R. Transient Folding Intermediates Characterized by Protein Engineering. *Nature* **1990**, *346*(6283), 440–445. https://doi.org/10.1038/346440a0.
55. Schymkowitz, J. W.; Rousseau, F.; Irvine, L. R.; Itzhaki, L. S. The Folding Pathway of the Cell-Cycle Regulatory Protein p13suc1: Clues for the Mechanism of Domain Swapping. *Structure* **2000**, *8*(1), 89–100. https://doi.org/10.1016/S0969-2126(00)00084-8.
56. Teilum, K.; Thormann, T.; Caterer, N. R.; Poulsen, H. I.; Jensen, P. H.; Knudsen, J.; Kragelund, B. B.; Poulsen, F. M. Different Secondary Structure Elements as Scaffolds for Protein Folding Transition States of Two Homologous Four-Helix Bundles. *Proteins* **2005**, *59*(1), 80–90. https://doi.org/10.1002/prot.20340.

57. Fowler, S. B.; Clarke, J. Mapping the Folding Pathway of an Immunoglobulin Domain: Structural Detail From phi Value Analysis and Movement of the Transition State. *Structure* **2001**, *9*(5), 355–366. https://doi.org/10.1016/S0969-2126(01)00596-2.
58. Cota, E.; Clarke, J. Folding of Beta-Sandwich Proteins: Three-State Transition of a Fibronectin Type III Module. *Protein Sci.* **2000**, *9*(1), 112–120. https://doi.org/10.1110/ps.9.1.112.
59. Jemth, P.; Day, R.; Gianni, S.; Khan, F.; Allen, M.; Daggett, V.; Fersht, A. R. The Structure of the Major Transition State for Folding of an FF Domain From Experiment and Simulation. *J. Mol. Biol.* **2005**, *350*(2), 363–378. https://doi.org/10.1016/j.jmb.2005.04.067.
60. Melnik, B. S.; Marchenkov, V. V.; Evdokimov, S. R.; Samatova, E. N.; Kotova, N. V. Multy-State Protein: Determination of Carbonic Anhydrase Free-Energy Landscape. *Biochem. Biophys. Res. Commun.* **2008**, *369*(2), 701–706. https://doi.org/10.1016/j.bbrc.2008.02.096.
61. Tang, K. S.; Guralnick, B. J.; Wang, W. K.; Fersht, A. R.; Itzhaki, L. S. Stability and Folding of the Tumour Suppressor Protein p16. *J. Mol. Biol.* **1999**, *285*(4), 1869–1886. https://doi.org/10.1006/jmbi.1998.2420.
62. Laurents, D. V.; Corrales, S.; Elías-Arnanz, M.; Sevilla, P.; Rico, M.; Padmanabhan, S. Folding Kinetics of Phage 434 Cro Protein. *Biochemistry* **2000**, *39*(45), 13963–13973. https://doi.org/10.1021/bi001388d.
63. Parker, M. J.; Marqusee, S. The Cooperativity of Burst Phase Reactions Explored. *J. Mol. Biol.* **1999**, *293*(5), 119–1210. https://doi.org/10.1006/jmbi.1999.3204.
64. Lowe, A. R.; Itzhaki, L. S. *Rational Redesign of the Folding Pathway of a Modular Protein. Proc. Natl. Acad. Sci. U. S. A.* **2007**, *104*(8), 2679–2684. https://doi.org/10.1073/pnas.0604653104. https://www.pnas.org/content/104/8/2679.
65. Choe, S. E.; Matsudaira, P. T.; Osterhout, J.; Wagner, G.; Shakhnovich, E. I. Folding Kinetics of Villin 14T, a Protein Domain With a Central β-Sheet and Two Hydrophobic Cores. *Biochemistry* **1998**, *37*(41), 14508–14518. https://doi.org/10.1021/bi980889k.
66. Mu noz, V.; Lopez, E. M.; Jäger, M.; Serrano, L. Kinetic Characterization of the Chemotactic Protein From Escherichia Coli, CheY. Kinetic Analysis of the Inverse Hydrophobic Effect. *Biochemistry* **1994**, *33*(19), 5858–5866. https://doi.org/10.1021/bi00185a025.
67. Stagg, L.; Samiotakis, A.; Homouz, D.; Cheung, M. S.; Wittung-Stafshede, P. Residue-Specific Analysis of Frustration in the Folding Landscape of Repeat β/α Protein Apoflavodoxin. *J. Mol. Biol.* **2010**, *396*(1), 75–89. https://doi.org/10.1016/j.jmb.2009.11.008.
68. Ratcliff, K.; Corn, J.; Marqusee, S. Structure, Stability, and Folding of Ribonuclease H1 From the Moderately Thermophilic Chlorobium Tepidum: Comparison With Thermophilic and Mesophilic Homologues. *Biochemistry* **2009**, *48*(25), 5890–5898. https://doi.org/10.1021/bi900305p.
69. Luo, L. F. Protein Photo-Folding and Quantum Folding Theory. *Sci. China Life Sci.* **2012**, *55*(6), 533–541. https://doi.org/10.1007/s11427-012-4316-9.
70. Fang, C.; Frontiera, R. R.; Tran, R.; Mathies, R. A. Mapping GFP Structure Evolution During Proton Transfer With Femtosecond Raman Spectroscopy. *Nature* **2009**, *462*(7270), 200–204. https://doi.org/10.1038/nature08527.
71. Zurek, W. H. Decoherence and the Transition From Quantum to Classical–Revisited. *Los Alamos Sci.* **2002**, *27*, 86–109.
72. Tegmark, M. Importance of Quantum Decoherence in Brain Processes. *Phys. Rev. E* **2000**, *61*(4), 4194–4206. https://doi.org/10.1103/PhysRevE.61.4194.
73. Luo, L. F.; Lv, J. Quantum Conformational Transition in Biological Macromolecule. *Quant. Biol.* **2017**, *5*(2), 143–158. https://doi.org/10.1021/bi961330s.

74. Hou, P.; Li, Y.; Zhang, X.; Liu, C.; Guan, J.; Li, H.; Zhao, T.; Ye, J.; Yang, W.; Liu, K.; Ge, J.; Xu, J.; Zhang, Q.; Zhao, Y.; Deng, H. Pluripotent Stem Cells Induced From Mouse Somatic Cells by Small-Molecule Compounds. *Science* **2013**, *341*(6146), 651–654. https://doi.org/10.1126/science.1239278.
75. Luo, L. F. Conformation Dynamics of Macromolecules. *Int. J. Quantum Chem.* **1987**, *32*(4), 435–450. https://doi.org/10.1002/qua.560320404.
76. Abreu, P. Bioenergetics Mechanisms Regulating Muscle Stem Cell Self-Renewal Commitment and Function. *Biomed. Pharmacother.* **2018**, *103*, 463–472. https://doi.org/10.1016/j.biopha.2018.04.036.
77. Ryall, J. G.; Lynch, G. S. The Molecular Signature of Muscle Stem Cells is Driven by Nutrient Availability and Innate Cell Metabolism. *Curr. Opin. Clin. Nutr. Metab. Care* **2018**, *21*(4), 240–245. https://doi.org/10.1097/MCO.0000000000000472.
78. Kogut, I.; McCarthy, S. M.; Pavlova, M.; Astling, D. P.; Chen, X.; Jakimenko, A.; Jones, K. L.; Getahun, A.; Cambier, J. C.; Pasmooij, A. M. G.; Jonkman, M. F.; Roop, D. R.; Bilousova, G. High-Efficiency RNA-Based Reprogramming of Human Primary Fibroblasts. *Nat. Commun.* **2018**, *9*(1), 745. https://doi.org/10.1038/s41467-018-03190-3.

CHAPTER SEVEN

Classical-quantum interfaces in living neural tissue supporting conscious functions

Alfredo Pereira Jr.*
Goldsmiths, University of London, London, United Kingdom
São Paulo State University (UNESP), Botucatu, São Paulo, Brazil
*Corresponding author: e-mail address: alfredo.pereira@unesp.br

Contents

Abstract

This chapter describes three classical-quantum interfaces in living neural tissue supporting conscious functions. The first is activation of the glutamatergic synapse, leading to memory formation, by means of the binding of calcium ions entering NMDA channels with calmodulin receptors and kinases. The second is the coupling of calcium ions with negative water (selected by membrane aquaporins) inside astrocytes, forming, by means of quantum spin configuration changes induced by Coulomb interactions, large-scale "hydro-ionic" waves that control the tissue's chemical homeostasis and support sentience. The third is the action of hydrogen protons (separated by aquaporins) on the extracellular fluid, generating by means of a Grotthuss-like effect, i.e., a coherent-dissipative "superconductive" medium, which impacts on the temporal patterning of action potentials of a neuronal population by means of Na—Ca ions exchange at distal parts of axons, as originally proposed by Tasaki. This process is adequate to account for the "conscious binding" of features processed in the parallel distributed architecture of the thalamocortical system.

Advances in Quantum Chemistry, Volume 82
ISSN 0065-3276
https://doi.org/10.1016/bs.aiq.2020.08.002

1. Introduction

Neural tissues are complex systems composed of neuronal and glial cells, ions, intersticial water, blood and several macromolecules (transmitters, proteins, hormones, peptides) forming the extracellular "chemical soup". In these systems, negentropic mechanisms involved in the dynamical control of homeostasis putatively generate transient quantum phases supporting conscious functions. Neural tissue function include electrochemical processes and biological signaling of exteroceptive and interoceptive information patterns. These functions involve micro, *meso* and macroscale mechanisms.[1] Understanding the complexity of brain and whole-body conscious functions requires an interdisciplinary approach to describe these mechanisms and to identify the respective interfaces responsible for the integration of multiscale processes in living neural tissue. How do processes in different scales and their interfaces compose *an unitary whole* with a first-person perpective of reality? Here I approach this complex issue in three steps.

First, I describe the mechanisms of glutamatergic information processing in neural tissue, involving both classical and quantum (ionic computing) processes, and their interface. Information transmission, in learning, memory formation and attentional focusing, is based on glutamatergic excitatory circuits balanced with GABAergic inhibition, determining patterns of calcium ion entry in neuron membranies and spines, where they interact with protein receptors. This interaction involves microscale processes of the binding of ions with protein sites, composing a first type of classical-quantum interface.

Second, I claim that sentience is putatively based on *Hydro-Ionic Waves*[2] induced by the neuronal release of several transmitters and modulators that impact on astroglial metabotropic receptors, activating signaling pathways that lead to the release of calcium ions from the cells' endoplasmic reticulum. The wave is composed of negative configurations of water induced by aquaporins and dynamical processes, involving mostly calcium, sodium and potassium cations. The Coulomb interactions of these ions and negative water imply changes in spin within the atoms, composing a second classical-quantum interface.

Third, I focus on the problem of *conscious binding*: how does distributed local information processing in the brain lead to the formation of unitary conscious episodes? I raise the hypothesis that the conscious "binding" of features processed in the parallel-distributed architecture of the neocortex can be explained by the Tasaki's concept of the action potential,[3] further involving the Grotthuss mechanism. Taking into consideration the interaction of hydrogen protons with the extracellular "chemical soup" in living neural tissue, I propose the existence of a "superconductive" classical-quantum interface, made possible by an increase of concentration of hydrogen protons that results from the selective action of aquaporins. The effects of quantum information patterns carried by spin configurations of calcium, sodium and potassium ions (previously formatted by the hydro-ionic wave) on the axon of large populations of neurons coordinates neuronal activity to form unitary conscious episodes and guide adaptive behavior in the environment. This "superconductive" process is claimed to propagate in living tissue according to the *vortex effect*.[4]

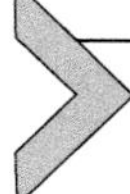

2. Glutamatergic information transmission and reception

Glutamate (Glu) is the main excitatory transmitter in the brain, being largely present in cortico-cortical networks and operating both on excitatory (as pyramidal cortical) and inhibitory neurons (as GABAergic interneurons). The Glu-induced excitation (i.e., membrane depolarization) of interneurons increase their inhibitory action (i.e., GABAergic transmission inducing the flow of chloride ions to hyperpolarize their membrane) on the excitatory ones. Glu transmission is a key component in the balance of excitation and inhibition that is a necessary condition for brain function.[5]

Glu also operates as an information carrier to thalamocortical and cortico-cortical synapses, a role that is crucial for the understanding of perceptual processing in the brain. All perceived information, from the environment is carried by neuronal spike trains, which are transduced by Glu at each synapse, reaching the sensory cortex, where they activate specialized feature detectors. The Central Nervous System (CNS) constructs conscious episodes using feature detectors activated by the information patterns carried by a series of neurons by means of spike trains and Glu transmission at each synapse (Fig. 1).

The role of Glu as information carrier in thalamocortical and cortico-cortical synapses is crucial for the understanding of how conscious perception is possible.[6] Glu is largely present in cortico-cortical networks, with a central role in the generation of conscious content in normal states, dreams and altered states. This role has been proved in experiments when the Glu

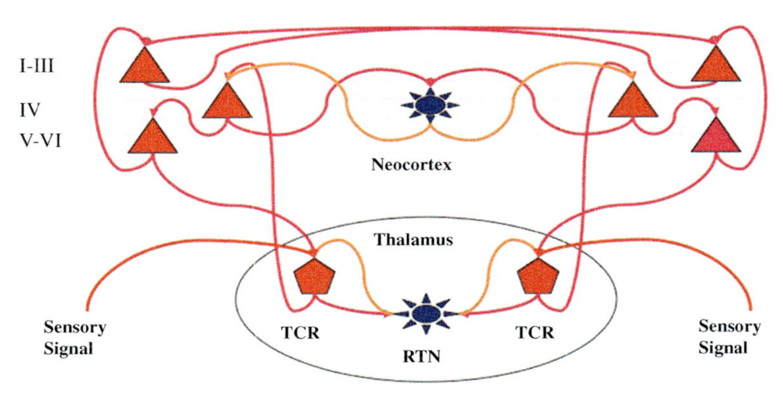

Fig. 1 Canonical circuit of the balance of excitation and inhibition in the neocortex: Excitatory sensory signals (red arrows) pass through thalamic relay cells (TCR) and reach the fourth layer of a column of the sensory cortex, which fires to neurons located in the fifth and sixth layer. These neurons fire back to the thalamus and send the excitatory signal to the superficial layers, where they are horizontally spread to other columns of the neocortex. This excitatory process is soon extinguished by habituation mechanisms, comprising the excitation of thalamic inhibitory interneurons (RTN) that in turn inhibit (brown arrows) the thalamic excitatory neuron soon after, and the neocortical inhibitory interneurons that inhibit the excitatory neurons that excited them. With this mechanism, the excitatory process is conceived as dynamically moving through neocortical columns, composing the flux of thought (with unconscious and conscious aspects). *Original figure by APJ.*

NMDA receptor is transiently blocked by subanesthetic doses of an antagonist (ketamine, PCP or MK-801), thus generating perceptual distortions and hallucinations.[7]

Spike trains encode information by means of frequency and phase in populations of axons. The CNS constructs conscious episodes from the ensemble of neuron firing patterns received within a temporal period of approximately 2–3 s.[8] The concepts of feature-detectors and population-rate coding can be combined in the idea of a *sparse population code*.[9] In this view, the detection of real-world objects would be made by a cooperative group of neurons, forming a Hebbian *cell assembly*. A cell assembly is a relatively small neuronal population, located in cortical columns, with strengthened connections elicited by previous learning.[10]

Glu membrane receptors control intracellular signaling pathways targeting the dendritic spine, where a *molecular device* is able to register the relevant afferent patterns, supporting conscious perceptual learning and selective triggering of memory formation, as well as unconscious priming. The mechanism involved in such a recording of sensory patterns has been studied as the early stage of LTP.[11] It involves biological molecular structures and functions, including the system of Glu receptors, and calcium-binding proteins as Calmodulin (CaM) and Calmodulin-Dependent Protein Kinase II (CamKII, a protein from the kinase family, having several receptor and effector active sites).

This mechanism operates in dendritic spines distributed over the whole neocortex. In the sensory cortex, exogenous patterns transmitted through thalamocortical glutamatergic projections are received and processed by post-synaptic mechanisms.[12–14] Activation of Glu receptors combined with voltage-dependent calcium channels (VDCCs) converge to the dendritic spine, where they control CaM/CaMKII signaling mechanisms. Glu released from the presynaptic neuron's axon terminal is spread in synaptic space and bind to three different kinds of receptors (AMPA, NMDA and Metabotropic Glu Receptors—MetGR) located at the post-synaptic neuron membrane. The three kinds of receptors activate signal transduction pathways that converge into the dendritic spine.[15]

Calcium cations (Ca^{++}) are largely employed biological ions with a flexible electronic structure able to encode information.[16] CaM and CamKII have several receptor and effector active sites, where Ca^{++} ions entering through NMDA and VDCCs are trapped. The ions trapped in CaM are transferred to the kinase and trigger regulatory functions. The informational state of CaM/CaMKII is dependent on the interaction with the Ca^{++} population passing through NMDA and VDCCs. In normal cases, most of Ca^{++} entry is made through NMDA channel, which is considered to be a coincidence-detector for both bottom-up (sensory afferent) and top-down (previously learned) patterns.[17] Because of this condition, the NMDA channel assures the reliability of percepts in regard to stimuli, since it is opened to Ca^{++} entry only if endogenous and afferent pulses reach the NMDA receptor together. When VDCCs (which are not coincidence detectors) assume the main role in glutamatergic transmission, perceptual distortions and hallucinations occur (see Table 1).

Table 1 Three different modes of functioning of the glutamatergic synapse.

Presynaptic	Postsynaptic	Ca^{++} entry	Ca^{++} channel
Inactive	Inactive	No	No transmission
Active	Inactive	Low	VDCC following AMPA activation
Inactive	Active	Low	VDCC following AMPA activation
Active	Active	High	VDCC and NMDA following AMPA

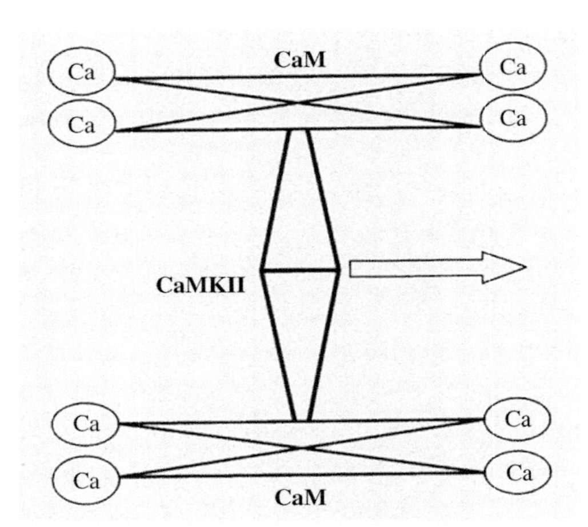

Fig. 2 The structure of Calcion Ion binding with CaM and CaMKII. The input is composed by the flux of calcium ions (Ca) entering the post-synaptic neuron and binding to several CaM units. On a second step, CaM binds to CaMKII, controlling its conformational states. The output of the computer is the action of CaMKII on other substrates (see Ref.[18]). *Original figure by APJ.*

The multimeric structure of CamKII, having binding sites for Ca^{++} and phosphatases that participate in the phosphorylation of other proteins, constitutes a *micro computing device* able to read quantized information from incoming Ca^{++}, to process this information, and to activate other proteins according to the results of the information processing (Fig. 2). A model of quantum computing with calcium ions in spines supporting the formation of conscious states was presented in two publications.[17,19]

The multimeric structure of CaMKII contains four sites that bind to CaM, determining the conformational state of the kinase and the resulting phosphorylation functions.[20] Such a micro computing device uses quantum information encoded in the electronic configuration of the ions. Besides binding with CaMKII, Ca^{++}-activated CaM can trigger other signaling pathways in the cell, some of them exerting feedback control on the state of the membrane and others reaching the nucleus and inducing the formation of long-term memory.

The above signaling pathways can be related to cognitive processes that depend on glutamatergic mechanisms responsible for conscious perception. For instance, one of the key proteins present in the converging glutamatergic and dopaminergic pathways is DARPP-32, which is found, among other places, in the striatum, controlling thalamocortical glutamatergic and cholinergic neurons with the participation of dopaminergic modulation.[21] Striatal signals processed by such converging molecular pathways convey information to the cortex as an "efferent copy" (i.e., signals sent from motor to perceptual areas when a voluntary action is initiated; for a review of such concepts, including the roles of the cerebellum and hippocampus, see[22]). DARPP-32 activation, as a converging route for intra-neuronal dopaminergic and Glu-activated signal transduction pathways, has been related to mental function. It is defective in schizophrenia, possibly participating in the generation of symptoms.[23,24] An explanation for the role of DARPP-32 is that the feedback from motor to sensory areas would have the role of *reinforcing* the perceptual pattern, in order to boost its learning and memorizing process.

3. The neuron-astrocyte interaction model

Glutamatergic synapses are mostly "tripartite" (composed of two neurons and one astrocyte). The Glu released by the presynaptic neuron reaches both the post-synaptic neuron and the astrocyte. By binding to the NMDA and AMPA receptors of the postsynaptic neuron, Glu induces membrane depolarization, and the opening of NMDA channels to fast calcium currents. This excitation decays in around 150 ms[25] but can be sustained by glial transmission (Glu released by the astrocyte, binding to extrasynaptic NMDA receptors), producing slow calcium currents through NMDA receptors. The second input of calcium may activate the path of calmodulin and its kinase (CaMKII), which phosphorylates or dephosphorylates AMPA receptors and thus induces synaptic potentiation or depression.

When the animal is awake, the astrocytic network is pre-activated by adrenergic, purinergic and cholinergic mechanisms, facilitating the generation of calcium waves at the time when Glu transmission occurs. In the astrocyte, the induction of local calcium waves by means of excitatory transmission (by another transmitter or Glu) may generate intercellular, global waves that broadcast information to many other parts of the brain.[26]

While in the "Neuron Doctrine" (formulated by Ramon y Cajal) neurons were considered the functional units of the mind, in the new emerging model of neuro-astroglial interactions the tripartite synapse becomes the functional unit, constituting the basis for psycho-physiological processes. Neural networks are mostly responsible for cognitive processes (such as the formation of representations and logical operations in processes of perception, attention, action, learning and memory formation), whereas the astrocyte network performs an appreciation of the information carried by the neural network (e.g., in terms of "pleasant" or "unpleasant", "attractive" or "disgusting"). The integration-and-appreciation process is achieved by means of intercellular, global calcium waves. Based on the appreciation, the astrocyte network modulates neuronal activity, reinforcing the patterns of information that have positive valence for the individual, and weakening those with negative valence.

According to the *Neuron Doctrine,* the neuronal network predictive computation in social contexts "represents" emotions. This view is limited in the sense that it does not address the phenomena of *experiencing* emotional feelings in the first-person perspective. A representation is not an emotional experience. The neuro-astroglial interaction model has theoretical resources to address not only the cognitive representation of emotions, but also the lived experience of emotional feelings. In the neuro-astroglial model, the generation of subjective feelings is attributed to amplitude-modulated ionic waves in astrocytes and extracellular fluid. The "Endogenous Feedback Network" hypothesis[27] states that conscious feeling involves the interaction of neuronal and astroglial networks, the first responsible for the processing of information and the constructing of representations, while the second would be responsible for the lived, direct experience of feeling.[28]

The functions of integration and appreciation of information are closely related, because the lived experience intrinsically contains an appraisal, by assigning positive, neutral or negative valence to certain aspects of the experience. The astroglial network participates in both processes of integration and appreciation of information patterns processed in a distributed manner in the thalamocortical system, interacting with somatic processes by means of signaling via blood flow and cerebral fluid, and uses the results to modulate the neuronal network, thus participating in processes of perception, attention, learning, training of semantic and episodic memory, emotion, consciousness and control of behavior.

As the astrocyte is in contact with blood and cerebrospinal fluid (while neurons are not), it receives the signals that come in the flow, forming a continuously updated reference of the state of the body in the world. The intracellular astrocyte signaling generate small calcium waves that interfere with each other in the whole astroglial network, resulting in larger waves build by means of constructive interference. The intrinsic patterns of the larger waves, to be studied experimentally, are claimed to correspond to the first-person experienced feelings.

Astrocytes are not connected to sensory transducers or muscle and endocrine effectors. All sensation and perception, as well as all actions in the world begin with neurons; the results of neuronal processes (information patterns embodied in local electromagnetic fields) reach the astrocytes and induce the larger waves (see the "domino" and "carousel" effects[26]). The larger waves in the astrocyte network require a coordinating action from neurons, by means of the balance of excitatory and inhibitory transmission:

(a) Local EM fields are formed by active neuronal assemblies;
(b) Neuronal large-scale synchronization occurs in theta to gamma frequencies (not delta; synchronization in the slowest frequencies imply unconsciousness);
(c) There are chemical and ephaptic (magnetic) transmissions of information from the neuronal local fields to astroglial waves; and.
(d) There is an interference of the smaller waves, leading to the formation of the larger ones in the astroglial network.

The temporal process in which neurons activate calcium waves in astrocytes and these waves modulate neurons in the following time interval is summarized in Fig. 3.

The existence of astroglial calcium waves in vitro as a response to glutamatergic excitation was optically registered since the 1990s, but the existence of the same phenomenon in vivo and its biological function was questioned until recently. The hypothesis was confirmed in three publications,[29–31] pointing towards a paradigm shift in brain science.

There are experimental indications for the existence of two kinds of astroglial calcium waves, the local and global ones.[32] For instance, six publications show promising results in the analysis of calcium waves in vivo: a new method to trace fast global calcium waves across relatively large distances in the brain[33]; wavelet analysis to identify spatiotemporal dynamic patterns of calcium waves[34]; detection of "rich information content" in astroglial calcium waves[35]; use of probabilistic methods to approach the possibility of decoding stimuli messages carried by calcium waves[36]; optogenetic methods combined with optical imaging to produce realistic videos that reveal the dynamics of the waves in freely behaving animals[37]; a study on the complexity of waves and their alterations in mouse models of neurogenerative diseases.[38]

The role of astrocytes in the control of brain functions is beyond any reasonable scientific doubt, but the question about the role of calcium waves remains nowadays. Animals with deficient expression of connexins have major behavioral disabilities.[39] Nevertheless, animals with damage in the

Neuro-Astroglial Interaction and Brain Function

Neuron **Graded Potentials** -> Local Field Potentials ("Oscillations") and Action Potentials -> Synchronization of Neural Assemblies -> Transmitter Release -> **Ionic Waves** in Astrocytes and Extracellular Fluid -> Feedback on Neural Graded Potentials -> Action Potentials to Muscles and Glands -> Interoceptive Cycles and External Stimulaton -> **Neuron Graded Potentials**...

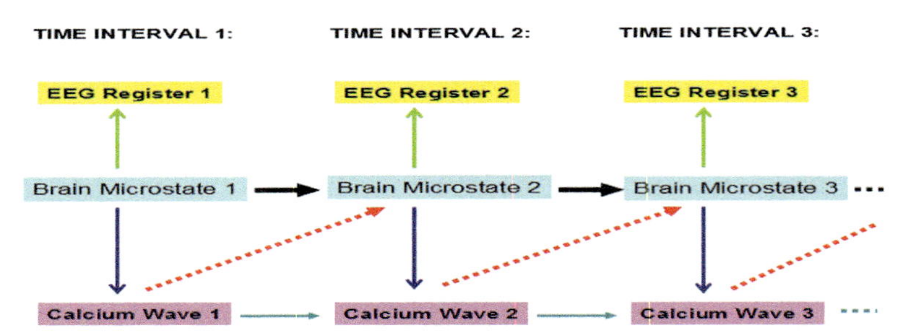

Fig. 3 Temporal process of neuro-astroglial interactions and the EEG. A given microstate generates both the EEG signal and changes in the previously existing standing calcium wave in astrocytes. In the next moment, the calcium wave modulates the following microstate, reinforcing patterns with systemic positive valence, and depressing patterns with systemic negative valence (by means of the molecular mechanisms reviewed in[26]). The second microstate generated both the next EEG register and changes in the existing calcium wave, and so on. Although the scalp EEG does not measure the calcium wave directly, the modulatory function exerted by the wave is implicit in the temporal evolution of the registered electromagnetic signal. *Figure originally published in Pereira, A., Jr.; Foz, F.B.; Rocha, A.F. Cortical Potentials and Quantum-like Waves in the Generation of Conscious Episodes.* Quantum Biosys. ***2015,*** 6, *10–21.*

inositol triphosphate (IT) signal transduction pathway apparently do not have notable losses of sensory, cognitive or motor functions.[40] A possible interpretation of this result is that such animals would compensate the shortage of IT glutamatergic transmission and signal transduction by means of other transmitters such as cholinergic and purinergic ones. There are two mechanisms of wave formation and propagation: from the endoplasmatic reticulum, and at distal processes of the cell, where the waves are induced by synaptic processes.[32] Only the first mechanism is impaired by IT genetic knock out. In this sense, it is important to emphasize that astrocytic calcium waves "in vivo" also involve cholinergic[30,31] and purinergic transmission,[41] which could hardly be blocked in transgenic animals without disturbing their vital processes.

It is known that both BOLD fMRI[42] and scalp electroencephalogram[43] measure neuronal and glial actions together. Although astrocytes maintain their membrane in hyperpolarized states, not generating action potentials, they exhibit Ca^{++} waves that modulate dendritic fields and exert vascular control,[25] determining the magnitude of the hemodynamic response measured by fMRI. Imaging of astrocytic calcium waves "in vitro" by optical means raised the interest in the phenomenon, leading experts to wonder about the functionality of the waves.[44] Images of waves using fluorescence microscopy ("two-photon microscopy", a technique in which a beam of light is projected on brain tissue, allowing the microscope to capture reflectance patterns generated by astrocytic calcium waves) proved the existence of the phenomenon "in vivo."[45] In the new field of Optogenetics,[45] fluorescent fish genes were inserted into the DNA of transgenic mice, in the sectors responsible for the expression of the calcium ion receptor proteins. The physiological activity of the brain of these animals display increased brightness when calcium ions bind to such receptors. Through a "window" opened on the skull of transgenic mice, it is possible to view (using a microscope) the calcium waves.

Recent results in brain physiology and pharmacology research (e.g.[46]) suggest that wakefulness is dependent on astroglial calcium waves. Astrocytes have also been implicated in the etiology and therapy of most psychiatric and neurological problems, for instance in Alzheimer's.[47] Experts in the morphology and physiology of astroglial cells have argued that the main function of the astroglial transport of macromolecules, related ionic currents and waves, is *the control of brain homeostasis*.[48] It has became increasingly clear that all psychiatric disorders involve some type of malfunction of astrocytes, impairing the dynamics of adaptive homeostasis. Our proposed scientific modelling[2,26,49–54] addresses the problem of how the astroglial network—interacting with distributed neuronal assemblies—integrates dynamic patterns embodied in local fields, taking into consideration that the same neurotransmitters that generate the electric fields can elicit—by a variety of pathways—calcium waves in astrocytes.

The development of the model is based on the current state of the art of research on cognitive and affective functions of the astrocyte network. Regarding the relationship between astrocyte activity and human consciousness, besides the fundamental discoveries[55,56] of types of astrocytes unique to our species, it is worth noting a recent discovery of Brazilian neuroanatomists[57] pointing towards the existence of a greater amount of glial cells in brain regions correlated with conscious activity, while a higher proportion of neurons is found in the cerebellum, which has little or no contribution to conscious processing.

We have proposed two mechanisms of formation of global calcium waves in the astrocyte network.[26] The *domino effect* explains signal propagation in astrocytic calcium waves by the transferring the vibrational energy from ion to ion, and signal amplification by ATP both in gap junctions and in extracellular mediums, without requiring displacement of the ions between the microdomains. The *carousel effect* explains how synchronized neuronal activity induces a large-scale calcium wave in astrocytic network, which in the next moment modulates neuronal activity. The "domino" and "carousel" effects conjointly explain how patterns embodied in activity of synchronized neuronal networks can be readily transferred to calcium waves in the astroglial network.

A parsimonious explanation of our capacity of operating both unconsciously at millisecond times and consciously at the scale of seconds is a combination of neuronal and astroglial processing in superposed time scales. The neuro-astroglial interaction mechanisms operate with these two time-scales, one neuronal (at the range of milliseconds) and other astroglial (in the range of seconds to minutes). Event-Related Potentials (ERPs) correlated with conscious events take from 100 to 1000 ms to occur. If their generation depended only on neuronal transmission through cortico-cortical axons, ERPs would take only 50 to 100 ms. The astroglial timing also corresponds to the Slow Cortical Potential[58] described for BOLD fMRI results, considering both positive correlations (between percepts and fMRI activations) and default networks.

Summarizing the evidence found in this field, a leading researcher reported that at the workshop *Glial Biology in Learning and Cognition*, organized by the USA National Science Foundation, "our unanimous conclusion was that neurons working alone provide only a partial explanation for complex cognitive processes, such as the formation of memories... The complex branching structure of glial cells and their relatively slow chemical (as opposed to electrical) signaling in fact make them better suited than neurons to certain cognitive processes. These include processes requiring the integration of information from spatially distinct parts of the brain, such as learning or the experiencing of emotions".[59]

The concept of "plasticity" as the strengthening of synapses applies mostly to neurons, because the astroglial network does not have chemical synapses (while astrocytes communicate by means of gap junctions); however, calcium waves generated by astrocytes *modulate* neural plasticity; for instance, a person is more likely to remember conscious experiences associated to strong feelings. This modulation has been proposed to be mediated by a tissue wave that can be studied using the retina as a model.[60]

4. The hydro-ionic wave

Besides taking into consideration molecular and chemical properties of macromolecules, also ions and water are considered important for brain information processing, cognitive and affective functions. In the context of classical chemistry, calcium waves are conceived as generated by changes in the concentration of ions, which are considered as material particles devoid of any intrinsic information processing capabilities. Such a framework makes it difficult or impossible to explain the optically registered dynamics of calcium waves. According to a biophysics expert, "A given ion...seems to be transported along a chain (cascade) of macromolecules...A signal (calcium, for example) occurring at the entry of the chain induces the liberation of the sequestered ion from the first element of the chain. And this one, in its turn, induces the liberation of the ion from the following element, etc. The ion appearing at the end of the chain is liberated by the last element of the chain. This type of transport differs deeply from diffusion. It is not a transport of matter but a transfer of a level of energy (transduction)".[61] In the framework adopted in this chapter, I claim that "transfer of energy" involves a quantum potential that unfolds itself as a spatiotemporal wave. In this section, I address the phenomenon from a biological perspective, and in the next sections I address the putative quantum computing processes that underpin the biological ones.

Calcium ions trapped in the astroglial endoplasmic reticulum and intracellular microdomains interact and form local correlations. These microdomains are connected to each other by means of gap junctions and communicate sequentially by means of extracellular ATP signaling. When a synchronized population of neurons activates—by means of Glu, other neurotransmitters, or electromagnetically—the astroglial network, there is constructive interference of the signal with previous correlations, inducing in the ionic population a vibrational mode, thus forming a global wave pattern, which feeds back on neuronal networks. The physical-chemical processes putatively involved in this phenomenon are the kinetic energy (vibrational states) of the ions, and related changes in the electronic structure, related to emission of photons and changes in the total spin of the ions; these issues are discussed in the next section of this chapter; this section is focused on the interactions of ions and water in neural tissue.

Calcium waves in astrocytes are part of larger *Hydro-Ionic Wave* in living tissue, involving a complex interaction of ions with water and proteins.

Water can interact with the ions in complex ways, making possible the explanation of observed phenomena. The negative "Exclusion Zone" (EZ) in structured interfacial water[62,63] can be regarded as adequate both to attract cations and to be attracted by positive protein sites. In the EZ, the molecules of water are tightly coupled, not allowing the presence of ions and molecules; this condition can have the effect of aligning positive ions *at the boundary surfaces* of the EZ, forming standing ionic waves. When the water changes to the unstructured phase, the line of cations move to another charge attractor. This is a possible mechanism for the formation of traveling ionic waves of energy, which is different from simple changes of concentration of the ion in solution.

Structured water is composed of the combination of hydrogen and oxygen, containing negative and positively charged regions. The negative region is the exclusion zone, having the texture of a gel. It is composed of H_3O_2, which results in a negative charge. The positive region is composed of positively charged molecules derived from water, as the proton. Positively charged electrolytes are attracted to the surface of the exclusion zone but cannot enter it because of its increased density. The result is that the vibrating cations are aligned to the surface of negative water and generate waves by means of the "domino effect"—the transference of kinetic energy from ion to ion in the chain, without spatial displacement of the ion.

A classic example is the binding of calcium ions with calmodulin we have mentioned before. Hyaluronic Acid,[64,65] which is abundant in the extracellular matrix, has negative sites that interact with cations, guiding them to targets at the neuronal membrane, such as ligand-gated calcium channels; another modality is the formation of a "combo" with negative water and positive protein sites (e.g., glycoproteins). The water negative zone bound to positive protein chains can compose a "tissue avenue" for the cationic wave to move from one astrocyte to another astrocyte, or towards the extracellular fluid, where it is conducted by structures of the extracellular matrix (Fig. 4), finally reaching the neuron membrane, where the cations enter ligand and voltage-gated channels, thus controlling neuronal activity and behavior. Astrocytes, but not neurons, accumulate polyamines such as spermine and spermidine.[66] The concentration of positively charged proteins such as spermine, inside astrocytes, can increase when the pH is shifted to acidity. These proteins can guide the hydro-ionic wave through gap junctions and towards neurons. Polyamine release in extracellular space can be a mechanism to guide the hydro-ionic wave towards targets (receptors, ion channels) located at the neuronal membrane.

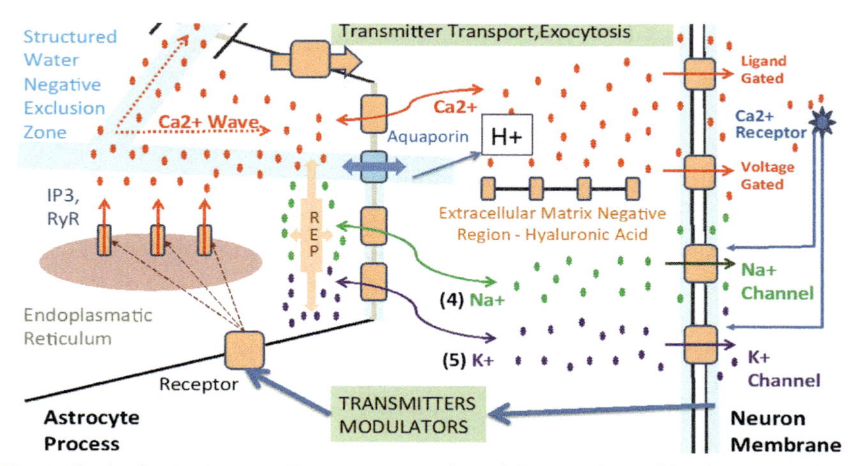

Fig. 4 The hydro-ionic wave. Transmitters and modulators released by the neuron bind to astroglial metabotropic receptors, activating signal transduction pathways that lead to the release of Ca^{2+} from the astrocyte endoplasmatic reticulum. The released ions repel each other, and also repel other cations present in the astrocyte intracellular compartments and processes. The small wave formed inside one astrocyte can propagate to other astrocytes through gap junctions, interfering with other astrocytes' waves and forming larger waves. Another action of the intracellular astroglial calcium wave is to prompt the release of gliotransmitters in the extracellular space, by means of vesicles released through hemichannels, or by means of molecular transporters. The gliotransmitters control neuronal activity and the timing of spike trains carried through the neuron's axon. *Figure by APJ; modified from Pereira, A.,Jr.; Astroglial Hydro-Ionic Waves Guided by the Extracellular Matrix: An Exploratory Model.* J. Integr. Neurosci. ***2017,*** 16, *1–16.*

The astrocyte calcium wave prompts the release of calcium ions in the extracellular space, where they are guided by proteins of the extracellular matrix to their respective neuronal targets, which may be voltage-gated or ligand-gated (NMDA) channels. The selective action of aquaporins result in the accumulation of hydrogen cations in the extracellular medium, possibly providing, by means of the Grotthuss mechanism (discussed in the next sections), a coherent-dissipative extension of the classical effect, which acts as quantum-like medium for long-range phase correlations for the possible "binding" of parallel distributed information patterns, forming integrated unitary conscious episodes.

5. Quantum computing with calcium ions

Calcium ions are involved with all conscious functions of the brain: perception, learning and memory formation, feeling and binding of features into integrated episodes. In the latter function, they are relevant to Tasaki's concept of the action potential,[3] because, as he claims, there is no suficient sodium buffering along the extracellular space around the neuron axon, and therefore, there would be a substitution (calcium in the place of sodium) in the generation of spike trains (as originally argued by Jacques Loeb long ago, in 1900). It is important to take into consideration the quantum computational properties of calcium ions, because the ion is the physical carrier of the patterns that appear in conscious experience.

In the last decades, we have seen the experimental realization of quantum computers composed of two or more calcium ions trapped in a magnetic quadripole. Information is transferred to and read from the ions by means of structured lasers that interact with the ions' vibration pattern, causing changes of energy distribution in their electronic structure. Departing from an initial state when the ions are cooled, the use of lasers modifies the internal state of one ion that is entangled with the others, then changing the collective states. In such quantum computers, some of the physically possible electronic states are avoided or not taken into consideration, forcing the system to work as a binary device.

In the classical Turing machine paradigm, computations are carried by means of symbolic operations implemented in physical devices, with the usage of a binary language. In this context, the property of *multiple realizability* of such machines (i.e., the possibility of implementing the computations in a variety of physically different systems; see[11]) is heavily dependent on the fact that a large variety of physical systems is susceptible of displaying binary states, e.g., being magnetized or not magnetized, transmitting or not transmitting an electrical current, reflecting or not reflecting light.

McCullough and Pitts conceived the neuron as a binary device, with two kinds of states relevant for computational processes: "firing" and "not-firing."[67] This simplification of neurobiological reality was useful for the development of the first models of neural networks, but does not correspond to current knowledge in neurobiology. Neuroscience evolved to understand that besides receiving and sending signals neurons perform important bio-molecular functions, in close interaction with glial cells and the extracellular "chemical soup." Neuronal firing occurs when the membrane electric potential reaches a threshold and transmits a signal along the axon, considering that the propagation of this signal implies a close interaction with the extracellular "chemical soup" from which, according to the Loeb/Tasaki conjecture, calcium ions are recruited to enter the axon and propagate the action potential. In the dendritic tree, under the threshold value for firing, there are also important functions controlled by neuronal excitation, as the activation of membrane receptors and voltage-gated ion channels, triggering a cascade of events in the cell.[68]

A "naturalistic" paradigm of computation, focusing on how computational processes spontaneously occur in biological, molecular and quantum systems, questions the binary assumption present in traditional approaches to computation—including the neural network approaches derived from the McCullough and Pitts modeling. In natural systems, the computations are determined by their respective physical and biological properties, instead of being determined by a "program" artificially introduced by the engineer.

Calcium ions are important for computational as well as well-known biological functions.[69,70] The computational importance derives from the fact that the ion is sufficiently complex to carry information in its electronic structure, while being sufficiently simple to be experimentally manipulated. One of the most successful and promising experimental realizations of quantum computers is the Ion-Trap Quantum computer (ITQC), using the calcium cation.[71,72] This kind of quantum computer is composed of two or more calcium ions trapped in a magnetic quadripole. Information is encoded in the computer and read from it by means of structured lasers. The lasers interact with the external vibrational activity of the ion, changing its "internal" state (i.e., the distribution of energy in the electronic structure). Departing from an initial state when all the ions are cooled to their ground state, the lasers modify one ion that is entangled with the others, then also changing their internal states.

The informational capacity of Ca^{++} is derived from its electronic structure. Calcium ions are endowed with a structure able to carry information both in external vibratory states and in the flexible arrangements of the electronic structure. The biological importance of this ion derives from the capacity of interacting with proteins like a hormone.[16] Regarding the informational capacity of Ca^{++}, "crystallographers have known for some time that calcium is a cut above the rest of the small inorganic ion crowd: it has a flexible crystal field—bond distances and angles are adjustable, with coordination numbers that can vary from six to ten—and it has higher ionization energies. The adaptable coordination sphere permits a wide variety of cooperative packings, giving the ion an advantage in the cross-linking of crystal structures."[16]

Ca^{+} has been used in experimental realizations of quantum computing, following the model of the ITQC, which is composed of two or more ions linearly trapped in a magnetic quadripole. Information is transferred to and read from the ions by means of structured lasers that interact with the ions' external vibrational pattern, causing changes of energy levels in their internal electronic structure. Departing from an initial state when the ions are cooled, the use of lasers modifies the internal state of one ion that is correlated with the others, then changing the collective states.

Some basic notions deserve a very brief review to highlight the physical basis of ion-trap quantum computing.[73] The electrons organize itself in shells, denoted *S*, *P*, *D*, etc., around the atom nucleus according to the aufbau principle. In calcium the configuration of the 20 electrons is [Ar]$4s^2$, where one follows the Madelung order in filling the shells outside the closest neutral atom, here argon, up to the *valence level*. By completing the shells commensurate with the Pauli exclusion principle, Hund's rule etc., each atom can be characterized by a term symbol according the convention $^{2s+1}L_J$ or $^1S_0$ for the calcium ground state and $^2S_{1/2}$ for the singly ionized ion Ca^+.

The Pauli exclusion principle imparts that the fermion, such as an electron, is equipped with an extra quantum number, $s=½$, which effects two directions in a magnetic field according to the quantum number $m_s=\pm½$. Integer spins characterize bosons. In an atom, the electron spin *S* may be combined with the nuclear spin *I* of the electrons giving $F=S+I$, eventually allowing different combinations.[74] This hyperfine interaction are defined by very small shifts and splittings, and are vanishingly small compared to the normothermic energy.

The calcium ion, $^{40}Ca^+$ with no nuclear spin, and one electron outside a closed shell exhibits different transitions for quantum logic operations. For instance if one defines a lower sublevel of a $^2S_{1/2}$ with $m_j=-½$ there are dipole transitions to upper sublevels $^2P_{1/2}$ and $^2D_{3/2}$, where the spin quantum number goes from 1/2 to 3/2. Given the evidence of Ca and its ions as an information carrier, knowledge about its electronic structure arrangements, the dynamics of the transitions and associated electromagnetic frequencies becomes important to understand how it processes information. Electrons can pass from an energy sublevel to another one, once receiving the right amount of energy. In ITQC, this energy is provided by the laser, which has photons with energy:

$$E = h\nu; \quad or \quad E = \frac{hc}{\lambda}$$

where h is Planck's constant and ν is the frequency, c the light velocity and λ the wave length. Quantum logic operations and readout the laser makes use of the $4\,^2S_{1/2} \rightarrow 4\,^2P_{1/2}$ and the $3\,^2D_{3/2} \rightarrow 4\,^2P_{1/2}$ transitions. Excitation to the $4P$ state and spontanous re-emission to the ground state is fast, about 7 ns, while during Doppler cooling, the ions may decay from the $4P$ to the $3D$ state without photon emission and with a life time of 1 s, for more details see.[75] The latter nonradiative process, involving the dark $3D$ state, is therefore useful for computational purposes, since it lasts for a time duration that is sufficient for the realization of logical operations. These techniques are, like those related to NMR techniques involving also hyperfine interactions and fine structure, extremely sensitive to the environment and will easily decohere unless the temperature is low enough. This is problematic and we will come back to it below below, but first some further comments on the quantum situation.

Considering the above physical properties of the ion, *we can attribute* the value $|0\rangle$ to the ground state, and the value $|1\rangle$ to the dark state. We will then know that an excited ion decays to the state $|0\rangle$ when it emits light, and to the state $|1\rangle$ when it is radiationless. Since the transition times to these states are extremely short, they are not considered in the attribution of states. It is important to note that such an attribution of binary states is a useful convention that allows quantum systems to operate with a quantum bit logic, cf. the language of classical computers and the execution of all classical logic gates with the fundamental qualities of the new qubit.[76]

The singly ionized calcium ion, discussed above will be considered a quantum computer with the capacity of 1 *quantum bit,* or *qubit* (for more on the basic unit of quantum information, see further below). The manipulation of the ionic levels is made by laser pulses that excite the ion from its ground state as input, while the read-off information is made by the detection, or not, of radiation. The present model can be made more powerful, using two or more ions.[72] In those cases, the number of processed *qubits* increase in proportion to the number of ions. For these multi-ion computers, *quantum entanglement* is essential for *quantum communication* between the ions, not to be confused with the classical information transmission.

Entanglement is the property of the quantum world that is of central importance to the operation of multi-particle quantum computers. Two particles, irrespective of their separation, are entangled when the quantum state of one particle cannot be described independently of the state of the other. The qubit is the basic unit of quantum information, realized as a two-state quantum system. Hence two qubits are entangled when one qubit cannot be described independently of the state of the other, which means, in contrast to a classical bit, that the unknown qubit state cannot, even in principle, be copied. However, measuring one qubit of any singlet state entangled pair, the other one, if measured, would always have been anti-correleted with the first one, but without knowing which is which.

In the operation of a multi-ion quantum computer, a message is encoded, utilizing e.g., calcium ions, represented as entangled qubits. In this configuration, the state obtained in one of these ions are correlated with all the others. This property makes quantum computers invariably much more powerful than the classical ones, since the reading, or binary choice, of one ion provides knowledge of the state of the other ions.

In the Los Alamos experiment[75] the ions were cooled to a temperature near the absolute zero, in order that the contribution of the spin to higher states would overcome the thermal effects. This again addresses the decoherence issue mentioned earlier. For instance, there appears some studies that seems to indicate that it is not necessary to cool the ions, as new modalities of ion-trap quantum computing have been proposed[77,78] to perform quantum computation with "hot" calcium ions. Nevertheless equipping biological matter with pulsing photo-ionization lazer beams is an arduous task, therefore alternative routes discussed in this volume with the theme "Quantum Boundaries of Life" might also be considered. Withal, if the modality of ITQC is somehow feasible, it will make possible artificial modeling of the fundamental biological role played by calcium cations.

6. Conscious "binding" in neural tissue

Conscious episodes are made of cognitive and emotional qualities ("qualia") bound to each other. Emotional feelings attribute a valence to episodes, giving motivation to action. *Qualia* derive from the inner workings of neural tissue and also from interactions with the physical and social environment, causing the activation of specialized feature detectors and circuits. Each subjective quality is initially formed as a psycho-physiological pattern[79] instantiated in neural tissue; in the human brain, each perceptual quale is believed to correspond to a specific activation of a neocortical column in the neocortex.

A conscious episode requires the binding of each quality with the others, under the valence given by the dominant emotional feeling. There are several hypotheses of mechanisms for conscious binding, such as gamma synchrony,[80] microtubule-based quantum operators,[81] and correlations described by Information Integration Theory.[82–84]

Binding phenomena are omnipresent in biological systems: proteins bind to effectors, organisms bind for reproduction, and in human language verbs bind to predicates. In cognitive neuroscience, a similar "binding" problem has become famous: the problem of explaining how sensations from different modalities bind to produce a unified perception of the world.[85–87] There are three inter-related, but different problems, referred by "binding":

(a) The problem of thalamocortical integration of patterns processed in the brain's parallel distributed architecture;

(b) The problem of inter-modal integration, that is relatively independent of cortical mechanisms, since in many species inter-modality is performed by the superior colliculus[88] . In humans this subcortical structure has a relatively smaller size, having a more limited function of controlling eye movements.[89] This fact justifies the association between cortical integration and perceptual binding for humans, but the generalization to non-primates is not adequate;

(c) The problem of psychological binding, i.e., how different aspects of perception (e.g., in visual perception, form, color and movement), presumably processed by distributed cortical systems, are inter-correlated, influence each other, and become unified in a single phenomenal world.[87]

A solution of the "binding problem" has to account for the three above issues. The approaches to the first two are necessary, but not suficient to explain conscious binding. I have claimed that the solution of the third issue (psychological binding) involves some type of quantum or quantum-like hypothesis about the superposition and entanglement of perceptual patterns.[54] The recovery of quantum coherence in consciousness was called "recoherence."[79]

In the conscious first person perspective of reality, we are always in the presence of an integrated and unitary phenomenal world; we cannot distinguish between perceptual objects before being integrated and after being integrated. There is no report of cases of neuronal tissue lesion where the subjects fail to perform "binding". Reported cases[87] show patients that are not able to perform one step of the "binding" process, or are not able to perceive some type of stimulus or location. However, the remaining perceptual capabilities of these patients always preserve diverse—if not all—modalities of "binding". It seems to be an essential part of the continuous process that goes from sensation to perception, and not a separate intermediate stage "between" sensation and perception. In this sense, any elementary sensation that is consciously perceived already displays some modality of "binding."

The consequence for scientific research is that we cannot introspectively identify the steps of binding processes, in contradiction to Treisman's optimistic statement that "the strongest evidence will come when changes in neural activity are found to coincide with perceived changes in binding, perhaps in ambiguous figures or in attentional capture."[87] If there is no such a thing as "perceived changes in binding", even when neuroscientists come to discover all mechanisms of neuronal integration a problem will remain, about which of them support unconscious psychological binding.

In the human brain, primary and associative perceptual areas form an integrated network, combining forward and feedback connections, simultaneously processing features of the stimuli and the respective integrated percept. There are currently some hypotheses consistent with this approach, based on the concepts of "synchrony", "spatiotemporal coherence" and "resonance". They are supported by the analysis of experimental data on electrical patterns of neuronal activity, measured by EEG or arrays of invasive electrodes. As synchrony entails informational redundancy, it is difficult to understand how synchronous activity would support a variety of conscious experiences. Spatio-temporal coherence[90] is an interesting and powerful hypothesis, which includes and goes beyond synchrony. One possibility of generating coherence is the temporal autocorrelation of electric patterns. There are good neurophysiological evidences that temporally structured patterns correlate with the experience of "qualia."[91]

The resonance hypothesis[92] goes one step further, accounting for processes of "reciprocal causality" (when two neurons or two assemblies resonate, each one amplifies the activity of the other). If reciprocal causality has a special role in physics (e.g., non-linearity), then the resonance hypothesis may have an explanatory role in neuroscience (it has already proven to be powerful in artificial network modeling). The concept of inter-columnar resonance is able to account for the cognitively relevant kinds of firing synchrony,[93] but the converse is not true. Oscillatory synchrony may be produced by subcortical mechanisms without any direct causal relation between the cortical neurons that synchronize; resonance, on the other hand, implies reciprocal causality between cortical neurons.

The "temporal synchrony" hypothesis[94] proposing that evoked potentials "bind" the activity of different feature detectors into a unified perception is a classical reference in the discussion of the problem. However, how could synchronic oscillations encode informational patterns[95]? The difficulty is that such oscillations are informationally redundant; they may be considered *carriers* for the non-synchronous patterns that modulate them.

One solution for this problem would be to distinguish between synchrony coding and oscillatory activity.[10] In this case, the neurons in an assembly have different temporal firing patterns that synchronize only partially.

The idea of "localization of function" in the neocortex, or the existence of a "spatial code", is frequently combined with populational theories of neuron encoding of information. The resulting hypotheses may be summarized in two statements: first, the assumption that sensors produce (spatial) *mappings* of the stimulus (e.g., retinotopic mapping); second, that properties of the stimulus are represented by spatially distributed activity (including synaptic activity) among a cortical sensory neuronal population. The mapping assumption seems to be supported by the physiology of peripheral sensors (specially of the eye, frequently considered to be paradigmatic), as well as the structure of the transmission pathways of the signals to the CNS. Study of afferent fibers (e.g., in the visual and somatic sensory systems) revealed a multi-channel structure, suggesting that the encoding of properties from the stimulus could be based on the distribution of nerve pulses across the population of fibers (i.e., axons of peripheral neurons). A criticism that has been made to the spatial/populational hypotheses is that they cannot account for data about response times. Cortical sensory neurons seem to respond faster to stimulation than models based on the spatially distributed firing rate predict.

A sophisticated version of the spatial-populational hypotheses is the connectionist theory that *patterns of connectivity* (not exactly synaptic weight but *recurrent* excitatory and inhibitory connections; see Ref. 96,97) between neurons *represent* information from the stimulus. The specification of patterns of connectivity has been considered sufficient to account for the specification of properties of the stimulus. A usual objection to this view is that the formation of such patterns would require *supervised* learning, but results seem to indicate that also unsupervised models can do it.[98] Possibly the "trick" is that the notion of *recurrent patterns* is richer than purely spatial-populational codes. If the patterns display a temporal structure, sophisticated connectionist models would be classified as a hybrid *spatio/populational/temporal* one.

What is not taken into consideration in the classical discussion of the binding problem is that electric patterns of neuronal activity are a part of a larger causal chain. The patterns are directly controlled by the homeostasis of the "chemical soup" at the synapses (transmitters coming from the axon of the presynaptic neuron, receptors produced in the post-synaptic neuron, calcium from the surrounding environment, etc.). The three models above mentioned fail to consider the role of the *Hydro-Ionic Wave* on the control of the "chemical soup" homeostasis, and the intrinsic properties of this system, including quantum effects, that may boost the process of communication between parallel distributed circuits of neuronal computation.

The presence of two classes of factors, internal and external to the brain, generating coherent activity of neurons, suggests that "binding" should be supported by a complex mechanism, including, besides proper neuronal computation, also the afferent patterns, reafferent patterns (the "efference copy" or "corollary discharge"), and, last but not the least, the control exerted by Hydro-Ionic Waves on the extracellular medium, determining temporal patterns of neuron firing, and consequently, influencing overt behavior and psychosomatic responses.

Astrocytes receive somatic signals carried by blood flow and cerebrospinal fluid, as well as sensory and cognitive information carried by neuronal assemblies. Their position, hub-like structure and intrinsic processing capabilities suggest that they integrate spatially distributed information. Oscillatory synchrony and constructive wave interference have been proposed to constitute mechanisms of neuro-astroglial interaction. The astrocyte network is stimulated by neuronal dendritic graded potentials and feed back on the neuronal network, modulating neuronal activity (and then the behavior), possibly according to valences attributed to them.

The processing of sensory information patterns and their combination into temporal conscious episodes is a complex process with several phases. In humans and possibly other animal species, the integration of dynamic ionic patterns processed by neuronal assemblies into "gestalts" (conscious episodes) requires the broadcasting and integration of the patterns carried by feed-forward/feedback neuronal circuits. The pattern of a stimulus (a three-dimensional object and/or invariants in a dynamical interaction process) is transduced by sensory receptors (e.g., in the cells of the retina and/or the olfactory bulb) in electric patterns. Signals from peripheral sensors (or central sensors, in the case of the retina) to the central nervous system (CNS) are carried by nerves, using a population frequency code. In the CNS, perceptual processing begins with single-neuron feature detection, filtering of salient features, local broadcasting and activation of specialized circuits. This initial sensory processing phase leads to global feed-forward broadcasting by means of axonal vertical (cortico-thalamic) and horizontal (cortico-cortical) connections, and feedback from associative to sensory areas. The feedback has an adaptive function of making the sensory pattern salient in the respective context.

The form of the stimulus is decomposed into an ensemble of signals, each one presumably reproducing an aspect of the object and/or process being perceived. In the CNS, these patterns are embodied in receptive fields, by means of an activation of the neuronal dendritic graded potential. These potentials are generated mostly by ligand-gated ion channels (e.g., AMPA glutamatergic receptors) that control the movements of ions through neuronal membranes. At this stage, the sensory message about aspects of the stimulus elicits the formation of an ensemble of Local Field Potentials (LFP) located at several cortical areas. Considering the spatial distribution of LFPs and their effects on the astroglial network, the latter integrate and instantiate feelings about the content of the information.[26]

A sketch of the proposed process leading to the formation of a conscious percept is illustrated in Fig. 5.

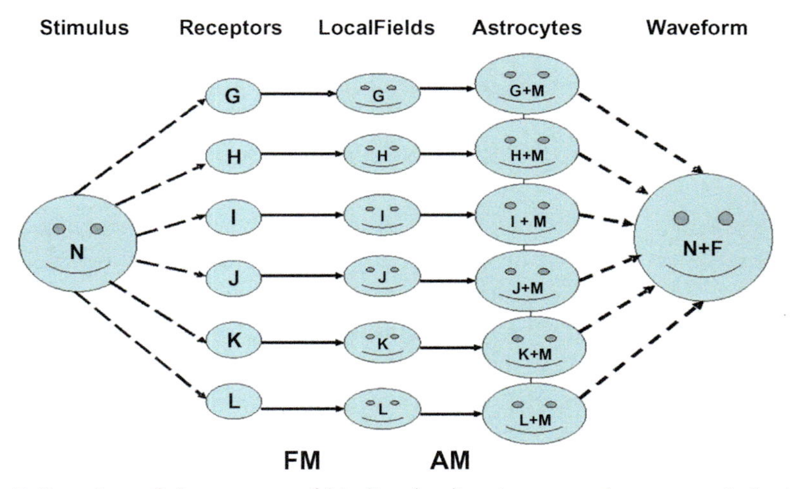

Fig. 5 Overview of the process of binding leading to a conscious percept. A given stimulus has N properties (including G, H, I, J, K, L), separately detected and processed by different sensory circuits in the brain, leading to the formation of a spatial ensemble of LFPs, each one reproducing one aspect of the stimulus (G, H, I, J, K, L). Upon transmission of these patterns to astrocytes, these cells attach to the information pattern a feeling that expresses the first-person perspective living individual. These feelings physiologically correspond to large Hydro-Ionic Waves generated from smaller calcium waves in astrocytes. These waves control the homeostasis of the extracellular "chemical soup" that determine the global state of neural tissue, thus reproducing the N processed qualitative properties (qualia) in the perspective of the experiencer or "self"—the person who feels (F). The information is frequency modulated (FM) in neuronal axonal transmissions and amplitude modulated (AM) in neuronal dendritic potentials and Hydro-Ionic Waves. *Figure made by APJ.*

The final waveform that corresponds to the conscious feeling of the whole episode has dynamical frequency, amplitude and phase properties, and possibly involves superposition and entanglement, as discussed in the next sections.

7. Negentropy, recoherence and the formation of conscious episodes

The binding of qualitative features (*qualia*) processed in a parallel-distributed manner in the neural tissue requires more than ordinary inter-neuronal signaling and astroglial calcium waves. Even departing from an initial stage of quantum coherence, at the micro level, the transition to macro level implies the process of decoherence, partially destroying the original superposition/entanglement and leading to macroscopic phenomena known as Local Field Potentials. In other words, the decoherence process is closely related to the process of thermodynamic entropy increase; recoherence implies some type of reversal of entropy, or negentropy action, in far from equilibrium systems, generating chaotic metastable dynamics, in which a transient phase instantiates conscious episodes.

The actualization of conscious states requires the satisfaction of necessary conditions, the most salient of which possibly consists of distance from thermodynamic equilibrium.[99] This condition is provided by means of the action of local entropy reducers, without violating the Second Law of Thermodynamics. When Non-Equilibrium Thermodynamics and Quantum Theory are combined, a plausible interpretation suggests a connection between the processes of decoherence and irreversibility.[100] If quantum decoherence—understood as loss of correlation between particles that have interacted—is related to entropy increase, a mechanism that locally decreases entropy—the Biological Maxwell Demon[101]—can allow microscopic potentialities of superposition and entanglement to re-emerge at the *meso* and macroscopic levels, supporting conscious binding in the first-person perspective. In this regard, Schrödinger argued that a local decrease of entropy in living systems occurs at the cost of increasing it in environment, thus implying that complex organization in living tissue could be generated by means of mechanisms that convert external into internal low entropy.[99] The idea historically derives from conjectures made by Maxwell[102] and resumed by Loschmidt[103] in his objection to the H-Theorem (an attempt to mechanically demonstrate the tendency to entropy increase) presented by Boltzmann.[104,105]

When considered in a contemporary perspective, it is possible to conjecture that Boltzmann's concept of *molecular disorder* relates to the *common cause* of *entropy increase* and *decoherence*. Since superposition and entanglement are correlations between microscopic particles, the elimination of correlations would be also a loss of quantum coherence.[106]

Could *molecular order* allow entropy *decrease*? From a historical perspective, the idea that coherent states supporting life and cognition can be generated from micro/mesoscopic mechanisms that locally decrease entropy was anticipated in a series of converging speculations by brilliant scientists, beginning with James Clerk Maxwell. Boltzmann's work advanced the idea that entropy increase depends on a loss of microscopic correlations. This idea is coherent with Maxwell's conjecture years before, and possibly an influence for Boltzmann's mature thought. In fact, when the small entity manipulates the gas particles to separate the slower from the faster molecules, *it creates molecular order*, a difference in concentration of molecules with different kinetic energy that can be used to produce work (e.g., moving a cylinder of an engine). The Demon uses information and also a small amount of energy to generate molecular order, thus decreasing the entropy of the system. However, as the operation of Demons requires consumption of useful energy from the environment, it is not able to decrease the total entropy of the universe.

Some approximations are needed to relate Maxwell's scheme with biological systems. Biological systems are not isolated, but depend on the assimilation of low-entropy parts of the environment (i.e., food) to stay alive. The work of Ilya Prigogine and his group in the field of Non-Equilibrium Thermodynamics has showed that—under adequate conditions—dissipative processes could lead to macroscopic organization. Prigogine argued that an open system's internal production of entropy could be balanced by an external energy flux generating "dissipative structures," by means of an "order from fluctuation" mechanism.[106,107] In this explanation, the emerging organization is conceived as the result of a pressure towards equilibrium in open systems receiving a flux of free energy from the outside. Consumption and usage of external free energy makes possible the organization of a macro system into low entropy states, when some of the system's recessive potentialities are amplified, making possible the existence of superposition and entanglement within the n-dimensional state space that encompasses the whole of reality, inclusing, of course, proper dimensions for conscious *qualia*.[79]

Leon Brillouin,[108] inspired by the similarity of the Boltzmann mathematical expression of entropy and the expression for average information transmitted between a source and a receiver,[109] proposed the identification of information and negentropy (the entropy that a system exports to keep its own entropy low). However, in order to undergo a negentropic trajectory, it is necessary for a system to possess adequate mechanisms to absorb low entropy energy from the environment. As indicated by Leo Szilard's classical analysis of the Maxwell proposal, this kind of operation requires the use of information, which must exist previously to the operation of the Demon.[110]

In biological systems, the DNA is the main source of structural information. Proteins operate as Maxwellian mechanisms using molecular and other available information to reduce entropy in metabolic systems that coordinate biological functions.[111] Information is a requisite for, not the product of emerging organization; previous information patterns stored in the system are used to detect signals and to act upon the underlying processes to decrease the entropy of the system.

In an engine, the product of entropy reduction is movement (work), while for living beings movement is only part of the output. The initial step that leads to several functions, including locomotion, is metabolism, a set of physio-chemical transformations. Two processes are central to understanding of the work of proteins in producing metabolic interactions: enzymatic reactions (catalysis: enzyme is not modified) and allosteric interactions (structure and activity of the enzyme are modified by the binding of a metabolic molecule). Since the pioneering work on the allosteric model,[111] we know that proteins have a flexible structure limited to mutually exclusive states: relaxed and tense. The lock-and-key binding of effectors with one or more active microsites of the protein generates a transition of states, and then the protein can itself act as an effector to produce a change in another molecule. Allosteric proteins can form sequential chains to transport information, control gating mechanisms (like those present in membrane ion channels), and other functions. These processes, which are low-energy and cybernetic, coexist with biochemical processes that use energy from glucose to sustain such operations.

Qualitative states in neural tissues may be chemical and/or electric. The action of proteins depends on spatial immediacy of effectors and substrates. In overcoming this limitation, evolution developed physiological processes whereby some proteins encode messages in ionic populations, which broadcast and carry signals to distant places in the biological tissue. The evolution of perceptual, emotional and cognitive systems of the brain was based on the progressive development of neuro-glial connectivity, by means of bioelectric activity (electrical currents and corresponding magnetic fields generated by the movement of ions). In synapses, the bioelectric activity in a preceding neuronal fiber causes release of chemical neurotransmitters, which induces another wave of bioelectric activity in subsequent neurons throughout a signal transduction pathway. In the neuronal axon, electric currents are generated by the movement of ions in and out the membrane, while in the astrocytes electric currents are produced by the movement of ions through cellular microdomains and gap junctions.

Signal transduction pathways are complex mechanism used not only for transmitting information between cells but also to connect two independent systems such as the environment and the genome. The neuronal ligand gated ion channel is a typical example of a *biological* Maxwell Demon, which uses information and low energy to control the movements of ions. The emergence of brain affective and cognitive functions can be explained by means of mechanisms of ionic control in neurons and astrocytes, inducing the formation of brain-wide coherent states. An example is the NMDA (N-Methyl-D-Aspartate) neuronal receptors, containing ion channels that cross the membrane. It is an allosteric protein with three or more binding sites, controlling Ca^{++} entry in the post-synaptic neuron by means of a gate that, in its resting state, is blocked by magnesium ions (Mg^{++}). The opening of the NMDA gate depends on the activation of at least two sites in a hundred-millisecond temporal window. Such an opening depends on activation of recurrent neuronal networks by excitatory Glutamate (Glu) and Glycine (Gly) inputs to distal (NR2) and proximal (NR1) receptor sections, respectively, of apical dendrites of pyramidal neurons in neocortex.[17,19]

In open non-equilibrium systems, when entropy decreases the mechanism of Order through Fluctuations[112] allows the formation of complex structures and functions. Under the operation of adequate mechanisms, the system can evolve to a "coherent macrostate," which has been called *recoherent*.[113–115]

Conscious recoherence *is not the reversal of decoherence*, leading back to the micro domain, because the information processing in neural tissues involves the *meso* and macro levels. Recoherence should be understood as *the superposition and entanglement of patterns instantiated in the parallel-distributed architecture of neural tissues*. The recoherent states resulting from the above operation exist in proper dimensions of reality, possibly fractal dimensions[116] nested in the portion of the 4D spacetime contained in the brain, or in the living body as a whole. This statement does *not* imply fractality in the sense that these states display self-similarity across spatial or temporal scales; it does mean that *qualia inhabit fractional dimensions additional to the classical four*, to be represented numerically as 4,XXX, where XXX refer to the additional dimensions proper to *qualia*. This structure is intermediary between the n-dimensional Hilbert space used in quantum theory and the classical 4D spacetime of relativity theory.

Brain/body recoherent macrostates result from the activity of entropy reducers, as ion channels and proteins composing intracellular signal transduction pathways. The recoherent macrostate as a whole would correspond to the binding of qualitative features in one integrated conscious episode available in the first-person perspective for the living individual.[8] Contents of the conscious episode are superposed and entangled in a quantum-like pattern that is experienced in the first person perspective and involves simultaneous processes in micro, *meso* and macro scales. Considering the metastable character of activity in neural tissues,[117] recoherent states have a very brief temporal duration. These short periods are interspersed with decoherent states; in the first-person perspective, we do not consciously experience the decoherent states, but only the recoherent phases of the whole process, which appear to us in a continuous "flow."[79]

8. The Tasaki action potential and conscious binding

In the this section I present my new hypothesis of quantum-like recoherence on the basis of the action of hydrogen cations in the "chemcial soup" by means of the Grotthuss effect, transforming the extracellular intersticial water into a "superconductive" medium that operates as the generator of the Tasaki action potential.[3,60]

In the context of the "Neuron Doctrine," the neuron is considered to be the structural and functional unit of the bran/mind. The main activity of the neuron is the action potential, a process dependent on the dendritic tree activation reaching a threshold, converting to spike trains at the axon hillock. The state of the "chemical soup" around the myelinated (or non-myelinated) axon, as well as the modulation exerted by glial cells on the extracellular medium, are not taken into account. An alternative view of the generation of action potentials was proposed by Tasaki. Considering that the propagation of the action potential through the axon also depends on the state of the "chemical soup" of the extracellular medium, Tasaki highlighted the control of ionic homeostasis of the "soup" by astrocytes, and the relevant changes in the configuration of intersticial water, both shaping the spatiotemporal dynamics of spike trains, and, therefore, determining mental and behavioral processes. Tasaki states that: "The cortical gel layer of nerve fibers has the properties of a cation exchanger. Hence, this layer can, and actually does, undergo a reversible abrupt structural change when monovalent cations (e.g., $Na+$) are substituted for the divalent counter-ions (e.g., $Ca2+$). This structural change brings about a sudden rise in the water content of the layer which in turn produces a large enhancement of cation mobilities accompanied by a shift of ion-selectivity in favor of hydrophilic cations. Based on these grounds, it is argued that the electrophysiological processes known as "nerve excitation and conduction" are, basically, manifestations of abrupt structural changes in the cortical gel layer."[9]

There are two types of abrup structural changes predicted by Tasaki: "The polypeptide chains in solutions can be reversibly converted, as is well known, from the random coil to the helical form. Hydrogen bonds formed between different groups in one long polypeptide chain lead the whole chain into the helical form. This structural transformation is very sharp; that is, a change of a few degrees in temperature or a few percent of solvent composition is sufficient to complete the transformation. Hence, the term "phase transition" has been employed to describe this reversible structural change… Negatively charged polyelectrolyte gels in salt solutions can be converted from the swollen state to the compact state when the monovalent counter-ions are replaced with divalent cations. This structural transformation is also sharp; that is, it can be initiated and completed by a small change in the salt composition of the surrounding solution."[3]

Additionally to the structural change reported by Tasaki, the Hydro-Ionic Wave model also predicts a rise in hydrogen cations within the "chemical soup," impacting on all processes that take place in the extracellular milieu of neural tissues. The rise in hydrogen cations is a consequence of the selective action of aquaporins, blocking the entry of these molecules in astrocytes and neurons, by means of a positive site located inside the protein's channel. This feature of the activity of neural tissues, by means of the *Grotthuss mechanism,*[118] increases the general conductivity of the "chemical soup" and amplifies the effect of polyelectrolytes on proteins. Interfacial water contains a narrow zone of protons, as positively charged hydrogen ions, which can move very rapidly in water from one water molecule to the next: "This principle of protonic conduction in interfacial water by way of *Grotthuss mechanism* can transfer quantum and quantized information through proton nuclear spin ensembles in polarized proteins."[119]

There is huge physical and chemical complexity of water dynamics in living systems (see for instance[118,120]) that I will not attempt to cover here. The Grotthuss mechanism is probably the oldest and best known attempt to conceptualize this complex dynamics, but there are surely other possibilities deserving a close examination. The Grotthuss mechanism can also be combined with quantum theory. There are some interesting results that suggest some kind of microscopic self-organization (although related to superconductivity in the sense of entailing long-range phase correlations) providing steady state dynamics and time-temperature dependencies, in a transition state density matrix formulation.[121]

The structural changes in "chemical soup" impact on protein conformation changes, which ultimately depend on quantum fluctuations, as argued in the context of the research program of Stuart Kaufman and colleagues.[122] This type of modeling is complemantary to the work of Sisir Roy, Gustav Bernroider and colleagues, arguing for a role of quantum fluctuations and entanglement in the coordinated generation of spike trains.[123–125]

The effects of ionic waves on proteins and neural tissue activity correspond to predictions made by Katchalsky[126,127] and recent experimental work on electrochemical patterns in the brain[4]: "Wave propagation would show coherent flow of matter or energy. An almost horizontal dipole standing around a mechanical stimulus zone [is expected to—APJ] produce a pattern that shimmers and shifts in place, suggesting a dynamic pattern of space/time energy oscillations", and "Coherent flow was demonstrated beyond any doubt. The propagating wavelengths were in the order of centimeters".[60] The results can be generalized: "The intrinsic electric field of the neuropil membranes and the polyelectrolytes between them compose a vibrating, self-organizing system. Macroscopic patterns can be formed."[60]

The connection of this concept of a self-organizing macrosystem with the issue of conscious binding is direct, although not uncontroversial. As the function of the heart is pumping blood, the function of the brain is producing the experience of conscious episodes. Conscious episodes require the binding of features processed in a parallel-distributed computing architecture, corresponding to the activity of specialized receptors, circuits and neocortical columns in the human brain.

On the one hand, conventional views of neuron signaling, on the basis of the Neural Doctrine and the Hodkins-Huxley equation, is not sufficient to account for the integration of the parallel-distributed patterns that correspond to the pre-conscious processing of *qualia*. On the other hand, the Hydro-Ionic Wave is a powerful mechanism of integration, but does not contain mechanisms for precise information register and computation, as neurons do. The Hydro-Ionic Wave has a central role in the formation of conscious episodes, because it is the motivational factor recruiting neural assembiles that will compose each episode. However, it is not sufficient to account for them, because its structure is fuzzy. It determines the focus of attention and selects the patterns that will enter the "Global Workspace"[128] in which conscious episodes are formed and experienced, but it is not the system that encompasses all these patterns, before and/or after they are bound together.

If each neuron, and the spike trains they generate, operated like a Leibinizian Monad, the conscious binding of features would be or impossible or instantiated in single neurons, with the (strange) consequence that each brain would contains billions of conscious entities.[129] The Tasaki concept of the action potential allows the interpretation of the activity of an ensemble of neurons as being determined by structural changes in the "chemical soup," under the control of the Hydro-Ionic Wave. In this view, each neuron's activity is not of a Leibinizian Monad, but belongs to a larger whole that interacts with the environment, composing dynamical functional cycles of action-perception and formation of the inner world of conscious experience with the format of integrated episodes.

9. Final remarks

In this chapter three classical-quantum interfaces were identified, departing from experimental results about brain processes in the micro, *meso* and macroscopic scales. The work of the whole nervous system interacting with the environment to generate conscious experience is very complex. An ensemble of negentropic mechanisms involved in the dynamical control of homeostasis generate transient "recoherent" phases with quantum features of superposition and entanglement, composing conscious episodes. Although the recoherent phases are very short for systems at biological temperature, their discontinuity in time may not be an obstacle to conscious experience, because we cannot perceive the lapses.[79]

Two large networks, neuronal and astroglial, contribute to this result. Neurons integrate information generating cognitive representations, while astrocytes integrate temporal waveforms composing sensations and emotional feelings. The intercellular 'chemical soup" is important for the whole integration process, mediating neuron-astrocyte interactions and executing operations necessary for conscious binding.

A final word is needed about how superposition and entanglement are features of conscious binding. In Fig. 5, the architecture of the binding process was described. The superposition of features means that different properties of objects and processes are included in the conscious episode; for instance, the statements "the zebra is black" and "the zebra is white" are not contradictory, but imply that "the zebra is black and white". Entanglement means that the features of an episode are not statistically independent (see discussion in Ref. 8). In this regard, there is a whole area of study about "quantum cognition" that was not reviewed here, but may be of interest for the reader (see, for instance, Ref. 130,131).

Acknowledgment

FAPESP (Brazilian Funding Agency) for the support to this research; Roman Poznanski for the invitation to write this chapter; Erkki Brändas for scientific support to the work; Chris Nunn, Alexej Verkhratsky, James Robertson, Bernhard Mitterauer for discussion and criticism about the ideas, and Vera Maura Fernandes de Lima for introducing me to the framework used in the final section.

References

1. Bullock, T. H.; Bennett, M. V.; Johnston, D.; Josephson, R.; Marder, E.; Fields, R. D. The Neuron Doctrine, Redux. *Science* **2005**, *310*(5749), 791–793.
2. Pereira, A., Jr. Astroglial Hydro-Ionic Waves Guided by the Extracellular Matrix: An Exploratory Model. *J. Integr. Neurosci.* **2017**, *16*, 1–16.
3. Tasaki, I. On the reversible abrupt structural changes in nerve fibers underlying their excitation and conduction processes. In *Phase Transitions in Cell Biology*; Pollack, G. H., Chin, W.-C., Eds.; Dordrecht, The Netherlands: Springer, 2008; pp. 1–21.
4. Fernandes de Lima, V. M.; Silva, G. E.; Pereira, A., Jr. *The Spreading Depression Propagation. Preprint Posted in Research Gate*, 2020; Available at: https://www.researchgate.net/publication/342246867_THE_SPREADING_DEPRESSION_PROPAGATION_HOW_ELECTROCHEMICAL_PATTERNS_DISTORT_OR_CREATE_PERCEPTION.
5. Marino, J.; Schummers, J.; Lyon, D. C.; Schwabe, L.; Beck, O.; Wiesing, P.; Obermeyer, K.; Sur, M. Invariant Computations in Local Cortical Networks with Balanced Excitation and Inhibition. *Nat. Neurosci.* **2005**, *8*(2), 194–201.
6. Jones, B. E. Arousal Systems. *Front. Biosci.* **2003**, *8*, s438–s451.
7. Pereira, A., Jr.; Johnson, G. Towards an Understanding of the Genesis of Ketamine-Induced Perceptual Distortions and Hallucinations. *Brain Mind* **2003**, *4*(3), 307–326.
8. Pereira, A., Jr.; Foz, F. B.; Rocha, A. F. The Dynamical Signature of Conscious Processing: From Modality-Specific Percepts to Complex Episodes. *Psychol. Conscious* **2017**, *4*(2), 230–247.
9. Fotheringhame, D. K.; Young, M. P. Neural Coding Schemes for Sensory Representation: Theoretical Proposals and Empirical Evidence. In *Cognitive Neuroscience*; Rugg, M. D., Ed.; MIT Press: Cambridge, 1997.
10. Phillips, W. A. Theories of Cortical Computation. In *Cognitive Neuroscience*; Rugg, M. A., Ed.; MIT Press: Cambridge, 1997.

11. Bickle, J. *Philosophy and Neuroscience: A Ruthless Reductive Account*; Kluwer: Dordrecht, 2003.
12. Sabatini, B. L.; Maravall, M.; Svoboda, K. Ca2+ Signaling in Dendritic Spines. *Curr. Opin. Neurobiol.* **2001**, *11*(3), 349–356.
13. Sabatini, B. L.; Oertner, T. G.; Svoboda, K. The Life Cycle of Ca2+ Ions in Dendritic Spines. *Neuron* **2002**, *33*(3), 439–452.
14. Holtoff, K.; Tsay, D.; Yuste, R. Calcium Dynamics of Spines Depend on their Dendritic Location. *Neuron* **2002**, *33*, 425–437.
15. Krystal, J. H.; Belger, A.; D'Souza, C.; Anand, A.; Charney, D.; Aghajanian, G. K.; Moghaddam, R. Therapeutic Implications of the Hyperglutamatergic Effects of NMDA Antagonists. *Neuropsychopharmacology* **1999**, *21*(56), S133–S157.
16. Loewenstein, W. *The Touchstone of Life: Molecular Information, Cell Communication and the Foundations of Life*; Oxford University Press: New York, 1999.
17. Rocha, A. F.; Pereira, J. R.; A. and Coutinho, F.A.B. N-Methyl-d-Aspartate Channel and Consciousness: From Signal Coincidence Detection to Quantum Computing. *Prog. Neurobiol.* **2001**, *64*(6), 555–573.
18. Squire, L.; Bloom, F.; Mcconnell, S.; Roberts, J.; Spitzer, N.; Zigmond, M. *Fundamental Neuroscience*; Academic Press: New York, 1999.
19. Rocha, A. F.; Massad, E.; Pereira, A., Jr. *The Brain: From Fuzzy Arithmetics to Quantum Computing*; Springer: Berlin/Heidelberg, 2005; p. 227.
20. Wilson, M. A.; Brunger, A. T. The 1.0 A Crystal Structure of Ca2+−Bound Calmodulin: An Analysis of Disorder and Implications for Functionally Relevant Plasticity. *J. Mol. Biol.* **2000**, *301*, 1237–1265.
21. Gilbert, P. F. C. An Outline of Brain Function. *Cogn. Brain Res.* **2001**, *12*, 61–74.
22. Silkis, I. The Cortico-Basal Ganglia-Thalamocortical Circuit with Synaptic Plasticity I. Modification rules for excitatory and inhibitory synapses in the striatum. *Biosystems* **2000**, *57*(3), 187–196.
23. Svenningsson, P.; Tzavara, E. T.; Carruthers, R.; Rachleff, I.; Wattler, S.; Nehls, M.; Mckinzie, D. L.; Fienberg, A. A.; Nomikos, G. G.; Greengard, P. Diverse Psychotomimetics Act Through a Common Signaling Pathway. *Science* **2003**, *302*, 1412–1415.
24. Javitt, D.C. and Coyle, J.T. (2004) Decoding Schizophrenia. Sci. Am. 290:48-55 Dec. 15. Available at: http://www.sciam.com/print_version.cfm?articleID=000EE239-6805-1FD5-A23683414B7F0000.
25. Haydon, P. G.; Carmignoto, G. Astrocyte Control of Synaptic Transmission and Neurovascular Coupling. *Physiol. Rev.* **2006**, *86*, 1009–1031.
26. Pereira, A., Jr.; Furlan, F. A. Astrocytes and Human Cognition: Modeling Information Integration and Modulation of Neuronal Activity. *Prog. Neurobiol.* **2010**, *92*, 405–420.
27. Carrara-Augustenborg, C.; Pereira, A., Jr. Brain Endogenous Feedback and Degrees of Consciousness. In *Consciousness: States, Mechanisms and Disorders*; Cavanna, A. E., Nani, A., Eds.; Nova Science Publishers, Inc.: New York, 2012.
28. Almada, L. F.; Jr, P.; Carrara-Augustenborg, C. What Affective Neuroscience Means for a Science of Consciousness. *Mens Sana Monogr.* **2013**, *11*(1), 253–273.
29. Kuga, N.; Sasaki, T.; Takahara, Y.; Matsuki, N.; Ikegaya, Y. Large-Scale Calcium Waves Traveling through Astrocytic Networks In Vivo. *J. Neurosci.* **2011**, *31*(7), 2607–2614.
30. Takata, N.; Mishima, T.; Hisatsune, C.; Nagai, T.; Ebisui, E.; Mikoshiba, K.; Hirase, H. Astrocyte Calcium Signaling Transforms Cholinergic Modulation to Cortical Plasticity In Vivo. *J Neurosci.* **2011**, *31*(49), 18155–18165.
31. Navarrete, M.; Perea, G.; de Sevilla, D. F.; Gómez-Gonzalo, M.; Núñez, A.; et al. Astrocytes Mediate in Vivo Cholinergic-Induced SynapticPlasticity. *PLoS Biol.* **2012**, *10*(2), e1001259.

32. Bazargani, N.; Attwell, D. Astrocyte Calcium Signaling: The Third Wave. *Nat. Neurosci.* **2016**, *19*, 182–189.
33. Smeal, R. M.; Economo, M. N.; Lillis, K. P.; Wilcox, K. S.; White, J. A. Targeted Path Scanning: An Emerging Method for Recording Fast Changing Network Dynamics across Large Distances. *J. Bioengi. Biomed. Sci.* **2012**, S5.
34. Brazhe, A.; Mathiesen, C.; Lind, B.; Rubin, A.; Lauritzen, M. Multiscale Vision Model for Event Detection and Reconstruction in Two-Photon Imaging Data. *Neurophotonics* **2014**, *1*, 011012.
35. Zheng, K.; Bard, L.; Reynolds, J. P.; King, C.; Jensen, T. P.; Gourine, A. V.; Rusakov, D. A. Time-Resolved Imaging Reveals Heterogeneous Landscapes of Nanomolar ca(2+) in Neurons and Astroglia. *Neuron* **2015**, *88*, 277–288.
36. Croft, W.; Reusch, K.; Tilunaite, A.; Russell, N. A.; Thul, R.; Bellamy, T. C. Probabilistic Encoding of Stimulus Strength in Astrocyte Global Calcium Signals. *Glia* **2015**, *64*, 537–552.
37. Resendez, S.; Jennings, J. H.; Ung, R. L.; Namboodiri, V. M.; Zhou, Z. C.; Otis, J. M.; Nomura, H.; McHenry, J. A.; Kosyk, O.; Stuber, G. D. Visualization of Cortical, Subcortical and Deep Brain Neural Circuit Dynamics during Naturalistic Mammalian Behavior with Head-Mounted Microscopes and Chronically Implanted Lenses. *Nat. Protoc.* **2016**, *11*(3), 566–597.
38. Shigetomi, E.; Patel, S.; Khakh, B. S. Probing the Complexities of Astrocyte Calcium Signaling. *Trends Cell Biol.* **2016**, *26*, 300–312.
39. Robertson, J. M. Astrocyte Domains and the Three-Dimensional and Seamless Expression of Consciousness and Explicit Memories. *Med. Hypotheses* **2016**, *81*, 1017–1024. https://doi.org/10.1016/j.mehy.2013.09.021.
40. Smith, K. Settling the Great Glia Debate. *Nature* **2010**, *468*, 160–162.
41. Verderio, C.; Matteoli, M. ATP in Neuron-Glia Bidirectional Signalling. *Brain Res. Rev.* **2011**, *66*(1–2), 106–114.
42. Schummers, J.; Yu, H.; Sur, M. Tuned Responses of Astrocytes and their Influence on Hemodynamic Signals in the Visual Cortex. *Science* **2008**, *320*, 1638–1643.
43. Banaclocha, M. A. M. Neuromagnetic Dialogue between Neuronal Minicol- Umns and Astroglial Network: A New Approach for Memory and Cerebral Computation. *Brain Res. Bull.* **2007**, *73*, 21–27.
44. Agulhon, C.; Petravicz, J.; McMullen, A. B.; Sweger, E. J.; Minton, S. K.; Taves, S. R.; Casper, K. B.; Fiacco, T. A.; KD, M. C. What Is the Role of Astrocyte Calcium in Neurophysiology? *Neuron* **2008**, *59*, 932–946.
45. Perea, G.; Yang, A.; Boyden, E. S.; Sur, M. Optogenetic Astrocyte Activation Modulates Response Selectivity of Visual Cortex Neurons In Vivo. *Nat. Commun.* **2014**, *5*, 3262.
46. Thrane, A. S.; Thrane, V. R.; Zeppenfeld, D.; Lou, N.; Xu, Q.; Nagelhus, E. A.; Nedergaard, M. General Anesthesia Selectively Disrupts Astrocyte Calcium Signaling in the Awake Mouse Cortex. *Proc. Natl. Acad. Sci. U. S. A.* **2012**, *109*, 18974–18979.
47. Furman, J. L.; Sama, D. M.; Gant, J. C.; Beckett, T. L.; Murphy, M. P.; Bachstetter, A. D.; Van Eldik, L. J.; Norris, C. M. Targeting Astrocytes Ameliorates Neurologic Changes in a Mouse Model of Alzheimer's Disease. *J. Neurosci.* **2012**, *32*(46), 16129–16140.
48. Verkhratsky, A.; Nedergaard, M. Physiology of Astroglia. *Physiol. Rev.* **2018**, *98*(1), 239–389.
49. Pereira, A., Jr.; Foz, F. B.; Rocha, A. F. Cortical Potentials and Quantum-like Waves in the Generation of Conscious Episodes. *Quantum Biosys.* **2015**, *6*, 10–21.
50. Pereira, A., Jr.; Furlan, F. A. On the Role of Synchrony for Neuron-Astrocyte Interactions and Perceptual Conscious Processing. *J. Biol. Phys.* **2009**, *35*(4), 465–481.

51. Pereira, A., Jr.; Almada, L. F. Conceptual Spaces and Consciousness: Integrating Cognitive and Affective Processes. *Int. J. Mach. Conscious.* **2011**, *3*(1), 127–143.
52. Pereira, A., Jr.; Furlan, F. A.; Pereira, M. A. O. Recent Advances in Brain Physiology and Cognitive Processing. *Mens Sana Monogr.* **2011**, *9*, 183–192.
53. Pereira, A., Jr.; Furlan, F. A. Analog Modeling of Human Cognitive Functions with Tripartite Synapses. *Stud. Comp. Int. Dev.* **2011**, *314*, 623–635.
54. Pereira, A., Jr. Perceptual Information Integration: Hypothetical Role of Astrocytes. *Cogn. Comput.* **2012**, *4*(1), 51–62.
55. Oberheim, N. A.; Wang, X.; Goldman, S. A.; Nedergaard, M. Astrocytic Complexity Distinguishes the Human Brain. *Trends Neurosci.* **2006**, *29*, 547–553.
56. Oberheim, N. A.; Takano, T.; Han, X.; He, W.; Lin, J. H. C.; Wang, F.; Xu, Q.; Wyatt, J. D.; Pilcher, W.; Ojemann, J.; Ransom, B. R.; Goldman, S. A.; Nedergaard, M. Uniquely Hominid Features of Adult Human Astrocytes. *J. Neurosci.* **2009**, *29*, 3276–3287.
57. Lent, R.; Azevedo, F. A.; Andrade-Moraes, C. H.; Pinto, A. V. How Many Neurons Do you Have? Some Dogmas of Quantitative Neuroscience under Revision. *Eur. J. Neurol.* **2012**, *35*(1), 1–9.
58. He, B. J.; Raichle, M. E. The fMRI Signal, Slow Cortical Potential and Consciousness. *Trends Cogn. Sci.* **2009**, *13*, 302–309.
59. Douglas Fields, R. Neuroscience: Map the Other Brain. *Nature* **2013**, *501*, 25–27.
60. Fernandes de Lima, V. M.; Pereira, A., Jr. The Plastic Glial-Synaptic Dynamics within the Neuropil: A Self-Organizing System Composed of Polyelectrolytes in Phase Transition. *Neural Plast.* **2016**, *2016*, 7192427. https://doi.org/10.1155/2016/7192427.
61. Mentré, P. Water in the Orchestration of the Cell Machinery Some Misunderstandings: A Short Review. *J. Biol. Phys.* **2012**, *38*, 13–26.
62. Pollack, G. H. *Cells, Gels, and the Engines of Life*; Ebner and Sons: Seattle, USA, 2001.
63. Pollack, G. H. Water, Energy and Life: Fresh Views from the Water's Edge. *Int J. Des. Nat. Ecodyn.* **2010**, *5*, 27–29.
64. Dityatev, A.; Seidenbecher, C. I.; Schachner, M. Compartmentalization from the Outside: The Extracellular Matrix and Functional Microdomains in the Brain. *Trends Neurosci.* **2010**, *33*, 503–512.
65. Dityatev, A.; Rusakov, D. A. Molecular Signals of Plasticity at the Tetrapartite Synapse. *Curr. Opin. Neurobiol.* **2011**, *21*, 353–359.
66. Benedikt, J.; Inyushin, M.; Kucheryavykh, Y. V.; Rivera, Y.; Kucheryavykh, L. Y.; Nichols, C. G.; Eaton, M. J.; Skatchkov, S. N. Intracellular Polyamines Enhance Astrocytic Coupling. *Neuroreport* **2012**, *23*, 1021–1025.
67. McCullough, W. S.; Pitts, E. A Logical Calculus of the Ideas Immanent in Nervous Activity. *Bull. Math. Biophys.* **1943**, *5*, 115–133.
68. Pereira, A., Jr.; Lungarzo, C. A. A Framework for the Computational Approach to Cellular Metabolism Supporting Neuronal Activity. *Int. J. Comput. Cogn.* **2005**, *3*(3), 87–92.
69. Carafoli, E. Calcium Signaling: A Tale for All Seasons. *Proc. Natl. Acad. Sci.* **2002**, *99*(3), 1115–1122.
70. Jaiswal, J. K. Calcium—How and Why? *J. Biosci.* **2001**, *26*(3), 357–363.
71. Cirac, J. I.; Zoller, P. A. Scalable Quantum Computer with Ions in an Array of Microtraps. *Nature* **2000**, *404*, 579–581.
72. Kielpinski, D.; Monroe, C.; Wineland, D. J. Architecture for a Large-Scale Ion-Trap Quantum Computer. *Nature* **2002**, *417*, 709.
73. Pereira Jr. A. e Polli, R. S. Trapped Ion Quantum Computing and the Principles of Logic. *Manuscrito* **2006**, *28*, 559–573.
74. Nielsen, M. A.; Chuang, I. L. *Quantum Computation and Quantum Information*; Cambridge Univiversity Press: Cambridge, 2000.

75. Hughes, R. J.; et al. The Los Alamos Trapped Ion Quantum Experiment. *Fortschritte der Physik* **1998**, *46*, 329–361.
76. Schmidt-Kaler, F.; et al. Realization of the Cirac-Zoller Controlled-NOT Quantum Gate. *Nature* **2003**, *422*, 408–411.
77. Milburn, G. J.; Schneider, S.; James, D. F. V. Ion Trap Quantum Computing with Warm Ions. *Fortschritte der Physik* **2000**, *48*, 801–810.
78. Schneider, S.; James, D. F. V.; Milburn, G. J. *Method of Quantum Computation with "Hot" Trapped Ions*; 2004. ArXiv:quant-ph/980801.
79. Pereira, A., Jr.; Vimal, R. L. P.; Pregnolato, M. Can Qualitative Physics Solve the Hard roblem? In *Biophysics of Consciousness: A Foundational Approach*; Poznanski, R., Tuszynski, J., Feinberg, T., Eds.; World Scientific: Singapore, 2017.
80. Engel, A. K.; Singer, W. Temporal Binding and the Neural Correlates of Sensory Awareness. *Trends Cogn. Sci.* **2001**, *5*(1), 16–25.
81. Hameroff, S.; Penrose, R. Quantum Computation in Brain Microtubules? The Penrose–Hameroff 'Orch OR' Model of Consciousness. *Philos. Trans. R. Soc. A* **1998**, *356*, 1869–1896.
82. Tononi, G. An Information Integration Theory of Consciousness. *BMC Neurosci.* **2004**, *5*(1), 42.
83. Tononi, G. Consciousness as Integrated Information: A Provisional Manifesto. *Biol. Bull.* **2008**, *215*(3), 216–242.
84. Tononi, G. Integrated Information Theory of Consciousness: An Updated Account. *Arch. Ital. Biol.* **2012**, *150*(4), 293–329. http://www.ncbi.nlm.nih.gov/pubmed/23802335.
85. Crick, F. *The Astonishing Hypothesis*; Charles Scribner's/Maxwell McMillan: New York, 1994.
86. Hardcastle, V. G. Psychology's Binding Problem and Possible Neurobiological Solutions. *Int. J. Consum. Stud.* **1994**, *1*(1), 66–90.
87. Treisman, A. The Binding Problem. *Curr. Opin. Neurobiol.* **1996**, *6*(2), 171–178.
88. Stein, B. E.; Meredith, M. A. *The Merging of the Senses*; MIT Press: Cambridge, 1993.
89. Schiller, P. H. Innate Motor Action as a Basis for Learning. In *Instinctive Behavior*; Schiller, C. H., Lashley, K. S., Eds.; International Universities Press: New York, 1949.
90. Roy John, E.; Easton, P.; Isenhart, P. Consciousness and Cognition May be Mediated by Multiple Independent Coherent Ensembles. *Conscious. Cogn.* **1997**, *6*, 3–39.
91. Cariani, P. As time really mattered: Temporal strategies for neural coding of sensory information. In *Origins: Brain and Self-Organization*; Pribram, K., Ed.; Erlbaum: Hillsdale, 1994.
92. Grossberg, S. The Link between Brain Learning, Attention and Consciousness. *Conscious. Cogn.* **1999**, *8*(1), 1–44.
93. Grossberg, S.; Somers, D. Synchronized Oscillations for Binding Spatially Distributed Feature Codes into Coherent Spatial Patterns. In *Neural Networks for Vision and Image Processing*; Carpenter, G., Grossberg, S., Eds.; MIT PRess: Cambridge, 1992.
94. Gray, C.; Singer, W. Stimulus-Specific Neuronal Oscillations in Orientation Columns of Cat Visual Cortex. *Proc. Natl Acad. Sci. U. S. A.* **1989**, *86*(p), 1698–1702.
95. Pereira, A., Jr.; Rocha, A. F. Temporal Aspects of Neuronal Binding. In *(Orgs.) Studies in the Structure of Time: From Physics to Psychopathology*; Buccheri, R., Soniga, M. E., Gesu, V., Eds.; Kluwer: New York, 2000.
96. Edelman, G. M. *Neural Darwinism*; Basic Books: New York, 1987.
97. Edelman, G. M. *The Remembered Present: A Biological Theory of Consciousness*; Basic Books: New York, 1989.
98. Pizzi, R.; Meccoci, A.; Favalli, L. *A Neural Model for Real-Time Inductive Learning*; 1998 (Preprint. IEEE Transactions on Neural Networks).

99. Schrödinger, E. *What Is Life? The Physical Aspect of the Living Cell*; Cambridge University Press: Cambridge, 1944.
100. Halliwell, J. J.; Pérez-Mercader, J.; Zurek, W. H., Eds. *Physical Origins of Time Asymmetry*; Cambridge University Press: Cambridge (UK), 1994.
101. Monod, J. *Le Hasard et la Nécessité: Essai Sur la Philosophie Naturelle de la Biologie Moderne (Chance and Necessity: An Essay on the Natural Philosophy of Modern Biology)*; Editions du Seuil: Paris, 1970.
102. Maxwell, J. C. On the dynamical theory of gases. In *Kinetic Theory*; Brush, S., Ed.; Vol. 2; Pergamon Press: Oxford, London, 1866/1965; pp. 23–87.
103. Loschmidt, J. J. Sitzungsberichte der Akademie der Wissenschaften Zu Wien. *Proceedings of the Academy of Sciences in Vienna* **1876**, *73*, 128–336.
104. Boltzmann, L. Further studies in the thermal equilibrium of gas molecules. In *Kinetic Theory*; Brush, S., Ed.; Vol. 1; Pergamon Press: Oxford, London, 1872/1965; pp. 88–175.
105. Boltzmann, L. *Lectures on Gas Theory*; S. Brush, Trans. University of California Press: Berkeley, Los Angeles, 1896/1964.
106. Prigogine, I. *Time, Structure And Fluctuations. Nobel Lecture*, 1977; 8 December http://www.nobelprize.org/nobel_prizes/chemistry/laureates/1977/prigogine-lecture.pdf.
107. Prigogine, I.; Stengers, I. *Order out of Chaos. Man's Dialogue with Nature*; Bantam Books: New York, 1984.
108. Brillouin, L. *Science and Information Theory*; Academic Press: New York, 1956.
109. Weaver, W.; Shannon, C. E. *The Mathematical Theory of Communication*; University of Illinois Press: Urbana, 1949.
110. Szilard, L. Uber Die Entropieverminderung in Einem Thermodynamischen System Bei Eingriffen Intelligenter Wesen [about the Decrease of Entropy in a Thermodynamic System by the Intervention of Intelligent Beings]. *Z. Physik* **1929**, *53*, 840–856.
111. Monod, J.; Changeux, J. P.; Jacob, F. Allosteric Proteins and Cellular Control Systems. *J. Mol. Biol.* **1963**, *6*, 306–329.
112. Nicolis, G.; Prigogine, I. *Self-Organization in Nonequilibrium Systems—from Dissipative Structures to Order through Flucutations*; Wiley: New York, 1977.
113. Chin, A. W.; Prior, J.; Rosenbach, R.; Caycedo-Soler, F.; Huelga, S. F.; Plenio, M. B. The Role of Non-equilibrium Vibrational Structures in Electronic Coherence and Recoherence in Pigment-Protein Complexes. *Nat. Phys.* **2013**, *9*, 113–118.
114. de Ponte, M. A.; Cacheffo, A.; Villas-Boas, C. J.; Mizrahi, S. S.; Moussa, M. H. Y. Spontaneous Recoherence of Quantum States After Decoherence. *Eur. Phys. J. D* **2010**, *59*, 487–496.
115. Hsiang, J.-T.; Ford, L. H. *Decoherence and Recoherence in Model Quantum Systems*; 2008. http://arxiv.org/abs/0811.1596v1.
116. Pereira, A., Jr.; Bresciani, F.; Ilario, E.; E. Modelando o Monismo de Triplo Aspecto em um Espaço de Estados com Dimensões Fracionárias. *Complexitas* **2016**, *1*(1), 78–100.
117. Kelso, J. A. Multistability and Metastability: Understanding Dynamic Coordination in the Brain. *Philos .Trans. R. Soc. Lond. B Biol. Sci.* **2012**, *367*(1591), 906–918. https://doi.org/10.1098/rstb.2011.0351.
118. Agmon, N. The Grotthuss Mechanism. *Chem. Phys. Lett.* **1995**, *244*(5–6), 456–462. https://doi.org/10.1016/0009-2614(95)00905-J.
119. Poznanski, R., Tuszynski, J. and Feinberg, T. (Eds.) Biophysics of Consciousness: A Foundational Approach. Singapore: World Scientific.
120. Bai, C.; Herzfeld, J. Special Pairs Are Decisive in the Autoionization and Recombination of Water. *J. Phys. Chem. B.* **2017**, *121*(16), 4213–4219. https://doi.org/10.1021/acs.jpcb.7b02110.

121. Chatzidimitriou-Dreismann, C. A.; Brändas, E. J. Proton Delocalization and Thermally Activated Quantum Correlations in Water: Complex Scaling and New Experimental Results. *Ber. Bunsenges. phys. chem.* **1991**, *95*(3), 264–272.
122. Vattay, G.; Kauffman, S.; Niiranen, S. Quantum Biology on The Edge of Quantum Chaos. *PLoS One* **2014**, *9*(3). https://doi.org/10.1371/journal.pone.0089017.eCollection.
123. Bernroider, G.; Roy, S. Quantum Entanglement of K+ Ions, Multiple Channel States, and the Role of Noise in the Brain. *Proc. SPIE* **2005**, *5841*(29), 205–214. https://doi.org/10.1117/12.609227.
124. Bernroider, G.; Summhammer, J. Can Quantum Entanglement between Ion Transition States EffectAction Potential Initiation? *Cogn. Comput.* **2012**, *4*(1), 29–37. https://doi.org/10.1007/s12559-012-9126-7.
125. Roy, S.; Llinás, R. Relevance of Quantum Mechanics on some Aspects of Ion Channel Function. *C. R. Biol.* **2009**, *332*(6), 517–522. https://doi.org/10.1016/j.crvi.2008.11.009.
126. Katchalsky, A. Carriers and Specificity in Membranes VI. Biological flow structure and their relation to chemodiffusion coupling. *Neurosci. Res. Program Bull.* **1971**, *9*(3), 397–413.
127. Katchalsky, A. Polyelectrolytes and their Biological Interactions. *Biophys. J.* **1971**, *4*(1), 9–41.
128. Baars, B. *A Cognitive Theory of Consciousness*; Cambridge University Press: New York, 1988.
129. Edwards, J. C. W. Is Consciousness Only a Property of Individual Cells? *J. Conscious. Stud.* **2005**, *12*, 60–76.
130. Aerts, D.; Broekaert, J.; Gabora, L.; Sozzo, S. Generalizing Prototype Theory: A Formal Quantum Framework. *Front. Psychol.* **2016**, 7, 418. https://doi.org/10.3389/fpsyg.2016.00418.
131. Gabora, L.; Kitto, K. Towards a Quantum Theory of Humour. *Front. Phys.* **2017**, *4*(53). https://doi.org/10.3389/fphy.2016.00053.

CHAPTER EIGHT

Quantum transport and utilization of free energy in protein α-helices

Danko D. Georgiev[a,*] and James F. Glazebrook[b]
[a]Institute for Advanced Study, Varna, Bulgaria
[b]Department of Mathematics and Computer Science, Eastern Illinois University, Charleston, IL, United States
*Corresponding author: e-mail address: danko.georgiev@mail.bg

Contents

Abstract

The essential biological processes that sustain life are catalyzed by protein nanoengines, which maintain living systems in far-from-equilibrium ordered states. To investigate energetic processes in proteins, we have analyzed the system of generalized Davydov equations that govern the quantum dynamics of multiple amide I exciton quanta propagating along the hydrogen-bonded peptide groups in α-helices. Computational simulations have confirmed the generation of moving Davydov solitons by applied pulses of amide I energy for protein α-helices of varying length. The stability and mobility of these solitons depended on the uniformity of dipole–dipole coupling between amide I oscillators, and the isotropy of the exciton–phonon interaction. Davydov solitons were also able to quantum tunnel through massive barriers, or to quantum interfere at collision sites. The results presented here support a nontrivial role of quantum effects in biological systems that lies beyond the mechanistic support of covalent bonds as binding agents of macromolecular structures. Quantum tunneling and interference of Davydov solitons provide catalytically active macromolecular protein complexes with a physical mechanism allowing highly efficient transport, delivery, and utilization of free energy, besides the evolutionary mandate of biological order that supports the existence of such genuine quantum phenomena, and may indeed demarcate the quantum boundaries of life.

Advances in Quantum Chemistry, Volume 82
ISSN 0065-3276
https://doi.org/10.1016/bs.aiq.2020.02.001

1. Introduction

Living organisms are considered as open physical systems which utilize the availability of free energy to maintain homeostasis, respond to stimuli, adapt to their environment, grow, reproduce, and to evolve.[1–3] All of these biological functions are implemented by the large-scale participation and interaction of proteins. The high versatility of protein functions is achieved by linear polymerization of 20 different standard amino acids into polypeptide chains[4] (noteworthy, nonstandard amino acids also exist, cf. Ref. 5). The linear sequence of amino acids in polypeptides is often referred to as the primary structure. The primary structure folds into two main types of hydrogen-bonded secondary structures, α-helices and β-sheets.[6,7] Other, special helical secondary structures, such as the 3_{10}-helix and the π-helix, are also found in proteins (for review of their functional importance see Refs. 8, 9). The ensuing organization of secondary structures through hydrogen bonding, ionic bonding, dipole–dipole interactions, London dispersion forces, or through covalent disulfide bonds, provides each protein with its own tertiary structure that is uniquely shaped. Further quaternary assembly of multiple protein subunits provides the means for evolutionary design of nano-engines equipped with multiple active sites, including an adenosine triphosphate (ATP) hydrolytic site for release of free energy, an active catalytic site for converting the released energy into biologically useful work, and a number of allosteric sites for regulation of protein activity. This permits utilizing biochemical energy stored in high-energy ATP pyrophosphate bonds to fuel bioprocesses such as protein-assisted directed motion, synthesis of biomolecules, or for transporting biochemical substances across lipid membranes.

The primary importance of proteins for life is reflected in their name: the term "protein" comes from the Greek words "protos," meaning "first," or "proteos," meaning "first of all." In addition to being major catalysts in biological processes, proteins may indeed have been the constituents of the very first physical systems of life. While in modern organisms genetic information for protein production is stored in nucleic acids, recent stochastic simulations support the protein foldamer hypothesis for abiotic origin of life.[10] In particular, short protein sequences composed of hydrophobic and polar amino acids were found to collapse into relatively compact structures with exposed hydrophobic surfaces, which in turn catalyze the elongation of other such hydrophobic/polar protein polymers.[10] Once the first living replicators were stochastically assembled, natural selection would have kicked in and selected those replicators that exploit available energy sources in the most efficient way for the purposes of reproduction. Eventually, after 3.5 billion years of evolution,[11] modern day organisms are capable of the effective use of energy as released by single ATP molecules so as to execute highly specialized biological processes, including single steps of the kinesin motor on cytoskeletal protein railways,[12,13] or phosphorylation of single amino acid residues in voltage-gated ion channels for modulation of the channel electric conductance.[14,15]

In living systems, energy is transferred in minute quantities because higher energy densities are detrimental to the delicate and fragile structures.[16] While the behavior of all molecules is fundamentally described by quantum mechanics, the highly efficient utilization of energy by proteins suggests that quantum effects may play a nontrivial role, and one that lies beyond the deterministic, mechanistic structural support of covalent bonds that bind macromolecules together.[17–24] In this present paper we undertake studying the quantum dynamics resulting from the generalized Davydov Hamiltonian for the transport of energy by multiple amide I quanta in protein α-helices. We show that the resulting system of differential equations admits soliton solutions, for which the corresponding waveforms preserve their shapes during propagation, reflect from protein ends, tunnel through massive barriers, and interfere quantum mechanically at collision points to produce sharply focused peaks of concentrated energy. These instances of quantum phenomena appear to be instrumental in delivering biochemical energy to active catalytic sites, and moreover, are expected to be indispensable for the theoretical basis supporting the quantum boundaries of life. Indeed, in making this contribution to the currently burgeoning science of quantum biology, we are indebted to Schrödinger's astonishing insight which predicted quantum mechanical effects as ubiquitous within living systems.[25] Presently there is evidence of these effects occurring in several cases: photosynthesis (coherence), avian navigation (entanglement), olfaction (tunneling), the kinetic isotope effect in enzymatic reactions (tunneling), as particular instances (reviewed in Refs. 26, 27; see also earlier works such as Refs. 28–31).

2. Protein α-helix structure and infrared spectra

Protein polypeptide chains are linear polymers that are assembled from a repertoire of 20 different standard amino acids joined together through peptide bonds from N-terminus to C-terminus. The identity of each amino acid is determined by its side chain, known as an R group. The chemical structure of a generic tripeptide is shown below.

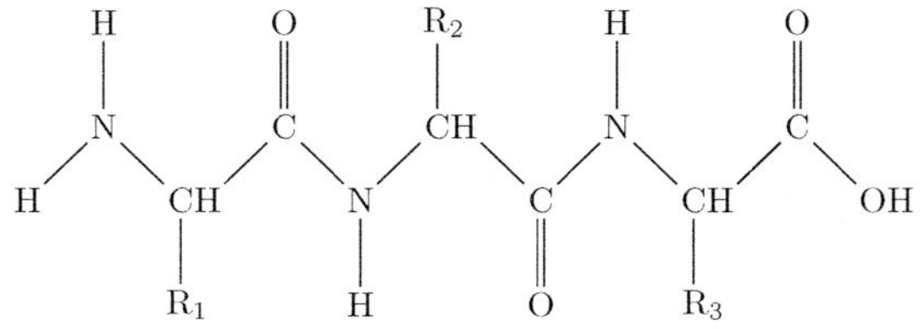

```
       H          O         R2         H          O
       |          ||        |          |          ||
       N          C         CH         N          C
     /   \      /   \     /    \     /   \      /   \
    H     CH          N            C        CH        OH
          |           |            ||       |
          R1          H            O        R3
```

For the construction of protein molecular machines, the flexibility of their polypeptide chains is essential.[32] This allows for organizing the polypeptide primary structure into ordered secondary structural elements such as the protein α-helix. Linus Pauling's prediction of the protein α-helix in 1951 is one of the greatest achievements in structural biology.[33] For his accomplishments in revealing the three-dimensional geometry of protein secondary structural elements, Pauling was awarded the 1954 Nobel Prize in Chemistry. Besides revealing the protein mysteries of life, it is perhaps not surprising that this distinguished fellow was also actively involved in the preservation of life itself, which subsequently earned him the Nobel Peace Prize in 1962, thus making Pauling the only person to have been awarded two unshared Nobel Prizes.[34]

Protein α-helices are right-handed spirals with 3.6 amino acids per turn, in which the N–H group of an amino acid is hydrogen-bonded with the C=O group of the amino acid that appears four residues earlier in the polypeptide chain (Fig. 1). In fact, for α-helices there is reliable evidence for the effect of hydrogen bonding on vibrational frequency.[35] The helical structure is supported by three chains of hydrogen-bonded peptide groups $\cdots$ H–N–C=O$\cdots$ H–N–C=O$\cdots$ referred to as α-helix spines. The typical bond lengths in the peptide group are as follows: N–H bond length is 101 pm, C–N bond length is 132 pm, and C=O bond length is 123 pm.[6] Because of the resonance between C=O and C–N bonds, their bond lengths are intermediate between a single and a double bond of the corresponding atoms, and the peptide bonds acquire a planar geometry. Full quantum simulations of electron molecular orbitals in the protein α-helix using Kohn–Sham density functional theory[36] are within the capabilities of modern quantum chemistry software applications, but this task would require thousands of hours of running time on a supercomputer.[37,38] Here, we will mainly rely on experimental data obtained from X-ray crystallography, or from the infrared spectroscopy of proteins.

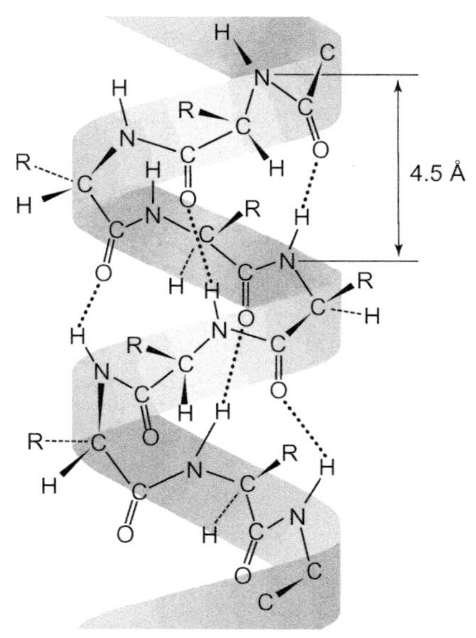

Fig. 1 A protein α-helix fragment with 3.6 amino acids per turn. The helical structure is supported by three chains of hydrogen-bonded peptide groups ··· H–N–C=O··· H–N–C=O··· referred to as α-helix spines. *Modified from Georgiev, D. D.; Glazebrook, J. F. On the Quantum Dynamics of Davydov Solitons in Protein α-Helices. Physica A 2019, 517, 257–269, https://doi.org/10.1016/j.physa.2018.11.026.*

Generally, hydrogen bonding is regarded as significantly instrumental for stabilizing the secondary structure of proteins in two ways: (i) through lowering the frequency of stretching vibrations by reducing restoration forces, and (ii) by increasing the frequency of bending through increased restoration.[35,39] Under these circumstances, hydrogen bonding can impact the amide functional class by stabilizing its $[^{-}O\text{–}C{=}N\text{–}H^{+}]$ structure over its [O=C–N–H] structure. Significantly, such bonding reliance can be influenced by the form of amide stretch vibrations. Amide I vibrations mainly arise from the C=O stretching vibrations, while supplemented by minor effects elsewhere.[35] As an example, Myshakina et al.[40] demonstrated that the frequency shifts of amide I and amide III bands (see below) may function as significant regulators for hydrogen bonding at the C=O and N–H sites of certain peptide bonds. Clearly, it is most significant that the proton in such hydrogen bonded systems is a quantum entity. Note that quantum nuclear effects may weaken relatively weak hydrogen bonds, while in contrast they may actually fortify the relatively strong ones.[41] Relevant in this case is the induced fit method for quantum H-bonding, which supports the molecular interactions for inducing conformational transitions in the binding sites of classes of enzymes.[42] In such a molecular recognition event, the dynamics of tunneling of electrons of proton-acceptor atoms or protons of hydrogen atoms is also significantly instrumental (see below). The tunneling of electrons of proton-acceptor atoms or protons of the hydrogen atoms generating quantum correlations have been studied in Refs. 42, 43. Further, in the thermal states of hydrogen bonds, substantial tunneling assisted quantum entanglement can be detected. In particular, if covalent bonding accompanies ionic associates, as is the case for a covalent bond created between electronegative and hydrogen atoms, then quantum entanglement may be subsequently hypothesized for various instances of ligand binding.[43]

Fourier-transform infrared (FTIR) spectroscopy allows for experimental measurement, and for plotting the absorption of infrared light by sample material versus the wavelength of the absorbed light. The application of FTIR spectroscopy to proteins has revealed several absorption bands that correspond to polypeptide backbone vibrations.[35,44–46]

Amide A. Among all absorption bands, most energetic is the amide A band near 3300 cm^{-1} (0.41 eV), which is due to >95% N–H stretch. The amount of this energy correspond exactly to the free energy released from a single ATP molecule. As a result of its exclusive localization on the N–H group, however, the amide A band in proteins appears to be insensitive to the secondary structure of the polypeptide backbone.[35]

Amide I. Particularly sensitive to the protein secondary structure is the amide I band near 1650 cm^{-1} (0.2 eV), which is due to 70%–85% C=O stretch and 10%–20% C–N stretch.[44] The free energy released from a single ATP molecule is sufficient to excite two amide I quanta. Importantly, resonance interaction can occur between two C=O oscillators when one of them is in an excited state. For distances over 300 pm, the main contribution to the interaction energy is due to transition dipole coupling.[44] The fundamental mechanism that renders the amide I vibration sensitive to secondary structure is the transition dipole coupling, because the coupling between the oscillating dipoles of neighboring amide groups depends upon their relative orientation and their distance.[35] The energy absorbed by a given C=O oscillator is readily transferred to nearby oscillators, which leads to delocalized excited states.[45]

Amide II. Also sensitive to the secondary structure of proteins, albeit in a less straightforward way, is the amide II band near 1550 cm^{-1} (0.19 eV), which is due to 40%–60% N–H bend, 18%–40% C–N stretch, and 10% C–C stretch.

Amide III. With lowest energy is the amide III band near 1300 cm^{-1} (0.16 eV), which is due to in-phase combination of 40% C–N stretch and 30% N–H bend.

As an experimental technology, FTIR spectroscopy can be applied to monitor protein structure in the liquid and dried (lyophilized) state. However, water can interfere with FTIR measurements of protein samples because it is strongly absorbent in the amide I region (how water is distinguished and interpreted in protein chemistry is surveyed in Refs. 47, 48). Consequently, FTIR spectroscopy is best suited for lyophilized (freeze-dried) protein samples. Measurements can also be obtained for protein samples in solution, but a high (>3 mg/ml) protein concentration is required.[49] We also mention that femtosecond infrared pump-probe spectroscopy has proven to be highly effective for analyzing the amide I band in relationship to the N–H vibrations.[50–53]

3. The generalized Davydov model of protein α-helices

Based on the strong dependence of amide I energy on the protein secondary structure, Alexander Davydov developed a quantum model for the transport of energy in terms of quasiparticles, here referred to as "solitons,"[55–61] where the full atomic complexity of the protein α-helix was reduced to the nonlinear interaction between amide I vibrations (excitons), and deformations (phonons) of the lattice of hydrogen bonds that stabilize the helical structure.[62–81]

In this work, we model only a single hydrogen-bonded spine in the protein α-helix, rather than the 3-spine structure of the entire helix. In realistic protein α-helices, the quantum dynamics of amide I energy in the 3-spine structure will also depend on inter-spine interactions and may give rise to complicated multihump solitons,[82,83] which the current study does not consider. For a single α-helix spine of hydrogen-bonded peptide groups, the generalized Davydov Hamiltonian is a sum of three parts

$$\hat{H} = \hat{H}_{\text{ex}} + \hat{H}_{\text{ph}} + \hat{H}_{\text{int}} \tag{1}$$

respectively, for amide I excitons $\hat{H}_{\text{ex}}$, hydrogen-bonded lattice phonons $\hat{H}_{\text{ph}}$, and exciton–phonon interaction $\hat{H}_{\text{int}}$. The three parts of the Hamiltonian include only nearest neighbor interactions, and they are formally similar to those featuring in Holstein polaron theory[70,84,85]

$$\hat{H}_{\mathrm{ex}} = \sum_n \left[E_0 \hat{a}_n^\dagger \hat{a}_n - J_{n+1} \hat{a}_n^\dagger \hat{a}_{n+1} - J_n \hat{a}_n^\dagger \hat{a}_{n-1} \right] \tag{2}$$

$$\hat{H}_{\mathrm{ph}} = \frac{1}{2} \sum_n \left[\frac{\hat{p}_n^2}{M_n} + w(\hat{u}_{n+1} - \hat{u}_n)^2 \right] \tag{3}$$

$$\hat{H}_{\mathrm{int}} = \chi_r \sum_n (\hat{u}_{n+1} + (\xi - 1)\hat{u}_n - \xi \hat{u}_{n-1}) \hat{a}_n^\dagger \hat{a}_n \tag{4}$$

where the index n counts the peptide groups along the α-helix spine, $\hat{a}_n^\dagger$ and $\hat{a}_n$ are the boson creation and annihilation operators for the amide I excitons, E_0 is the amide I exciton energy, J_n is the dipole–dipole coupling energy between the nth and $(n - 1)$th amide I oscillator along the spine, $\hat{p}_n$ is the momentum operator, $\hat{u}_n$ is the displacement operator from the equilibrium position of the peptide group n, M_n is the mass of the peptide group n, w is the spring constant of the hydrogen bonds in the lattice,[56,57,78] χ_r and χ_l are anharmonic parameters arising from the coupling between the amide I exciton and the phonon lattice displacements, respectively, to the right or to the left, $\overline{\chi} = \frac{\chi_r + \chi_l}{2}$, and $\xi = \frac{\chi_l}{\chi_r}$ is the anisotropy parameter of the exciton–phonon interaction (by construction $\chi_r \neq 0$, and $0 \leq \chi_l \leq \chi_r$ so that ξ varies in the interval $[0, 1]$).[54,80,86]

The quantum equations of motion for multiquanta states of amide I energy can be derived from the Hamiltonian (1) with the use of the following generalized ansatz state vector[74]:

$$|\Psi(t)\rangle = |a(t)\rangle |b(t)\rangle = \frac{1}{\sqrt{Q!}} \left[\sum_n a_n(t) \hat{a}_n^\dagger \right]^Q |0_{\mathrm{ex}}\rangle e^{-\frac{i}{\hbar} \sum_j \left(b_j(t)\hat{p}_j - c_j(t)\hat{u}_j \right)} |0_{\mathrm{ph}}\rangle \tag{5}$$

where

$$|a(t)\rangle = \frac{1}{\sqrt{Q!}} \left[\sum_n a_n(t) \hat{a}_n^\dagger \right]^Q |0_{\mathrm{ex}}\rangle \tag{6}$$

$$|b(t)\rangle = e^{-\frac{i}{\hbar} \sum_j \left(b_j(t)\hat{p}_j - c_j(t)\hat{u}_j \right)} |0_{\mathrm{ph}}\rangle \tag{7}$$

For ease of notation, the time dependence of $a_n(t)$, $b_n(t)$, $c_n(t)$, $|\Psi(t)\rangle$, $|a(t)\rangle$ and $|b(t)\rangle$, will henceforth be implicitly understood. Taking the inner product of this generalized ansatz with itself gives

$$\langle \Psi | \Psi \rangle = \left[\sum_n |a_n|^2 \right]^Q \tag{8}$$

which implies

$$\sum_n |a_n|^2 = 1 \tag{9}$$

Verification of (8) requires the multinomial theorem[87]

$$\left[\sum_n a_n \right]^Q = \sum_{k_1+k_2+\cdots+k_n=Q} \binom{Q}{k_1, k_2, \ldots, k_n} a_1^{k_1} a_2^{k_2} \ldots a_n^{k_n} \tag{10}$$

where the multinomial coefficient is

$$\binom{Q}{k_1, k_2, \ldots, k_n} = \frac{Q!}{k_1!k_2!\ldots k_n!} \tag{11}$$

with $k_1 + k_2 + \cdots + k_n = Q$. Then the inner product of the ansatz is

$$\langle \Psi | \Psi \rangle = \frac{1}{Q!} \left\langle 0_{\text{ex}} \middle| \left[\sum_{n'} a_{n'}^*(t) \hat{a}_{n'} \right]^Q \left[\sum_n a_n(t) \hat{a}_n^\dagger \right]^Q \middle| 0_{\text{ex}} \right\rangle \langle b | b \rangle \tag{12}$$

Taking into account that a nonzero result is only possible for creation and annihilation operators with same powers at a given site, together with $\langle b|b \rangle = 1$ and

$$\left\langle 0_{\text{ex}} \right| \left(a_n^* \hat{a}_n \right)^{k_n} \left(a_n \hat{a}_n^\dagger \right)^{k_n} \left| 0_{\text{ex}} \right\rangle = \left(|a_n|^2 \right)^{k_n} k_n! \tag{13}$$

we obtain

$$\begin{aligned}
\langle \Psi | \Psi \rangle &= \frac{1}{Q!} \sum_{k_1+k_2+\cdots+k_n=Q} \left(\frac{Q!}{k_1!k_2!\ldots k_n!} \right)^2 \left(|a_1|^2\right)^{k_1} \left(|a_2|^2\right)^{k_2} \ldots \left(|a_n|^2\right)^{k_n} \\
&\quad \times \langle 0_{\text{ex}} | \hat{a}_1^{k_1} \hat{a}_2^{k_2} \ldots \hat{a}_n^{k_n} \left(\hat{a}_1^\dagger\right)^{k_1} \left(\hat{a}_2^\dagger\right)^{k_2} \ldots \left(\hat{a}_n^\dagger\right)^{k_n} | 0_{\text{ex}} \rangle \\
&= \sum_{k_1+k_2+\cdots+k_n=Q} \frac{Q!}{k_1!k_2!\ldots k_n!} \left(|a_1|^2\right)^{k_1} \left(|a_2|^2\right)^{k_2} \ldots \left(|a_n|^2\right)^{k_n} = \left[\sum_n |a_n|^2 \right]^Q
\end{aligned}$$

For $Q = 1$, the generalized ansatz $|\Psi(t)\rangle$ reduces to Davydov's original $|D_2(t)\rangle$ ansatz,[58,81] which was studied extensively in several previous works.[54,73,86]

Given that the ansatz (5) is a suitable approximation to the exact solution of the Schrödinger equation,[88] we obtain

$$\imath\hbar\frac{d}{dt}|\Psi\rangle = \hat{H}|\Psi\rangle \tag{14}$$

For a given peptide group n, the expectation values for the exciton number operator $\hat{N}_n = \hat{a}_n^\dagger\hat{a}_n$, phonon displacement operator $\hat{u}_n$ and phonon momentum operator $\hat{p}_n$ are

$$\langle\hat{N}_n\rangle = \langle\Psi|\hat{N}_n|\Psi\rangle = Q\,|a_n|^2 \tag{15}$$

$$\langle\hat{u}_n\rangle = \langle\Psi|\hat{u}_n|\Psi\rangle = b_n \tag{16}$$

$$\langle\hat{p}_n\rangle = \langle\Psi|\hat{p}_n|\Psi\rangle = c_n \tag{17}$$

The verification of (15) requires the use of

$$\langle 0_{\text{ex}}|\left(a_n^*\hat{a}_n\right)^{k_n}\hat{a}_n^\dagger\hat{a}_n\left(a_n\hat{a}_n^\dagger\right)^{k_n}|0_{\text{ex}}\rangle = \left(|a_n|^2\right)^{k_n}k_nk_n! \tag{18}$$

Substitution of the ansatz (5) in $\langle\Psi|\hat{N}_n|\Psi\rangle$, and using the multinomial theorem (10), gives

$$\begin{aligned}
\langle\Psi|\hat{N}_n|\Psi\rangle &= \frac{1}{Q!}\sum_{k_1+k_2+\cdots+k_n=Q}\left(\frac{Q!}{k_1!k_2!\ldots k_n!}\right)^2\left(|a_1|^2\right)^{k_1}\left(|a_2|^2\right)^{k_2}\ldots\left(|a_n|^2\right)^{k_n}k_nk_1!k_2!\ldots k_n! \\
&= \sum_{k_1+k_2+\cdots+(k_n-1)=Q-1}\frac{Q(Q-1)!}{k_1!k_2!\ldots(k_n-1)!}\left(|a_1|^2\right)^{k_1}\left(|a_2|^2\right)^{k_2}\ldots\left(|a_n|^2\right)^{(k_n-1+1)} \\
&= Q|a_n|^2\left[\sum_{n'}|a_{n'}|^2\right]^{Q-1} = Q|a_n|^2
\end{aligned}$$

Verification of (16) requires the result that the expectation values of the position and momentum operators of peptide groups with the vacuum are zero, $\langle 0_{\text{ph}}|\hat{u}_n|0_{\text{ph}}\rangle = 0$ and $\langle 0_{\text{ph}}|\hat{p}_n|0_{\text{ph}}\rangle = 0$, together with the Hadamard lemma:

$$\begin{aligned}
e^{\hat{A}}\hat{B}e^{-\hat{A}} &= \exp\left(\text{ad}_{\hat{A}}\right)\left(\hat{B}\right) = \sum_{k=0}^{\infty}\frac{1}{k!}\left(\text{ad}_{\hat{A}}\right)^k\left(\hat{B}\right) \\
&= \hat{B} + \left[\hat{A},\hat{B}\right] + \frac{1}{2!}\left[\hat{A},\left[\hat{A},\hat{B}\right]\right] + \frac{1}{3!}\left[\hat{A},\left[\hat{A},\left[\hat{A},\hat{B}\right]\right]\right] + \cdots
\end{aligned} \tag{19}$$

where $\text{ad}_{\hat{A}}(\hat{B}) \equiv \left[\hat{A},\hat{B}\right]$ is the adjoint operator. Substitution of the ansatz (5) in $\langle\Psi|\hat{u}_n|\Psi\rangle$, and application of the Hadamard lemma (19) with the standard quantum commutation relations $[\hat{u}_n,\hat{p}_n] = \imath\hbar$ and $[\hat{p}_n,\hat{u}_n] = -\imath\hbar$, gives

$$
\begin{aligned}
\langle \Psi|\hat{u}_n|\Psi\rangle &= \langle a|a\rangle \left\langle 0_{\text{ph}}\left| e^{\frac{\imath}{\hbar}\sum_j \left(b_j\hat{p}_j - c_j\hat{u}_j\right)} \hat{u}_n e^{-\frac{\imath}{\hbar}\sum_j \left(b_j\hat{p}_j - c_j\hat{u}_j\right)} \right| 0_{\text{ph}}\right\rangle \\
&= \left\langle 0_{\text{ph}}\left| e^{\frac{\imath}{\hbar}(b_n\hat{p}_n - c_n\hat{u}_n)} e^{\frac{\imath}{\hbar}\sum_{j\neq n}\left(b_j\hat{p}_j - c_j\hat{u}_j\right)} \hat{u}_n e^{-\frac{\imath}{\hbar}\sum_{j\neq n}\left(b_j\hat{p}_j - c_j\hat{u}_j\right)} e^{-\frac{\imath}{\hbar}(b_n\hat{p}_n - c_n\hat{u}_n)} \right| 0_{\text{ph}}\right\rangle \\
&= \left\langle 0_{\text{ph}}\left| e^{\frac{\imath}{\hbar}(b_n\hat{p}_n - c_n\hat{u}_n)} e^{\frac{\imath}{\hbar}\sum_{j\neq n}\left(b_j\hat{p}_j - c_j\hat{u}_j\right)} e^{-\frac{\imath}{\hbar}\sum_{j\neq n}\left(b_j\hat{p}_j - c_j\hat{u}_j\right)} \hat{u}_n e^{-\frac{\imath}{\hbar}(b_n\hat{p}_n - c_n\hat{u}_n)} \right| 0_{\text{ph}}\right\rangle \\
&= \left\langle 0_{\text{ph}}\left| e^{\frac{\imath}{\hbar}(b_n\hat{p}_n - c_n\hat{u}_n)} \hat{u}_n e^{-\frac{\imath}{\hbar}(b_n\hat{p}_n - c_n\hat{u}_n)} \right| 0_{\text{ph}}\right\rangle \\
&= \left\langle 0_{\text{ph}}\left| \hat{u}_n + \left[\frac{\imath}{\hbar}(b_n\hat{p}_n - c_n\hat{u}_n), \hat{u}_n\right] \right| 0_{\text{ph}}\right\rangle \\
&= \langle 0_{\text{ph}}|\hat{u}_n|0_{\text{ph}}\rangle + \langle 0_{\text{ph}}|b_n|0_{\text{ph}}\rangle = b_n
\end{aligned}
$$

where we used $\langle a|a\rangle = 1$ as implied from (8) and (9). Verification of (17) is analogous[89] (only the last two steps are shown):

$$
\begin{aligned}
\langle \Psi|\hat{p}_n|\Psi\rangle &= \left\langle 0_{\text{ph}}\left| \hat{p}_n + \left[\frac{\imath}{\hbar}(b_n\hat{p}_n - c_n\hat{u}_n), \hat{p}_n\right] \right| 0_{\text{ph}}\right\rangle \\
&= \langle 0_{\text{ph}}|\hat{p}_n|0_{\text{ph}}\rangle + \langle 0_{\text{ph}}|c_n|0_{\text{ph}}\rangle = c_n
\end{aligned}
$$

Identifying b_n and c_n as quantum expectation values is important for a further application of the Schrödinger equation and its complex conjugate in the form of the generalized Ehrenfest theorem,[54] which governs the quantum dynamics of the expectation values

$$
\frac{d}{dt}b_n = \frac{1}{\imath\hbar}\langle \Psi|\left[\hat{u}_n, \hat{H}\right]|\Psi\rangle \tag{20}
$$

$$
\frac{d}{dt}c_n = \frac{1}{\imath\hbar}\langle \Psi|\left[\hat{p}_n, \hat{H}\right]|\Psi\rangle \tag{21}
$$

From the Hamiltonian (1), together with $\left[\hat{u}_n, \hat{p}_n^2\right] = 2\imath\hbar\hat{p}_n$, and $\left[\hat{p}_n, \hat{u}_n^2\right] = -2\imath\hbar\hat{u}_n$, we obtain the two commutators to be

$$
\left[\hat{u}_n, \hat{H}\right] = \imath\hbar\frac{\hat{p}_n}{M_n} \tag{22}
$$

$$
\begin{aligned}
\left[\hat{p}_n, \hat{H}\right] &= \imath\hbar w(\hat{u}_{n-1} - 2\hat{u}_n + \hat{u}_{n+1}) \\
&\quad - \imath\hbar\chi_r\left(\hat{a}_{n-1}^\dagger\hat{a}_{n-1} + (\xi - 1)\hat{a}_n^\dagger\hat{a}_n - \xi\hat{a}_{n+1}^\dagger\hat{a}_{n+1}\right)
\end{aligned} \tag{23}
$$

Substitution in (20) and (21), followed by application of (15), (16) and (17), yields one of Davydov's equations

$$
\begin{aligned}
M_n\frac{d^2}{dt^2}b_n &= w(b_{n-1} - 2b_n + b_{n+1}) \\
&\quad - Q\chi_r\left(|a_{n-1}|^2 + (\xi - 1)|a_n|^2 - \xi|a_{n+1}|^2\right)
\end{aligned} \tag{24}
$$

The equation for the amide I probability amplitudes a_n can be derived by differentiating the $|\Psi\rangle$ ansatz using the product rule

$$\imath\hbar\frac{d}{dt}|\Psi\rangle = \left(\imath\hbar\frac{d}{dt}|a\rangle\right)|b\rangle+|a\rangle\left(\imath\hbar\frac{d}{dt}|b\rangle\right) \tag{25}$$

where (cf. Refs. 73, The equation for the amide I probability amplitudes 74)

$$\imath\hbar\frac{d}{dt}|a\rangle = \imath\hbar\sqrt{Q}\left(\sum_{n'}\frac{da_{n'}}{dt}\hat{a}^{\dagger}_{n'}\right)\frac{1}{\sqrt{(Q-1)!}}\left(\sum_{n}a_n\hat{a}^{\dagger}_n\right)^{Q-1}|0_{\text{ex}}\rangle \tag{26}$$

$$\imath\hbar\frac{d}{dt}|b\rangle = \sum_{n}\left[\frac{db_n}{dt}\hat{p}_n - \frac{dc_n}{dt}\hat{u}_n + \frac{1}{2}\left(b_n\frac{dc_n}{dt} - \frac{db_n}{dt}c_n\right)\right]|b\rangle \tag{27}$$

Next, we use the Schrödinger equation (14) and take the inner product with the state $\frac{1}{\sqrt{Q!}}\langle b|\langle 0_{\text{ex}}|(\hat{a}_n)^Q$ as follows:

$$\frac{1}{\sqrt{Q!}}\left\langle b|\langle 0_{\text{ex}}|(\hat{a}_n)^Q\imath\hbar\frac{d}{dt}|\Psi\right\rangle = \frac{1}{\sqrt{Q!}}\langle b|\langle 0_{\text{ex}}|(\hat{a}_n)^Q\hat{H}|\Psi\rangle \tag{28}$$

Again, a nonzero result is only possible if the creation and annihilation operators at a given site have the same power. Since the operators act at most on two different sites, the use of the multinomial theorem reduces to the binomial one (there are at most two sites with nonzero powers). Straightforward application of binomial coefficients $\binom{n}{k} = \frac{n!}{k!(n-k)!}$ allows the calculation of individual inner products

$$\begin{aligned}
\frac{1}{\sqrt{Q!}}\left\langle 0_{\text{ex}}\middle|(\hat{a}_n)^Q\frac{d}{dt}\middle|a\right\rangle &= \frac{1}{\sqrt{Q!}}\frac{Q}{\sqrt{Q}}\frac{da_n}{dt}\frac{1}{\sqrt{(Q-1)!}}\binom{Q-1}{Q-1}a_n^{Q-1}\left\langle 0_{\text{ex}}\middle|(\hat{a}_n)^Q(\hat{a}^{\dagger}_n)^Q\middle|0_{\text{ex}}\right\rangle\\
&= \frac{da_n}{dt}a_n^{Q-1}Q\\
\frac{1}{\sqrt{Q!}}\langle 0_{\text{ex}}|(\hat{a}_n)^Q|a\rangle &= \frac{1}{\sqrt{Q!}}\frac{1}{\sqrt{Q!}}\binom{Q}{Q}a_n^Q\left\langle 0_{\text{ex}}\middle|(\hat{a}_n)^Q(\hat{a}^{\dagger}_n)^Q\middle|0_{\text{ex}}\right\rangle\\
&= a_n^Q\\
\frac{1}{\sqrt{Q!}}\langle 0_{\text{ex}}|(\hat{a}_n)^Q\hat{a}^{\dagger}_n\hat{a}_n|a\rangle &= \frac{1}{\sqrt{Q!}}\frac{1}{\sqrt{Q!}}\binom{Q}{Q}a_n^Q\left\langle 0_{\text{ex}}\middle|(\hat{a}_n)^Q\hat{a}^{\dagger}_n\hat{a}_n(\hat{a}^{\dagger}_n)^Q\middle|0_{\text{ex}}\right\rangle\\
&= a_n^Q Q\\
\frac{1}{\sqrt{Q!}}\langle 0_{\text{ex}}|(\hat{a}_n)^Q\hat{a}^{\dagger}_n\hat{a}_{n\pm1}|a\rangle &= \frac{1}{\sqrt{Q!}}\frac{1}{\sqrt{Q!}}\binom{Q}{Q-1}a_{n\pm1}a_n^{Q-1}\left\langle 0_{\text{ex}}\middle|(\hat{a}_n)^Q(\hat{a}^{\dagger}_n)^Q\middle|0_{\text{ex}}\right\rangle\\
&= a_{n\pm1}a_n^{Q-1}Q
\end{aligned}$$

Substitution of (25) together with the Hamiltonian (1) into (28), together with the expectation values (16) and (17), after cancelation of common factor $a_n^{Q-1} Q$ yields Davydov's second equation

$$\imath\hbar\frac{d}{dt}a_n = \gamma(t)a_n - J_{n+1}a_{n+1} - J_n a_{n-1} + \chi_r\left(b_{n+1} + (\xi - 1)b_n - \xi b_{n-1}\right)a_n \tag{29}$$

where $\gamma(t)$ is a real-valued global term for all sites n given by

$$\gamma(t) = E_0 + \frac{1}{Q}\left[W(t) + \frac{1}{2}\sum_j\left(b_j\frac{dc_j}{dt} - \frac{db_j}{dt}c_j\right)\right] \tag{30}$$

The phonon energy $W(t)$ can be computed with the use of the Hadamard lemma (19) again utilizing the zero expectation values of the position and momentum operators of peptide groups with the vacuum, together with $c_j = M_j\frac{db_j}{dt}$, and the following commutators

$$\left[-\frac{\imath}{\hbar}c_j\hat{u}_j, \frac{1}{2}\frac{\hat{p}_j^2}{M_j}\right] = \frac{c_j}{M_j}\hat{p}_j$$

$$\left[-\frac{\imath}{\hbar}c_j\hat{u}_j, \frac{c_j}{M_j}\hat{p}_j\right] = \frac{c_j^2}{M_j}$$

$$\left[\frac{\imath}{\hbar}\left(b_j\hat{p}_j + b_{j+1}\hat{p}_{j+1}\right), \frac{1}{2}w\left(\hat{u}_{j+1}^2 - 2\hat{u}_j\hat{u}_{j+1} + \hat{u}_j^2\right)\right] = w\left(b_{j+1}\hat{u}_{j+1} - b_j\hat{u}_{j+1} - b_{j+1}\hat{u}_j + b_j\hat{u}_j\right)$$

$$\left[\frac{\imath}{\hbar}\left(b_j\hat{p}_j + b_{j+1}\hat{p}_{j+1}\right), w\left(b_{j+1}\hat{u}_{j+1} - b_j\hat{u}_{j+1} - b_{j+1}\hat{u}_j + b_j\hat{u}_j\right)\right] = w\left(b_{j+1}^2 - 2b_jb_{j+1} + b_j^2\right)$$

$$W(t) = \langle b|H_{\text{ph}}|b\rangle = W_0 + \frac{1}{2}\sum_j\left[M_j\left(\frac{db_j}{dt}\right)^2 + w\left(b_{j+1} - b_j\right)^2\right] \tag{31}$$

with $W_0 = \langle 0_{\text{ph}}|\hat{H}_{\text{ph}}|0_{\text{ph}}\rangle$ denoting the zero-point phonon energy.

In order to make further headway, we note that introduction of a global phase change on the exciton quantum probability amplitudes, namely $a_n \to \bar{a}_n e^{-\frac{\imath}{\hbar}\int\gamma(t)dt}$, will not change the quantum probabilities for finding the exciton at a given site

$$|a_n|^2 = e^{+\frac{\imath}{\hbar}\int\gamma(t)dt}\bar{a}_n^*\bar{a}_n e^{-\frac{\imath}{\hbar}\int\gamma(t)dt} \tag{32}$$

This is important because we will not need to transform the first Davydov Equation (24). After differentiation of the left-hand side of (29)

$$\imath\hbar\frac{d}{dt}\left(\bar{a}_n e^{-\frac{\imath}{\hbar}\int\gamma(t)dt}\right) = \imath\hbar e^{-\frac{\imath}{\hbar}\int\gamma(t)dt}\frac{d\bar{a}_n}{dt} + \bar{a}_n\gamma(t)e^{-\frac{\imath}{\hbar}\int\gamma(t)dt}$$

followed by cancelation of common terms and relabeling of $\bar{a}_n$ back to a_n, we obtain the following system of gauge transformed quantum equations of motion

$$\imath\hbar\frac{d}{dt}a_n = -J_{n+1}a_{n+1} - J_n a_{n-1} + \chi_r[b_{n+1} + (\xi - 1)b_n - \xi b_{n-1}]a_n \quad (33)$$

$$M_n\frac{d^2}{dt^2}b_n = w(b_{n-1} - 2b_n + b_{n+1}) - Q\chi_r\left(|a_{n-1}|^2 + (\xi - 1)|a_n|^2 - \xi|a_{n+1}|^2\right) \quad (34)$$

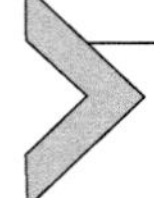

4. A computational study

4.1 Model parameters

The system of generalized Davydov equations (33) and (34) for multiple amide I quanta expands the scope of the original Davydov model (cf. Refs. 54, 56–64, 66–68, 73, 74, 86, 90–92) to the extent that it is now capable of answering a number of questions in regard to the quantum boundaries of life. In particular, anisotropy of the exciton–phonon coupling for different values of the parameter ξ, nonuniformity of amino acid masses M_n, and nonuniformity of amide I dipole–dipole coupling energies J_n could be directly incorporated in computer simulations and the resulting effects on quantum dynamics could be theoretically characterized. In order to improve on the comparison with a number of previous studies,[54,69,75–77,86] we here apply the following basic model parameters: spring constant of the hydrogen bonds in the lattice $w = 13$ N/m,[93] anharmonic parameter for the exciton–phonon coupling $\bar{\chi} = 35$ pN,[54] average mass of an amino acid inside the protein α-helix $M = 1.9 \times 10^{-25}$ kg,[69] average amide I dipole–dipole coupling energy $J = 0.155$ zJ,[94] initially unperturbed lattice of hydrogen bonds, and Q nonoverlapping Gaussian pulses of amide I energy each spread over five peptide groups with quantum probability amplitudes a_n given by $\sqrt{\frac{1}{Q}}\{\sqrt{0.099}, \sqrt{0.24}, \sqrt{0.322}, \sqrt{0.24}, \sqrt{0.099}\}$. For simulations with a double soliton, we used $Q - 1$ nonoverlapping Gaussian pulses, of which the double soliton had a factor of $\sqrt{\frac{2}{Q}}$ whereas the single solitons had a factor of $\sqrt{\frac{1}{Q}}$. For most of the simulations, we used an α-helix spine with length of $n_{\max} = 40$ peptide groups, which covers a distance of 18 nm.

4.2 Solitons in short protein α-helices

Davydov's original analysis aimed to establish the existence of soliton (quasiparticle) solutions via derivation of nonlinear Schrödinger equation (NLSE). To achieve that goal, the exciton–phonon interaction was taken to be completely isotropic $\xi = 1$, and all J_n and M_n were uniformly replaced with the corresponding average values J and M. Davydov further introduced the dimensionless variable $x = \frac{\tilde{x}}{a}$, where $\tilde{x}$ is the distance and a is the spacing between peptide groups.[56] He then approximated the system of discrete functions with continuous ones using the transformation $f(x \pm 1, t) \approx \left(1 \pm \frac{\partial}{\partial x} + \frac{1}{2}\frac{\partial^2}{\partial x^2}\right) f(x, t)$.[57] Lastly, the resulting system of partial differential equations (PDEs) was manipulated into NLSE, which is known to have analytic soliton solutions.[90] However, because a large number of mathematical assumptions enter at each step of Davydov's derivation, it is not known whether the soliton will persist in short protein α-helices in the presence of anisotropy and variability of various parameters.

In order to systematize the effect of the protein α-helix length upon these solitons, we have integrated numerically the system of Davydov equations with $Q = 1$ for completely isotropic exciton–phonon interaction $\xi = 1$ and uniform values J and M for all peptide groups n. The quantum dynamics of solitons launched from either the N-end or the C-end of the α-helix exhibited left–right mirror symmetry (Fig. 2). The solitons moved with velocity of 334 m/s, which is slower than 1 peptide group per picosecond (PG/ps), 1 PG/ps = 450 m/s. The solitons were capable of multiple reflections from the α-helix ends without disintegration. In fact, discreteness effects did not manifest even for ultrashort helices whose length was only $n_{\max} = 10$ peptide groups (Fig. 2C and F), even though the initial soliton width was already five peptide groups, which is half the length of the helix. This reveals that Davydov's intuition in regard to the approximations, was essentially correct; the continuum approximation is valid for $\xi = 1$ regardless of the actual protein α-helix length and the soliton width.

Next, we have integrated numerically the system of Davydov equations for completely anisotropic exciton–phonon interaction $\xi = 0$, which coincides with the model extensively studied by Scott and collaborators.[69,76–78] The speed of launched solitons was lower 218 m/s and wobbled in time

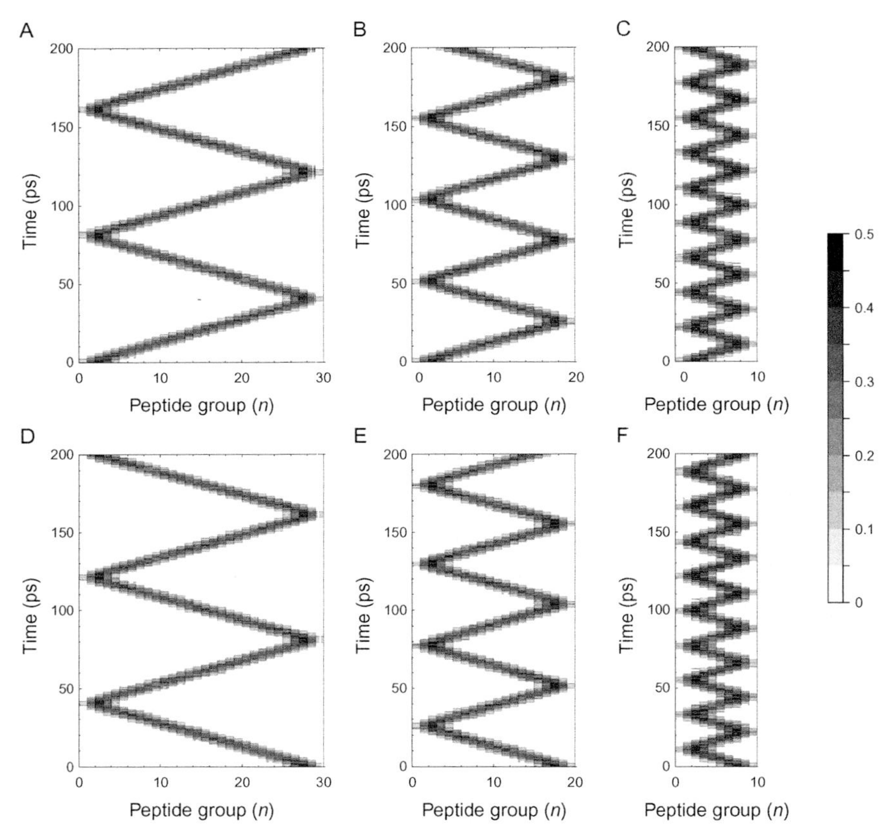

Fig. 2 The quantum dynamics of a moving Davydov soliton for completely isotropic exciton–phonon interaction $\xi = 1$, with $Q = 1$ quantum of amide I energy visualized through the expectation value of the exciton number operator $Q|a_n|^2$ at each peptide group n. (A)–(C) Gaussian pulse of amide I energy is applied at the N-end of a short protein α-helix spine with varying length, $n_{\max} = 30$ (A), $n_{\max} = 20$ (B), or $n_{\max} = 10$ (C). (D)–(F) Gaussian pulse of amide I energy is applied at the C-end of a short protein α-helix spine with varying length, $n_{\max} = 30$ (D), $n_{\max} = 20$ (E), or $n_{\max} = 10$ (F).

(Fig. 3). In fact, for ultrashort helices whose length was only $n_{\max} = 10$ peptide groups, the soliton reflected 6–7 times before settling in the middle of the helix within a timescale of 100 ps (Fig. 3C and F). Thus, the analytic soliton solution that moves at a constant speed fails to capture correctly the quantum dynamics of the system for $\xi = 0$.

In actual protein α-helices, the case of complete anisotropy of exciton–phonon interaction, $\xi = \frac{\chi_l}{\chi_r} = 0$, cannot be reached as this will require absolutely no interaction of the amide I oscillator with the hydrogen bond

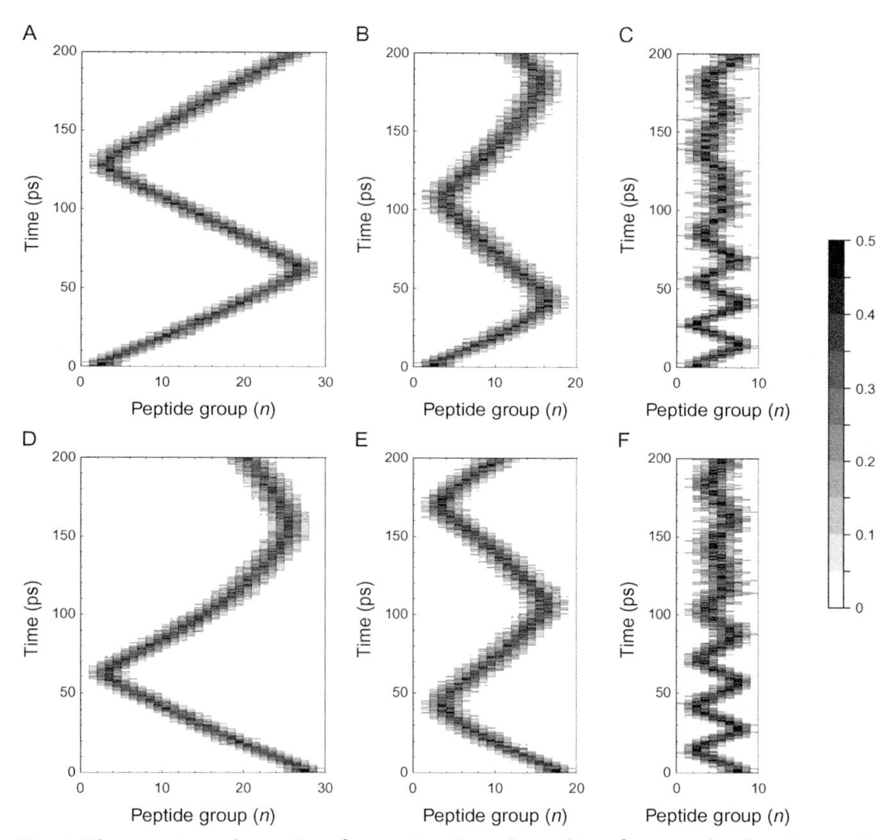

Fig. 3 The quantum dynamics of a moving Davydov soliton for completely anisotropic exciton–phonon interaction $\xi = 0$, with $Q = 1$ quantum of amide I energy visualized through the expectation value of the exciton number operator $Q|a_n|^2$ at each peptide group n. (A)–(C) The Gaussian pulse of amide I energy is applied at the N-end of a short protein α-helix spine with varying length, $n_{\max} = 30$ (A), $n_{\max} = 20$ (B) or $n_{\max} = 10$ (C). (D)–(F) The Gaussian pulse of amide I energy is applied at the C-end of a short protein α-helix spine with varying length, $n_{\max} = 30$ (D), $n_{\max} = 20$ (E), or $n_{\max} = 10$ (F).

to the left ($\chi_l = 0$). From a theoretical perspective, however, letting $\xi = 0$ provides a limiting case that may exhibit qualitatively different quantum behavior. Thus, varying ξ within the interval [0, 1] in systematic steps could provide a threshold for qualitative change in quantum dynamics. Interestingly, in the current context, such a threshold value exists and this could be set to $\xi = 0.02$. The presence of very slight isotropy of the exciton–phonon interaction $\xi = 0.02$ was able to restore the linear motion of the soliton increasing its speed to 241 m/s (Fig. 4). The soliton reflected

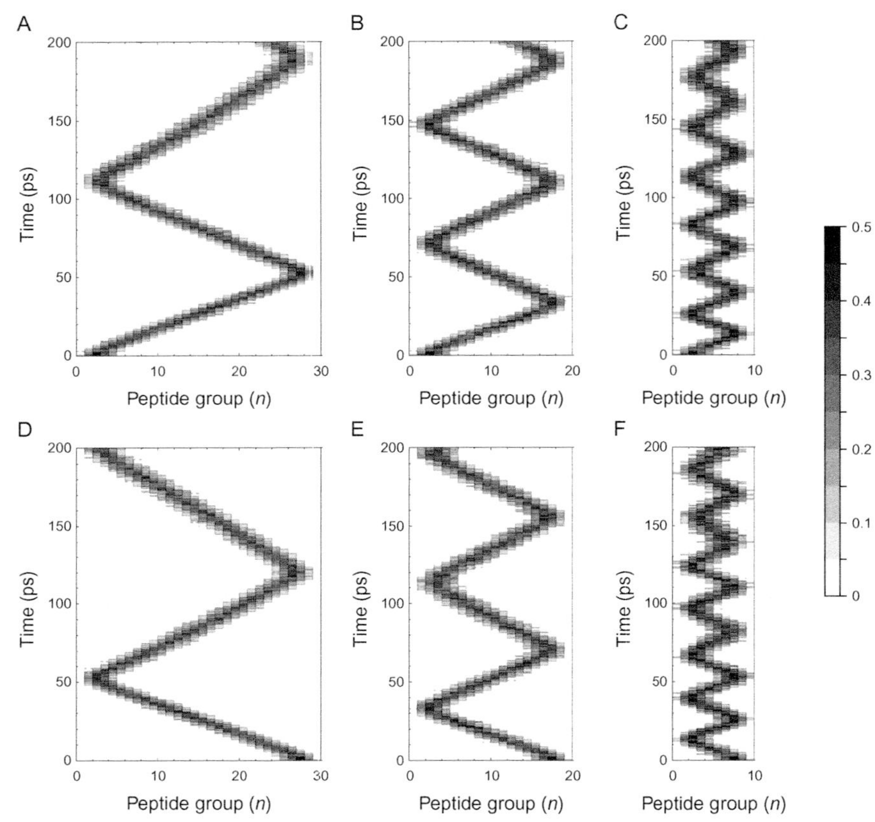

Fig. 4 The quantum dynamics of a moving Davydov soliton for very slightly isotropic exciton–phonon interaction $\xi = 0.02$, with $Q = 1$ quantum of amide I energy visualized through the expectation value of the exciton number operator $Q|a_n|^2$ at each peptide group n. (A)–(C) Gaussian pulse of amide I energy is applied at the N-end of a short protein α-helix spine with varying length, $n_{\max} = 30$ (A), $n_{\max} = 20$ (B) or $n_{\max} = 10$ (C). (D)–(F) Gaussian pulse of amide I energy is applied at the C-end of a short protein α-helix spine with varying length, $n_{\max} = 30$ (D), $n_{\max} = 20$ (E), or $n_{\max} = 10$ (F).

repeatedly in ultrashort helices with $n_{\max} = 10$ peptide groups for the whole time period of 200 ps of the simulation (Fig. 4C and F), similarly to the completely isotropic case with $\xi = 1$. In other words, for realistic protein α-helices the main effect of the anisotropy of exciton–phonon interaction could be to reduce the velocity of propagation.

Collectively, the above results indicate that the Davydov solitons are robust quasiparticles with respect to lattice discreteness, and in turn, with respect to the total length of the protein α-helix. Thus, in the process of evolutionary design, and optimization of protein functions in general, it is physically plausible that the lengths of various protein α-helices in fact gradually evolved through steps of a single amino acid residue toward achieving a certain optimal length for the delivery of free energy at a desired active site. For the remainder of the simulations, in order to ensure a sufficiently large arena with high spatial resolution of multisoliton dynamics with $Q > 1$, we will consider a protein α-helix spine with a fixed length of $n_{\max} = 40$ peptide groups. This length has been also used in our previous works[54,86] and allows direct comparison of the newly obtained results with the data that has already been published.

4.3 Multiquanta solitons

Previously we have shown that for $Q = 1$, pulses of Gaussian amide I energy are able to launch traveling solitons when applied to the protein α-helix ends, but generate pinned solitons if the amide I energy is applied in the interior of the protein α-helix.[54] To study the collision of several Davydov solitons for $Q > 1$, here we have considered a number of different scenarios in which the anisotropy of exciton–phonon interaction was systematically varied from $\xi = 1$ to $\xi = 0$ in steps of 0.1. Because the quantum dynamics was affected nonlinearly with variation of ξ, we have presented four panels per simulation in such a way that the exhibited changes between any two consecutive panels are the most prominent.

4.3.1 Double solitons

The simplest extension of the $Q = 1$ case is to apply a single Gaussian of amide I energy over five peptide groups while exciting $Q = 2$ amide I quanta. As noted by Kerr and Lomdahl,[74] the multiquantum property of the ansatz state (5) results in a stronger driving force on the phonon modes (34), but no modification of the equation for the amide I exciton probability amplitudes (33). Thus, it might be expected that double solitons may exhibit features similar to those resulting from an increased exciton–phonon coupling $\bar{\chi}$; namely, greater soliton stability, lower soliton velocity, and assisted soliton pinning.[54] Effectively, double Davydov solitons with $Q = 2$ launched from the N-end of the protein α-helix for completely isotropic exciton–phonon interaction $\xi = 1$ moved at lower velocity of 147 m/s (Fig. 5A) compared with velocity of 334 m/s for $Q = 1$ (Fig. 2A). Similarly, for completely anisotropic exciton–phonon interaction $\xi = 0$ the velocity of the soliton with $Q = 2$ was also lower, in fact zero due to soliton pinning (Fig. 5D), compared with velocity of 218 m/s for $Q = 1$

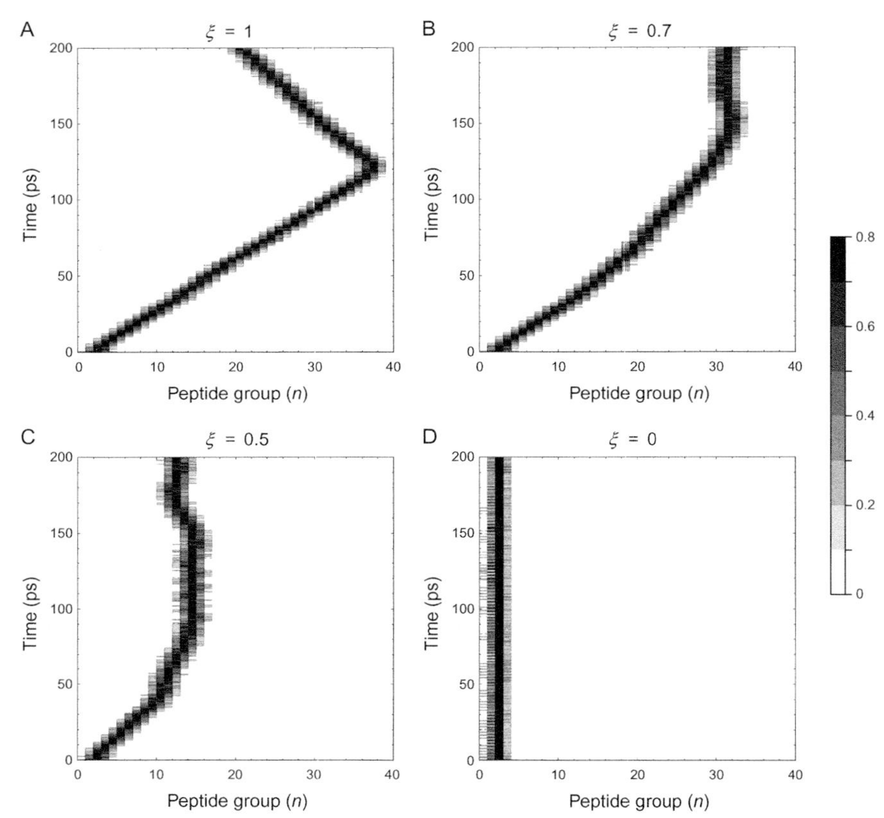

Fig. 5 The quantum dynamics of a double Davydov soliton with $Q = 2$ quanta of amide I energy delivered as a single Gaussian pulse over five peptide groups at the N-end of a protein α-helix visualized through the expectation value of the exciton number operator $Q|a_n|^2$. Decreasing the isotropy of exciton–phonon interaction leads to reduction in soliton velocity and eventual soliton pinning: $\xi = 1$ (A), $\xi = 0.7$ (B), $\xi = 0.5$ (C), and $\xi = 0$ (D).

(Fig. 3A). For intermediate values of ξ, the soliton migrated toward the interior of the protein α-helix where it started wobbling around some fixed interior point (Fig. 5B and C).

When the double soliton was generated in the interior of the protein α-helix, it was pinned and wobbled around its initial position (Fig. 6). When the isotropy of exciton–phonon interaction ξ was decreased, this led to reduction in the wobbling of the pinned soliton, which eventually came to a complete halt for $\xi = 0$ (Fig. 6D). Thus, increasing the number of amide I quanta increases the energy of the soliton, enhances soliton stability and in turn lowers the soliton propagation velocity.

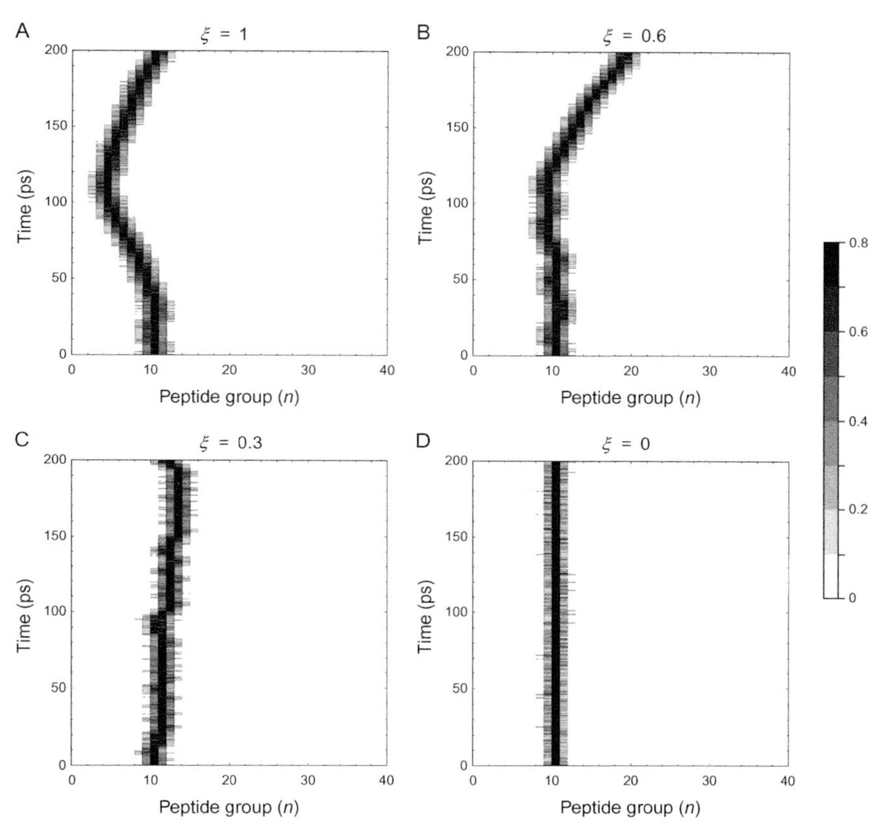

Fig. 6 The quantum dynamics of a double Davydov soliton with $Q = 2$ quanta of amide I energy delivered as a single Gaussian pulse over five peptide groups $n = 9 - 13$ in the interior of a protein α-helix visualized through the expectation value of the exciton number operator $Q|a_n|^2$. Decreasing the isotropy of exciton–phonon interaction leads to reduction in wobbling of the pinned soliton: $\xi = 1$ (A), $\xi = 0.6$ (B), $\xi = 0.3$ (C), and $\xi = 0$ (D).

4.3.2 Two-soliton collisions

The presence of multiple amide I quanta in the protein α-helix allows for the application of an initial multi-Gaussian distribution, which is a sum of several Gaussians, in order to generate several solitons whose eventual collision may lead to detecting either constructive, or destructive, quantum interference phenomena. This could be instrumental in how proteins utilize free energy for driving life-supporting bioprocesses.

The launching of two propagating solitons from the two ends of the protein α-helix, for $\xi = 1$, leads to significant constructive quantum interference

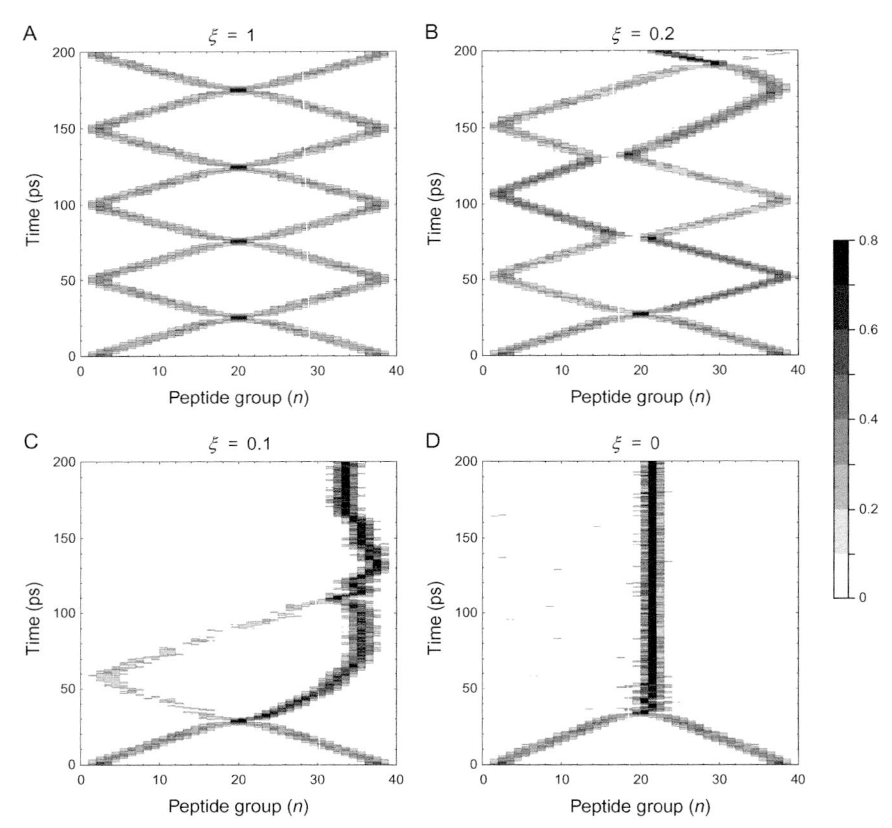

Fig. 7 The quantum collision of two moving Davydov solitons with $Q = 2$ quanta of amide I energy launched from the two ends of a protein α-helix visualized through the expectation value of the exciton number operator $Q|a_n|^2$. Constructive quantum interference may focus the amide I energy at the collision site to a width much narrower than each of the individual solitons. Decreasing the isotropy of exciton–phonon interaction may result in soliton pinning after the collision: $\xi = 1$ (A), $\xi = 0.2$ (B), $\xi = 0.1$ (C), and $\xi = 0$ (D).

of amide I quantum probability amplitudes a_n, which focuses the amide I energy at the collision site to a width much narrower than each of the individual solitons (Fig. 7A). Thus, constructive quantum interference may provide a mechanism for the brief focusing of energy at protein active centers for catalysis of biologically important reactions.

Interestingly, increasing the anisotropy of exciton–phonon interaction by lowering $\xi < 1$ is capable of inducing destructive quantum interference at sites of soliton collision, so that these solitons appear to bounce off each other without even touching (Fig. 7B). For $\xi < 0.2$, the soliton collision may lead to pinning of the soliton, which may wobble out of the collision site for $\xi = 0.1$ (Fig. 7C) or remain pinned at the collision site for $\xi = 0$ (Fig. 7D). Thus, another potentially useful mechanism for persistent pinning of energy at protein active centers may be the local modification of exciton–phonon interaction anisotropy toward lower ξ values.

Launching a propagating soliton from one end of the protein α-helix toward a pinned soliton in the interior of the helix, for $\xi = 1$, leads to destructive quantum interference of amide I quantum probability amplitudes a_n at the collision site with the moving and pinned soliton switching their roles between collisions (Fig. 8A). In such collision scenario, lowering of the isotropy of exciton–phonon interaction through the parameter ξ leads

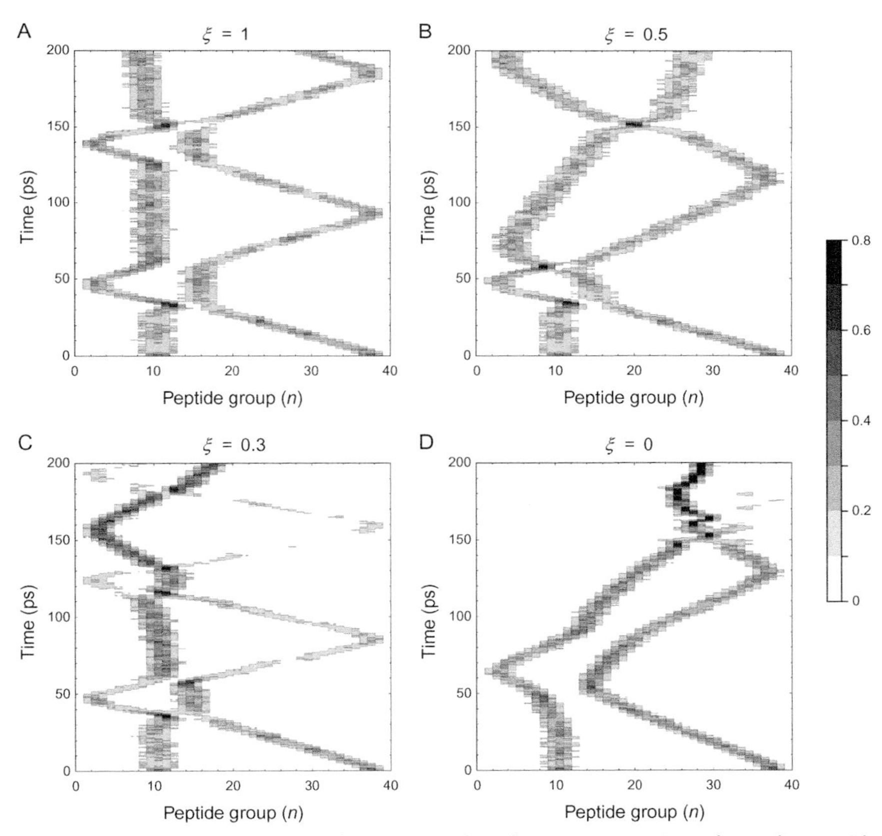

Fig. 8 The quantum collision of one pinned and one moving Davydov soliton with $Q = 2$ quanta of amide I energy visualized through the expectation value of the exciton number operator $Q|a_n|^2$. Decreasing the isotropy of exciton–phonon interaction leads to soliton pinning after the collision: $\xi = 1$ (A), $\xi = 0.5$ (B), $\xi = 0.3$ (C), and $\xi = 0$ (D).

to replacement of destructive with constructive quantum interference at the collision sites (Fig. 8B and C). For $\xi = 0$, the collision of the two solitons at a site with constructive quantum interference again leads to pinned soliton (Fig. 8D). The irregular trajectories of solitons in the computer simulations highlight the nonlinear dependence of the observed quantum dynamics on the boundary conditions and point toward the emergence of quantum chaos in regard to dynamics of quantum expectation values of amide I energy.

4.3.3 Three-soliton collisions

Increasing the number of amide I quanta to $Q = 3$ allows for collision of two moving solitons with one pinned soliton at a central position (Fig. 9) or at a noncentral position (Fig. 10). In both of these cases, the pinned soliton appears to act as a divider, which splits the protein α-helix into compartments. The two moving solitons, one launched from the N-end and the other launched from the C-end, then reflect forth and back within the left compartment or the right compartment, respectively (Fig. 9A). Because the soliton width is spread over five peptide groups, which is an odd number, in the simulations with central soliton the protein length was set to $n_{\max} = 41$ peptide groups in order to be able to perfectly center the soliton. Decreasing the isotropy of exciton–phonon interaction by lowering $\xi < 1$ introduces asymmetry in the quantum dynamics with accidental drifts to the right (Fig. 9B) or to the left (Figs. 9C and D) of the central pinned soliton and renders irregular the trajectories of the two propagating solitons (Figs. 9B–D). For $\xi \leq 0.1$, constructive quantum interference gives birth to a pinned soliton at the site of collision (Fig. 9C and D) qualitatively reproducing the behavior observed in two-soliton collisions (Figs. 7D and 8D).

The three-soliton collision between two moving solitons and one noncentral pinned soliton exhibited even richer quantum dynamics (Fig. 10). In such scenario, for $\xi = 1$ the moving soliton launched from the N-end of the protein α-helix collided first with the pinned soliton at $n = 11 - 15$ switching the roles of the moving and pinned soliton similarly to the two-soliton collision (cf. Fig. 10A vs Fig. 8A), whereas the moving soliton launched from the C-end remained compartmentalized and reflected forth and back without actual collision (cf. Fig. 10A vs Fig. 9A). Again, decreasing the isotropy of exciton–phonon interaction by lowering $\xi < 1$ led to irregularities in the soliton trajectories (Fig. 10C and D). For the completely anisotropic exciton–phonon interaction $\xi = 0$, the three solitons collided in the center of the protein α-helix producing a pinned soliton that did not wobble around (Fig. 10D).

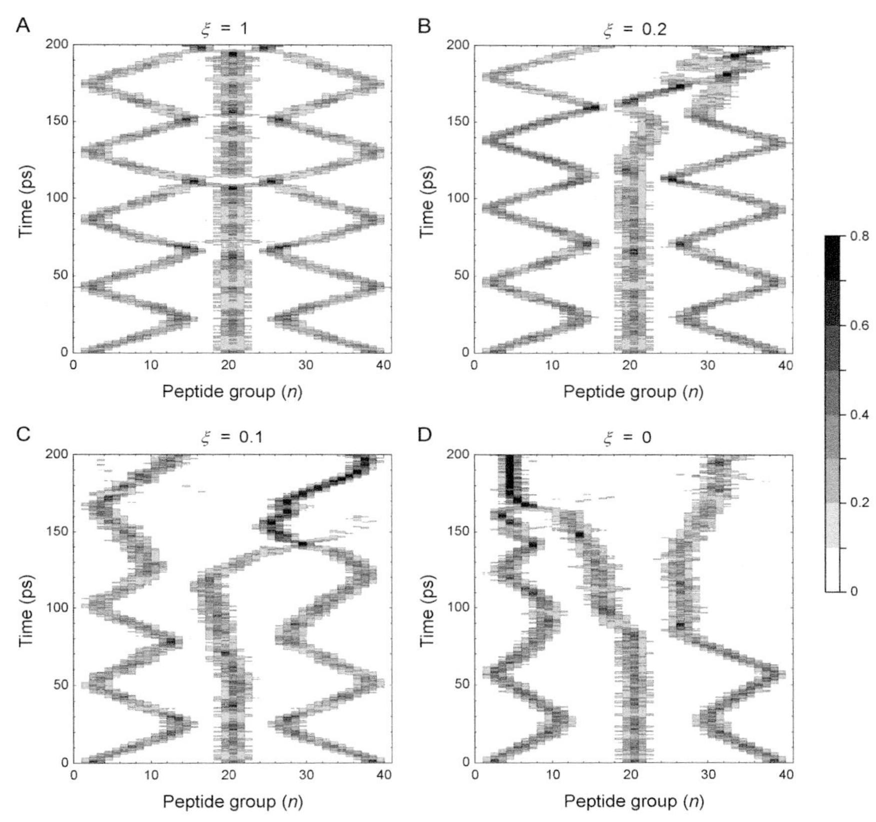

Fig. 9 The quantum collision of one central pinned soliton and two moving Davydov solitons with $Q = 3$ quanta of amide I energy visualized through the expectation value of the exciton number operator $Q|a_n|^2$. Destructive quantum interference prevents the solitons from colliding and it appears that they repel each other. Decreasing the isotropy of exciton–phonon interaction may lead to soliton pinning after the collision: $\xi = 1$ (A), $\xi = 0.2$ (B), $\xi = 0.1$ (C), and $\xi = 0$ (D).

To further test the effect on compartmentalization of the protein α-helix by a central pinned soliton, we have doubled the central soliton raising the total number of amide I quanta to $Q = 4$ (Fig. 11). Interestingly, for $\xi = 1$ the double central soliton stayed in the center where it devoured the single solitons feeding on their quantum probability amplitudes (Fig. 11A). Decreasing the isotropy of exciton–phonon interaction by lowering $\xi < 1$ introduced irregular wobbling of the central soliton (Fig. 11B–D). Thus, the collisions between single and double solitons in protein α-helices may have detrimental effects upon single solitons.

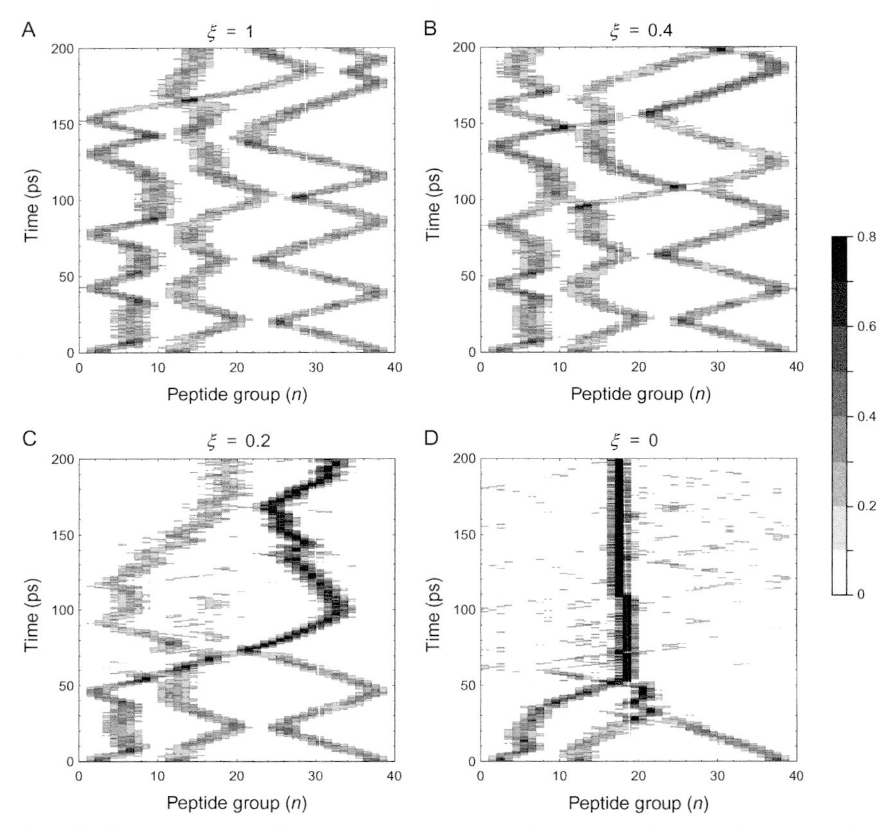

Fig. 10 The quantum collision of one noncentral pinned soliton and two moving Davydov solitons with $Q = 3$ quanta of amide I energy visualized through the expectation value of the exciton number operator $Q|a_n|^2$. Destructive quantum interference prevents the solitons from colliding and it appears that they repel each other. Decreasing the isotropy of exciton–phonon interaction may lead to soliton pinning after the collision: $\xi = 1$ (A), $\xi = 0.4$ (B), $\xi = 0.2$ (C), and $\xi = 0$ (D).

4.4 Quantum tunneling of solitons

Having verified the occurrence of quantum interference in soliton collisions, we have turned our attention to the possibility of quantum tunneling through massive barriers. Because the second of Davydov equations (34) depends on the mass of individual peptide groups M_n, in our previous work[86] we have studied whether external protein clamps could act as massive barriers by raising locally the effective mass of peptide groups inside a protein α-helix. In this latter case, we have shown that single solitons with $Q = 1$ that are wider

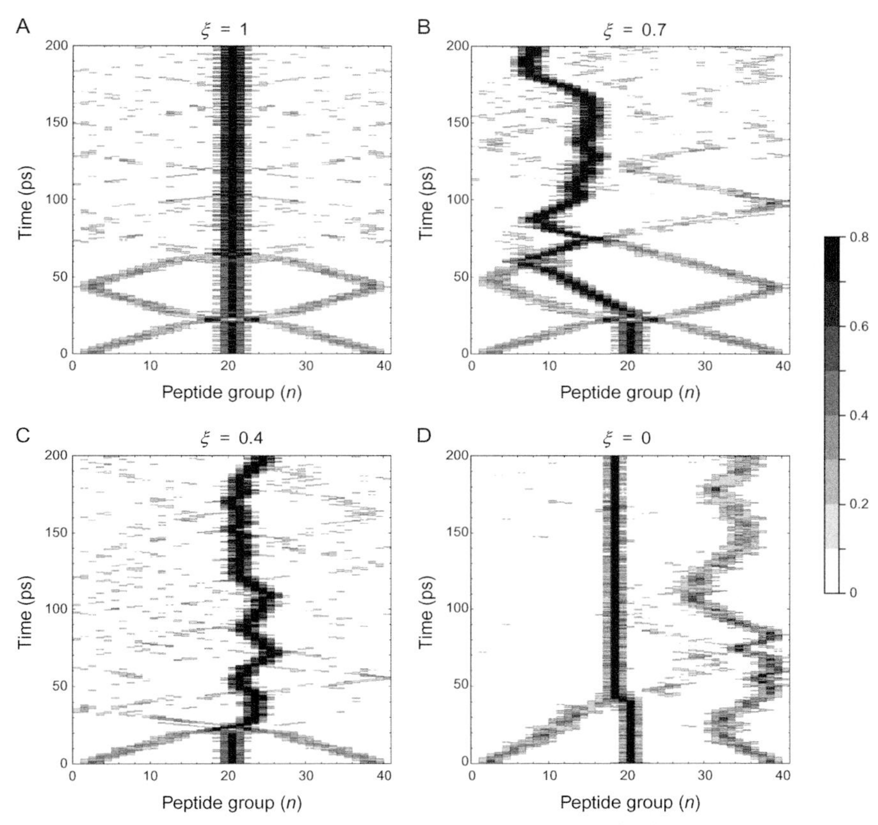

Fig. 11 The quantum collision of one double central pinned soliton and two moving Davydov solitons with $Q = 4$ quanta of amide I energy visualized through the expectation value of the exciton number operator $Q|a_n|^2$. Decreasing the isotropy of exciton–phonon interaction leads to reduction in the shift induced by the collision and suppresses the wobbling of the pinned soliton: $\xi = 1$ (A), $\xi = 0.7$ (B), $\xi = 0.4$ (C), and $\xi = 0$ (D).

behave as quasiparticles with higher energy and are capable of tunneling through heavier barriers in comparison with narrower solitons.[86]

Here, we have investigated the effect of increasing the number of amide I quanta for a fixed soliton width. To set a base for comparison, we have first launched a single Davydov soliton with $Q = 1$ by a Gaussian pulse of amide I energy distributed over five peptide groups at the N-end of the protein α-helix and aimed it at a massive barrier located over three peptide groups $n = 26 - 28$. This was repeated for the two limiting cases, $\xi = 1$ and $\xi = 0$, of exciton–phonon interaction isotropy (Figs. 12 and 13).

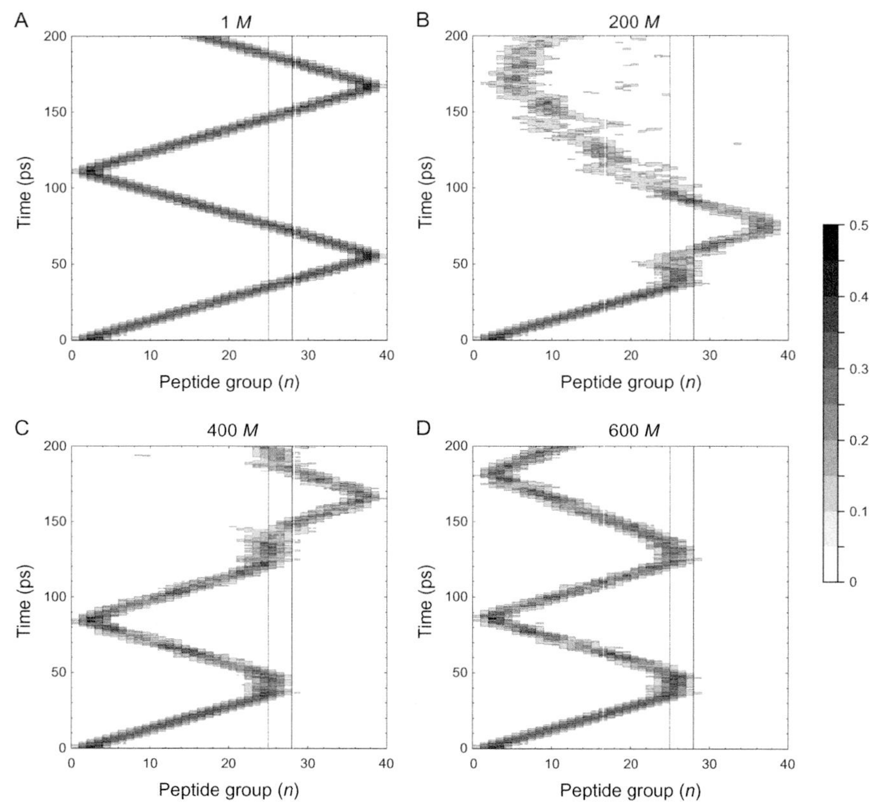

Fig. 12 The quantum dynamics of a Davydov soliton with $Q = 1$ quantum of amide I energy tunneling through or reflecting from a massive barrier located over three peptide groups $n = 26 - 28$ visualized through the expectation value of the exciton number operator $Q|a_n|^2$ for $\xi = 1$. Increasing the mass of the barrier decreases the probability of tunneling and increases the probability of reflection: no barrier 1 *M* (A), barrier in which each of the three peptide groups is with effective mass of 200 *M* (B), 400 *M* (C), and 600 *M* (D). The barrier location is indicated with *thin vertical lines*.

For $\xi = 1$, the soliton readily tunneled through the barrier in which each of the three peptide groups was with effective mass of 200 M (Fig. 12B), but got reflected from heavier barriers with 400 M (Fig. 12C) or 600 M (Fig. 12D). Thus, as it may be expected, increasing the mass of the barrier acts analogously to increasing the height of a potential barrier thereby reducing the probability of quantum tunneling of the soliton and increasing the probability of its reflection.

For $\xi = 0$, the soliton dynamics is not mirror symmetric with respect to launching from the N-end or the C-end of the protein α-helix.

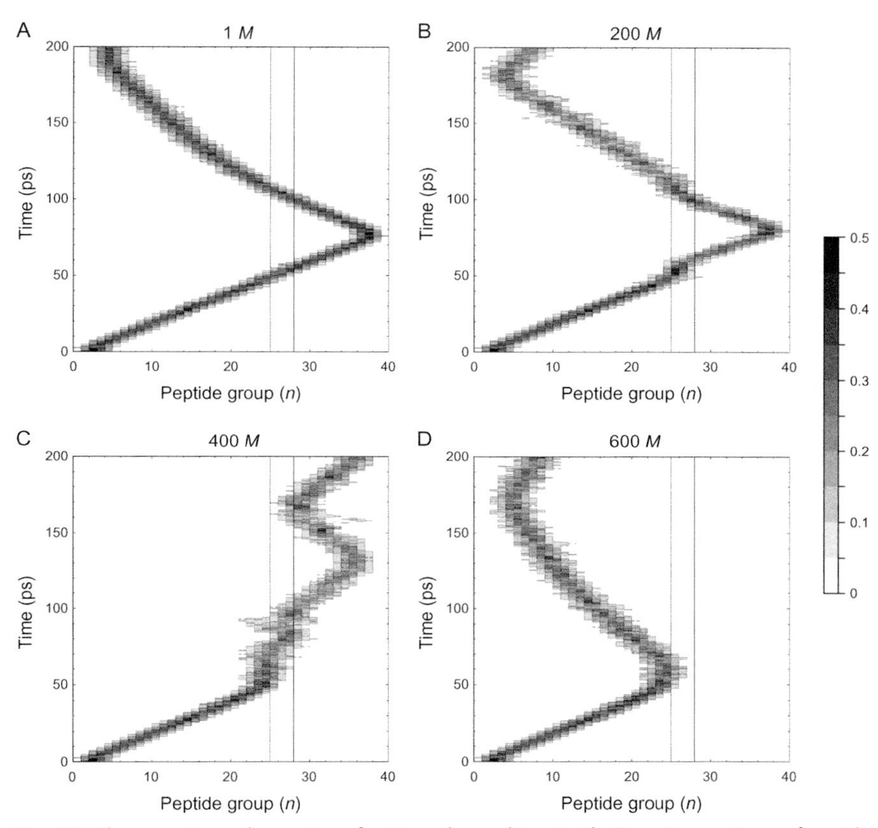

Fig. 13 The quantum dynamics of a Davydov soliton with $Q = 1$ quantum of amide I energy tunneling through or reflecting from a massive barrier located over three peptide groups $n = 26 - 28$ visualized through the expectation value of the exciton number operator $Q|a_n|^2$ for $\xi = 0$. Increasing the mass of the barrier decreases the probability of tunneling and increases the probability of reflection: no barrier 1 *M* (A), barrier in which each of the three peptide groups is with effective mass of 200 *M* (B), 400 *M* (C), and 600 *M* (D). The barrier location is indicated with *thin vertical lines.*

Despite the lack of complete mirror symmetry, qualitatively the soliton behavior was similar: it was able to tunnel through both 200 *M* barrier (Figs. 13B and 14B) and 400 *M* barrier (Figs. 13C and 14C), but reflected from the 600 *M* barrier (Figs. 13D and 14D). The soliton tunneling time through 200 *M* barrier was also faster for $\xi = 0$, 22.4 ps when launched from the N-end (Fig. 13B) and 29.8 ps when launched from the C-end (Fig. 14B), compared with 33.2 ps for $\xi = 1$ (Fig. 12B), which is the same for launching from either end of the α-helix. Thus, consistently with our

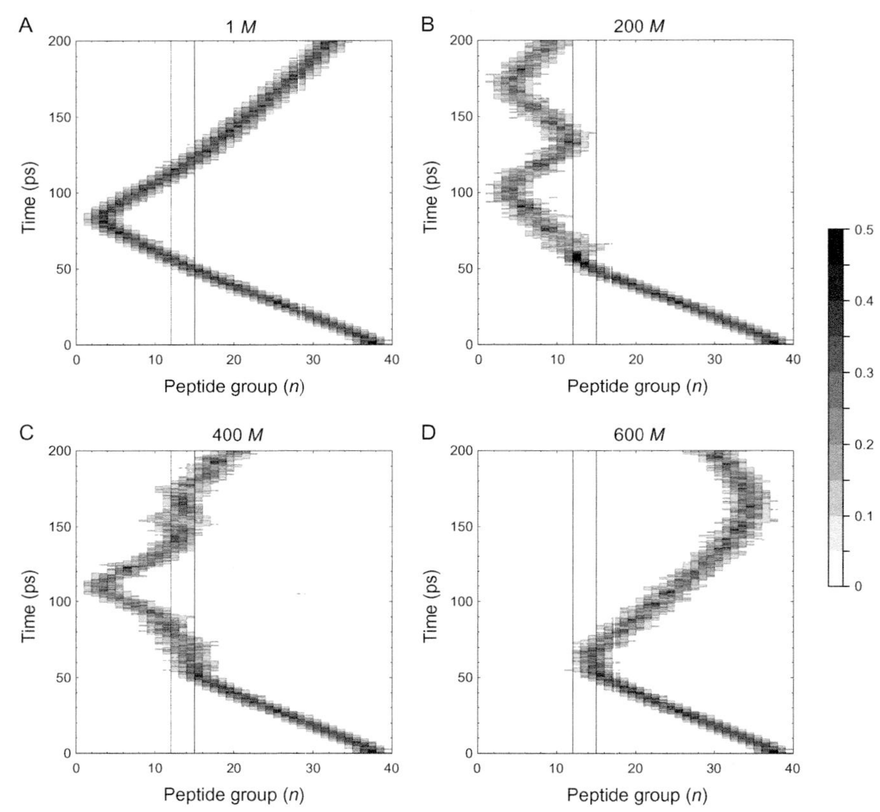

Fig. 14 The quantum dynamics of a Davydov soliton with $Q = 1$ quantum of amide I energy launched from the C-end of the α-helix tunneling through or reflecting from a massive barrier located over three peptide groups $n = 13 - 15$ visualized through the expectation value of the exciton number operator $Q|a_n|^2$ for $\xi = 0$. Increasing the mass of the barrier decreases the probability of tunneling and increases the probability of reflection: no barrier 1 M (A), barrier in which each of the three peptide groups is with effective mass of 200 M (B), 400 M (C), and 600 M (D). The barrier location is indicated with *thin vertical lines*.

previously reported results for $Q = 1$[86] decreasing the isotropy of exciton–phonon interaction by lowering ξ, decreases the probability of soliton reflection from the barrier, increases the probability of quantum tunneling of the soliton through the barrier and reduces the tunneling time in the event of successful tunneling.

It should be noted that the soliton width does not appear to be significantly affected by the anisotropy of exciton–phonon interaction with $\xi < 1$. The soliton width, defined by the spread of the exciton quantum probability amplitudes, is directly related to the expectation value of the exciton energy operator

$$\langle\Psi|\hat{H}_{\text{ex}}|\Psi\rangle = Q\sum_{n}\left[E_0|a_n|^2 - J_{n+1}a_n^* a_{n+1} - J_n a_n^* a_{n-1}\right] \tag{35}$$

The gauge transformation $a_n \rightarrow \bar{a}_n e^{-\frac{i}{\hbar}\int \gamma(t)dt}$ used to remove the highly oscillatory phase in Davydov's equations effectively sets $E_0 = 0$. For the case when all J_n are equal, the expectation of the exciton energy operator becomes

$$\langle\Psi|\hat{H}_{\text{ex}}|\Psi\rangle = -2QJ\sum_{n}[\text{Re}(a_n)\text{Re}(a_{n+1}) + \text{Im}(a_n)\text{Im}(a_{n+1})] \tag{36}$$

Thus, the soliton width is positively related to the absolute value of the exciton expectation energy. When the soliton is narrowly focused onto a single peptide group, then the expectation value is minimal $\left|\langle\Psi|\hat{H}_{\text{ex}}|\Psi\rangle\right| = 0$, whereas when the soliton is evenly spread over all peptide groups in the α-helix spine the expectation value is maximal $\left|\langle\Psi|\hat{H}_{\text{ex}}|\Psi\rangle\right| = 2QJ$.

To correctly determine the soliton width in the discrete lattice, it is necessary to consider the fact that, for most of the time, the soliton is in the process of transition between neighboring peptide groups. Thus, one would need a methodological rule that identifies time points when the soliton is best positioned over the peptide groups for measuring its width. Furthermore, on top of the soliton-induced deformation of the lattice of hydrogen bonds there are superposed small disturbances due to the phonon oscillations of the lattice, which may introduce some noise on the exciton envelope of the soliton. Therefore, to find out whether the soliton width changes in the course of the whole simulated time period of 200 ps, we have divided the α-helix spine into 35 overlapping local stretches of excition expectation energy given by 5 terms in the sum for $\left|\langle\Psi|\hat{H}_{\text{ex}}|\Psi\rangle\right|$ as follows

$$\mathcal{E}_i(t) = 2QJ\sum_{n=i-2}^{i+2}[\text{Re}(a_n)\text{Re}(a_{n+1}) + \text{Im}(a_n)\text{Im}(a_{n+1})] \tag{37}$$

The choice of $i \in [3, \ldots, 37]$ ensures that the local stretches do not extend outside the α-helix spine ends, whereas the sum over five terms ensures that exciton probability amplitudes over six peptide groups are captured. Then, the motion of the soliton leads to sequential peaks of neighboring

$\mathcal{E}_i(t)$ separated by the time period needed for the soliton to travel from a position perfectly centered on $\mathcal{E}_i(t)$, to a spatially translated position when the soliton is centered on $\mathcal{E}_{i+1}(t)$. Variation of the time intervals between peaks of neighboring local stretches $\mathcal{E}_i(t)$ indicates a variation in the soliton speed, whereas a nonzero slope of the trend line of exciton energy peaks will be an indication of changing soliton width. For example, a positive slope of the trend line of exciton energy peaks will indicate that the soliton becomes wider, whereas a negative trend line of exciton energy peaks will indicate that the soliton becomes narrower. This will hold true as long as the soliton width fits inside the stretches $\mathcal{E}_i(t)$ (hence no contributions to the exciton expectation energy will be trimmed) and explains why we have set the length of the stretches to be six peptide groups given that the initial soliton width is five peptide groups.

To assess the effects of ξ on soliton width and velocity, we have compared the simulations with $Q = 1$ reported in Figs. 12A, 13A, and 14A for $\xi = 1$ launched from the N-end, and for $\xi = 0$ launched from the N-end or C-end, respectively. The distances between consecutive $\mathcal{E}_i(t)$ peaks are shorter for $\xi = 1$ compared with $\xi = 0$ (Fig. 15A, C, and E) consistent with higher velocities of $\xi = 1$ solitons. While the soliton velocity is relatively constant for $\xi = 1$ for the whole simulation period of 200 ps (Fig. 15B), it appears to slow down for $t > 120$ ps for $\xi = 0$ (Fig. 15D and F). This retardation of the $\xi = 0$ solitons, however, is not accompanied by any significant spread in the envelope of exciton quantum probability amplitudes, since the trend lines for $\mathcal{E}_i(t)$ peaks remain horizontal at $1.6J$ (Fig. 15D and F). Thus, the mechanism behind the varying soliton velocity for $\xi = 0$ could be quantum interference within the finite length of the discrete lattice. Also, the ambient noise on the exciton envelope, resulting from the phonon lattice oscillations, is greater for $\xi = 0$, and is probably due to manifestly nonlinear effects dependent on χ_r, because to achieve the average $\bar{\chi} = 35\,\text{pN}$, the right exciton–phonon coupling becomes $\chi_r = 70$ pN given that $\chi_l = 0$ pN. In contrast, for $\xi = 1$ the average $\bar{\chi} = 35$ pN is obtained with $\chi_r = 35$ pN and $\chi_l = 35$ pN.

Doubling the launched soliton by increasing the amide I quanta to $Q = 2$, for $\xi = 1$, reveals a soliton reflection from 200 M barrier (Fig. 16B) or 400 M barrier (Fig. 16C), but remarkable tunneling phenomena through the much more massive 600 M barrier (Fig. 16D). This quantum behavior provides an indication that the tunneling of the Davydov soliton through the massive barrier could be analogous to a massive quantum particle tunneling through a potential barrier whose potential barrier height V_0 is lower than the energy E_0 of the particle.[95,96]

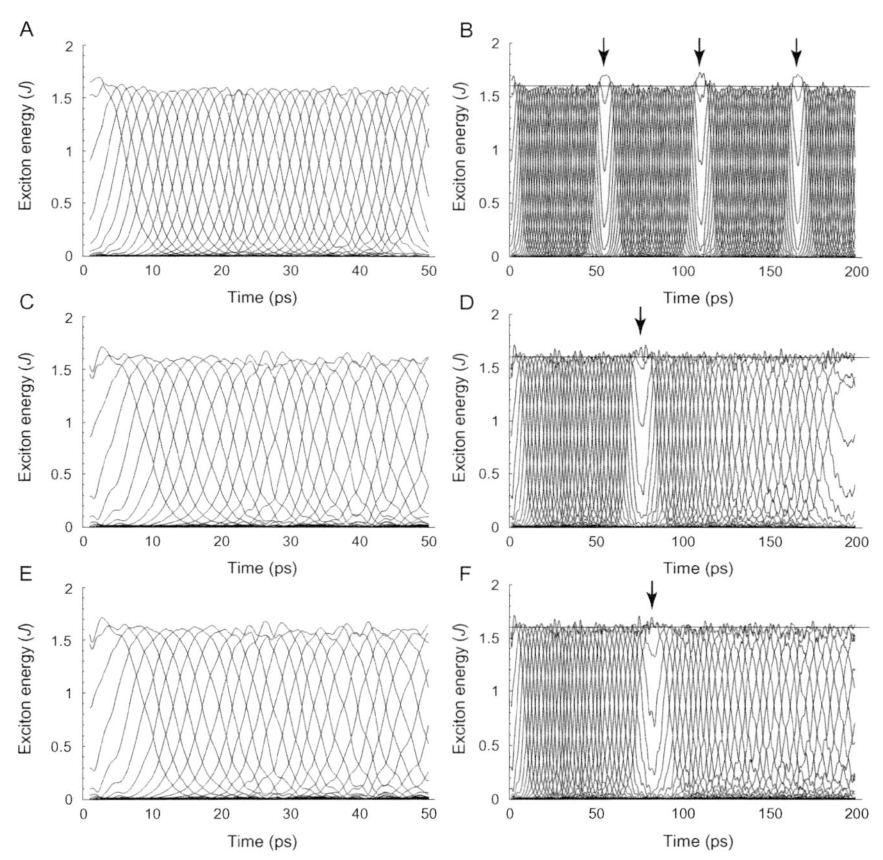

Fig. 15 The quantum dynamics of the local exciton energy (measured in units of $J = 0.155$ zJ) computed for stretches of five terms $\mathcal{E}_i(t)$ in the sum for $|\langle\Psi|\hat{H}_{\text{ex}}|\Psi\rangle|$ of a Davydov soliton with $Q = 1$ quantum of amide I energy for different values of exciton–phonon interaction isotropy ξ. (A) and (B) Exciton energy for $\xi = 1$ soliton launched from the N-end of the α-helix. (C) and (D) Exciton energy for $\xi = 0$ soliton launched from the N-end of the α-helix. (E) and (F) Exciton energy for $\xi = 0$ soliton launched from the C-end of the α-helix. The *horizontal trend lines* for $\mathcal{E}_i(t)$ peaks at 1.6J show that the changes in soliton velocity are not accompanied by spread in the envelope of exciton quantum probability amplitudes. *Arrows* indicate soliton reflection from the α-helix ends.

For a massive quantum particle with mass m and energy E_0 tunneling through rectangular potential barrier

$$V(x) = V_0[\Theta(x - x_1) - \Theta(x - x_2)] \tag{38}$$

with height V_0 and width $\Delta x = x_2 - x_1$, where $\Theta(x) = \frac{d}{dx}\max\{x, 0\}$ is the Heaviside step function, the transmission coefficient T is determined from

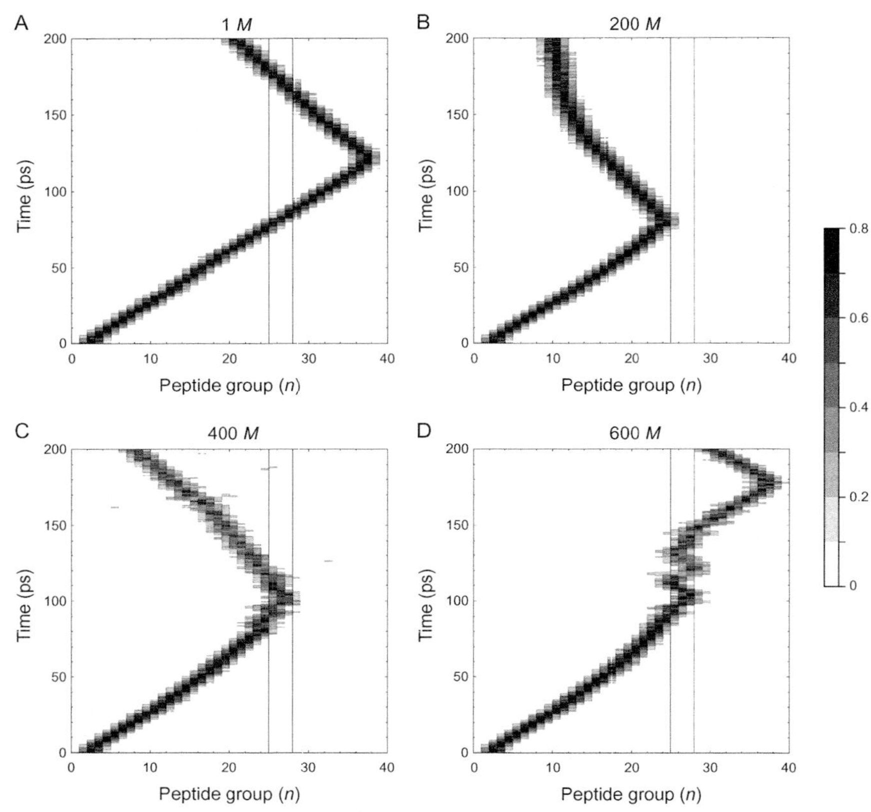

Fig. 16 The quantum dynamics of a Davydov soliton with $Q = 2$ quanta of amide I energy tunneling through or reflecting from a massive barrier located over three peptide groups $n = 26 - 28$ visualized through the expectation value of the exciton number operator $Q|a_n|^2$ for $\xi = 1$. Increasing the mass of the barrier may increase the interaction time with the barrier and increase the probability of tunneling through the barrier: no barrier 1 *M* (A), barrier in which each of the three peptide groups is with effective mass of 200 *M* (B), 400 *M* (C), and 600 *M* (D). The barrier location is indicated with thin *vertical lines*.

different quantum mechanical expressions depending on the magnitude of V_0 with respect to E_0, which is required to avoid the appearance of imaginary wavenumber k_2 as follows.

Case I: If $V_0 > E_0$, on setting $\hbar k_1 = \sqrt{2mE_0}$ and $\hbar k_2 = \sqrt{2m(V_0 - E_0)}$[95] pp. 75–80, the analytic derivation of the transmission coefficient gives

$$T = \frac{4k_1^2 k_2^2}{\left(k_1^2 + k_2^2\right)^2 \sinh^2(\Delta x k_2) + 4k_1^2 k_2^2} \tag{39}$$

Case II: If $V_0 < E_0$, on setting $\hbar k_1 = \sqrt{2mE_0}$ and $\hbar k_2 = \sqrt{2m(E_0 - V_0)}$[95] pp. 75–80, the analytic derivation of the transmission coefficient gives

$$T = \frac{4k_1^2 k_2^2}{\left(k_1^2 - k_2^2\right)^2 \sin^2(\Delta x k_2) + 4k_1^2 k_2^2} \tag{40}$$

In the latter case, for $\Delta x > \frac{1}{k_2}$, it is indeed possible to observe larger transmission coefficient T for larger V_0 due to the occurrence of quantum interference effects. Thus, the result of the simulation reported in Fig. 16, supports the conclusion that doubling the amide I quanta in the double soliton with $Q = 2$, is analogous to raising the energy E_0 of massive particle that now faces a potential barrier with height $V_0 < E_0$.

Because a particular effect of decreasing the isotropy of exciton–phonon interaction by lowering ξ, is to pin down the soliton, for the double soliton with $Q = 2$, we were unable to test quantum tunneling through the massive barrier for $\xi < 0.7$. In the case of $\xi = 0.7$, the double soliton reflected from 200 M barrier (Fig. 17B) or 400 M barrier (Fig. 17C), yet it tunneled through the much more massive 600 M barrier (Fig. 17D). This behavior was qualitatively similar to the completely isotropic case with $\xi = 1$. However, in comparison with the case $\xi = 1$, for which the tunneling time was 71.1 ps (Fig. 16D), the presence of some anisotropy for $\xi = 0.7$ delayed the passage through the barrier with a tunneling time of 92.9 ps (Fig. 17D). Thus, the faster tunneling time for higher ξ in double solitons with $Q = 2$ differs from the observed tunneling times for single solitons with $Q = 1$, and highlights the nonclassical nature of soliton transmission through a potential barrier with $V_0 < E_0$ due to manifested quantum interference effects in the transmission coefficient (40).

4.5 Disorder effects on soliton stability

The creation of soliton solutions by the system of Davydov equations (33) and (34) is made possible by the highly ordered protein α-helix structure, which is reflected in the construction of the Hamiltonian (1). Because both living and nonliving quantum physical systems are subject to the same fundamental quantum physical laws, we aimed at elucidating the importance of biological order[25,97,98] for outlining the quantum boundaries of life.

In most quantum chemistry applications, the Born–Oppenheimer approximation[99] allows the Schrödinger equation for a biomolecule to be separated it into two equations: (i) an electronic Schrödinger equation

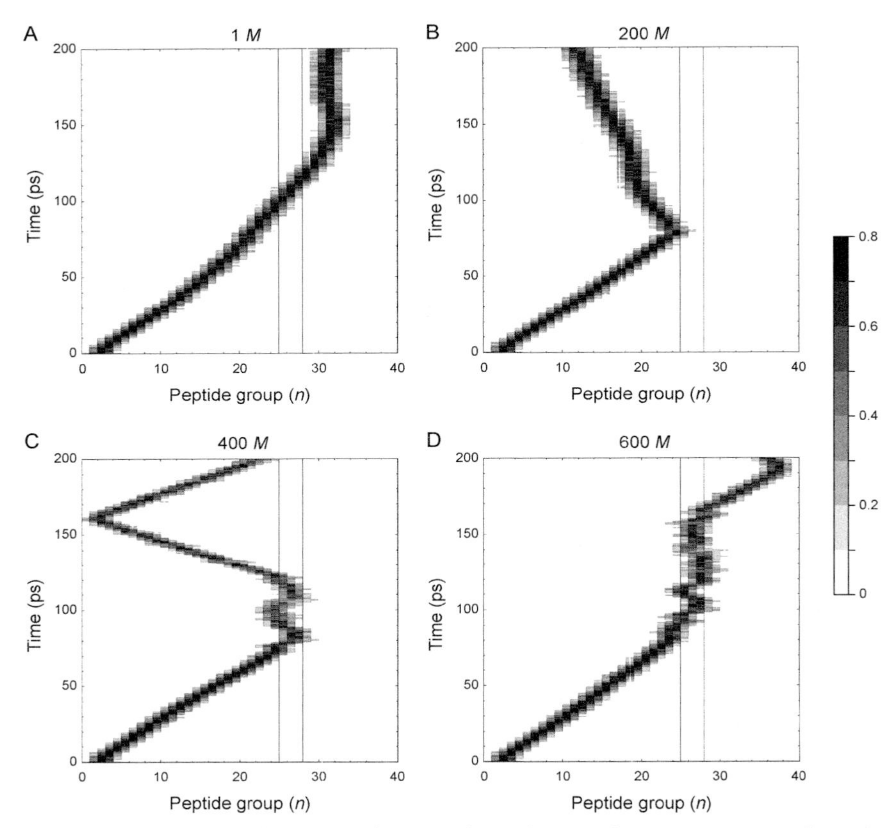

Fig. 17 The quantum dynamics of a Davydov soliton with $Q = 2$ quanta of amide I energy tunneling through or reflecting from a massive barrier located over three peptide groups $n = 26 - 28$ visualized through the expectation value of the exciton number operator $Q|a_n|^2$ for $\xi = 0.7$. Increasing the mass of the barrier may increase the interaction time with the barrier and increase the probability of tunneling through the barrier: no barrier 1 M (A), barrier in which each of the three peptide groups is with effective mass of 200 M (B), 400 M (C), and 600 M (D). The barrier location is indicated with thin *vertical lines*.

and (ii) a nuclear Schrödinger equation. Only the positions of the much heavier atomic nuclei (not their momenta) generate the potential that enters in the Hamiltonian for solving the Schrödinger equation for the electrons.[100] Once the solution ψ_e of the electronic Schrödinger equation is found, it is used to provide the potential energy function for the nuclear motion. The obtained vibrational nuclear wave functions ψ_n with corresponding energies solve the nuclear Schrödinger equation. The total wave function $\Psi = \psi_e \psi_n$ for the biomolecule is then composed as a product of the electronic wave function ψ_e and the nuclear wave function ψ_n. Further simplification of quantum mechanical calculations could be achieved with the Crude Born–Oppenheimer Approximation where the equilibrium separation of the nuclei is employed at all times.

The upshot of the Crude Born–Oppenheimer approximation manifests in the original Davydov Hamiltonian (1) in the guise of uniformity of dipole–dipole coupling energies J_n between neighboring amide I oscillators, which are all set to $J = 0.155$ zJ.[56–59] This introduces a high degree of stability in the biological system. In order to study the effects of instability on Davydov solitons, we have drawn randomly the value of each J_n from a Gaussian distribution with mean value J, and standard deviation σ_{J_n}, defined as a percentage of J. To avoid reporting an outlier quantum dynamics, we have simulated at least 5 randomly drawn distributions for a given σ_{J_n} and presented the outcome of a simulation run, which had been supported by another visually similar outcome.

For a protein α-helix with completely isotropic exciton–phonon interaction $\xi = 1$, the presence of a dipole–dipole coupling disorder $\sigma_{J_n} < 5\%$ was not detrimental for the moving soliton (Fig. 18A). When the disorder was increased between $\sigma_{J_n} = 5\%$ (Fig. 18B) and $\sigma_{J_n} = 10\%$ (Fig. 18C), the soliton propagated along the protein α-helix until it was trapped inside a region flanked by low J_n values. Visually, the quantum dynamics of the soliton resembled compartmentalization inside a short protein α-helix rather than pinning. For greater disorder $\sigma_{J_n} = 15\%$, features of soliton disintegration were observed even though part of the soliton still persisted (Fig. 18D).

For a protein α-helix with completely anisotropic exciton–phonon interaction $\xi = 0$, the presence of dipole–dipole coupling disorder had similar effects, however the soliton appeared to be pinned for $\sigma_{J_n} = 5\%$ (Fig. 19B) and $\sigma_{J_n} = 10\%$ (Fig. 19C). The destabilization of the soliton at $\sigma_{J_n} = 15\%$ was present (Fig. 19D), albeit a bit weaker compared with the completely isotropic $\xi = 1$ case. Thus, the order and uniformity of dipole–dipole coupling energies between neighboring amide I oscillators is pivotal for the highly efficient, dissipationless, solitonic transport of energy by protein α-helices. In addition, cooperative effects between backbone dipole–dipole interactions are instrumental for the formation

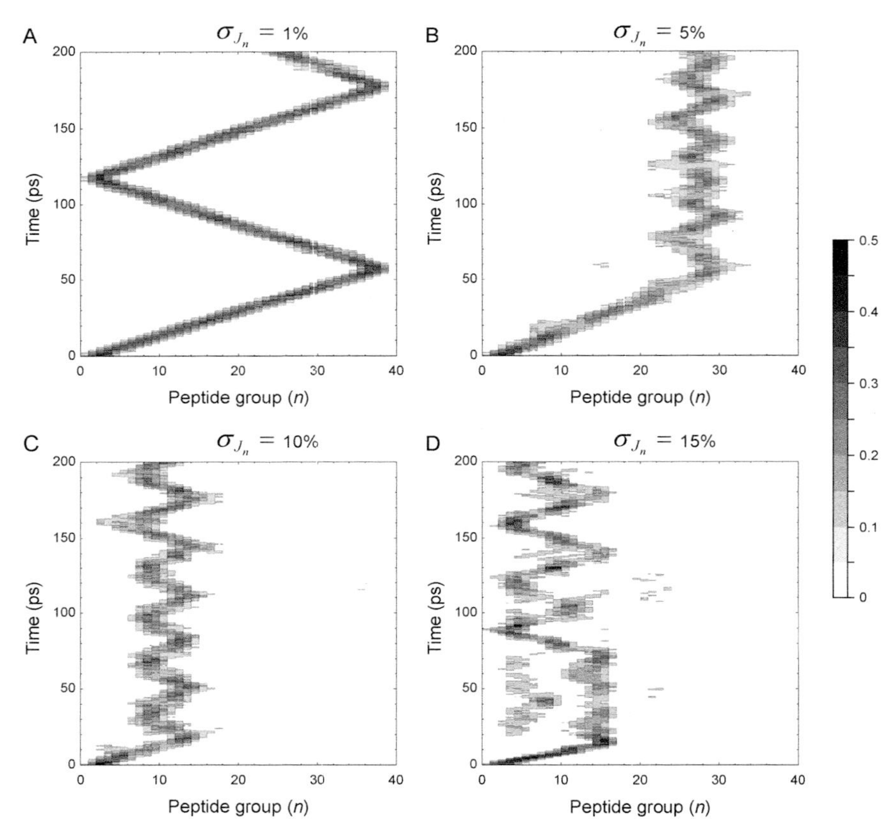

Fig. 18 The quantum dynamics of a Davydov soliton with $Q = 1$ quantum of amide I energy for protein α-helix with randomly variable J_n dipole–dipole coupling energy between amide I oscillators visualized through the expectation value of the exciton number operator $Q|a_n|^2$ for $\xi = 1$. Nonuniformity of J_n was quantified by the standard deviation σ_{J_n} expressed as percentage from the mean value J: $\sigma_{J_n} = 1\%$ (A), $\sigma_{J_n} = 5\%$ (B), $\sigma_{J_n} = 10\%$ (C), and $\sigma_{J_n} = 15\%$ (D). Increasing the nonuniformity of J_n destabilizes the soliton and traps it to a region flanked by low J_n values.

of secondary and supersecondary structures of proteins, and after successful folding keep proteins in their folded structural state,[101] while noting that protein conformational transitions, as microscopic biochemical processes, a fortiori, involve quantum states (see e.g., Ref. 102).

To illustrate the fact that not all types of disorder are equally adverse to soliton dynamics, we have also randomly varied the amino acid masses

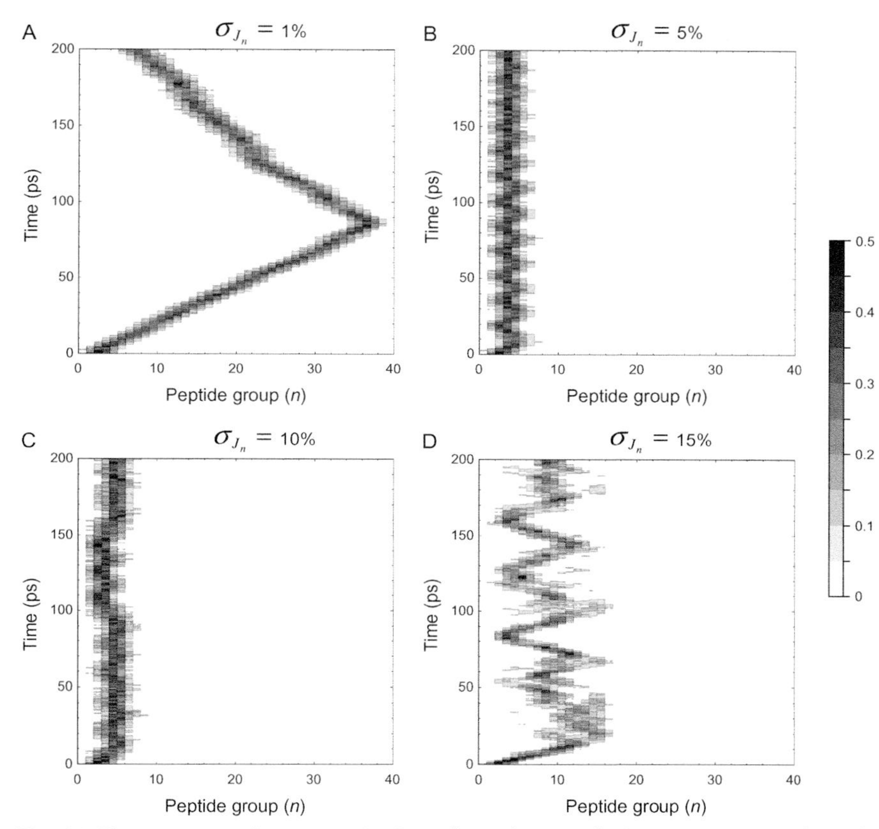

Fig. 19 The quantum dynamics of a Davydov soliton with $Q = 1$ quantum of amide I energy for protein α-helix with randomly variable J_n dipole–dipole coupling energy between amide I oscillators visualized through the expectation value of the exciton number operator $Q|a_n|^2$ for $\xi = 0$. Nonuniformity of J_n was quantified by the standard deviation σ_{J_n} expressed as percentage from the mean value J: $\sigma_{J_n} = 1\%$ (A), $\sigma_{J_n} = 5\%$ (B), $\sigma_{J_n} = 10\%$ (C), and $\sigma_{J_n} = 15\%$ (D). Increasing the nonuniformity of J_n destabilizes the soliton.

M_n within the maximal biochemical range of $\sigma_{M_n} = 25\%$ in a protein α-helix. For all values of ξ, the solitons readily propagated along the α-helix spine (Fig. 20) and were virtually indistinguishable from the case with uniform amino acid masses (cf. Fig. 12A vs Fig. 20A and Fig. 13A vs Fig. 20D).

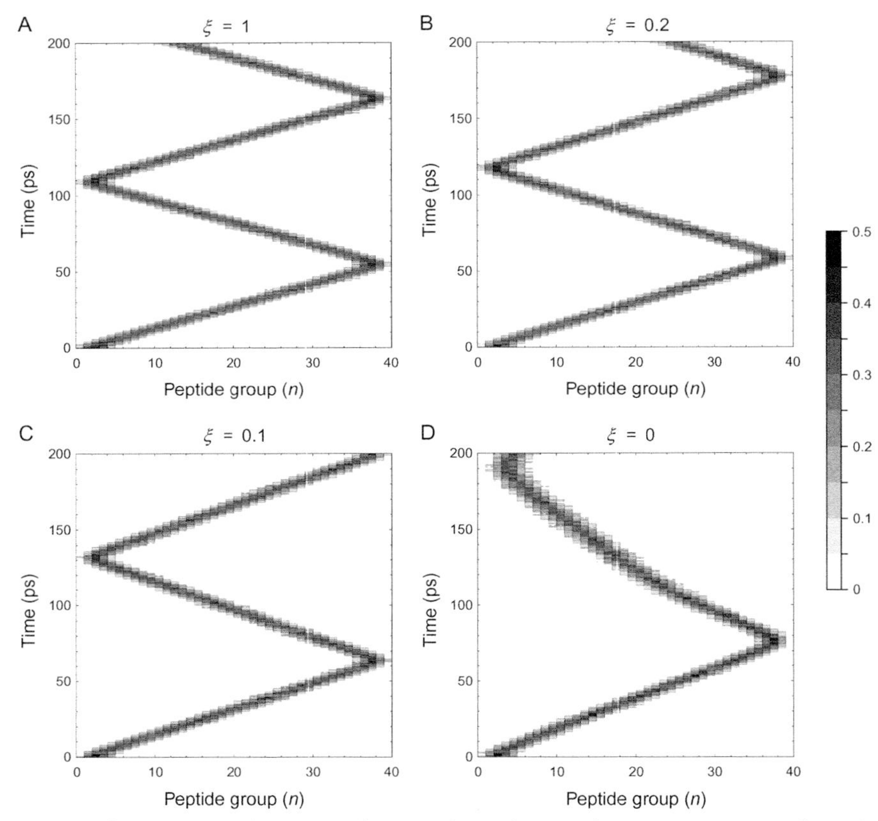

Fig. 20 The quantum dynamics of a Davydov soliton with $Q = 1$ quantum of amide I energy for protein α-helix with nonuniform M_n peptide group masses (standard deviation σ_{M_n} is 25% from the mean value $M = 1.9 \times 10^{-25}$ kg) visualized through the expectation value of the exciton number operator $Q|a_n|^2$ for different values of the isotropy of exciton–phonon interaction: $\xi = 1$ (A), $\xi = 0.2$ (B), $\xi = 0.1$ (C), and $\xi = 0$ (D).

5. Conclusions

The importance of proteins for maintaining life cannot be overstated.[103,104] The majority of biological processes are catalyzed or performed by proteins, and the main part of the coding DNA in the genome is dedicated to storing hereditary information for the production of proteins. The underlying mechanisms behind the versatility of protein function, however, remained elusive within the deterministic clockwork structures of classical physics.[25,97] In an attempt to address the perceived theoretical crisis in bioenergetics, Alexander Davydov, in 1973, proposed that the amide I quanta of peptide vibrational energy (C=O stretching) might become self-localized through interactions with lattice phonons in protein α-helices.[55] The original Davydov model was analytically studied with the use of the continuum approximation or simulated numerically,[56–68] and then further generalized to include explicit treatment of multiple amide I quanta and the possible anisotropies of various physical parameters in realistic protein models.[54,69–80,86,91,92]

To gain insights into the transport and utilization of energy by proteins, in this work we have studied the quantum dynamics of multiple amide I quanta, which arises from solving the Schrödinger equation for the generalized Davydov Hamiltonian (1). Utilizing only standard quantum mechanical commutators and the Schrödinger equation, we initially derived a discrete system of generalized Davydov equations (33) and (34) that govern the quantum dynamics of amide I excitons in protein α-helices (Section 3). Then, we performed computational simulations, which have revealed that the discrete system of Davydov equations supports the corresponding solitons, even for ultrashort protein α-helices whose length is only 10 peptide groups (Section 4.2). This suggests that natural evolution is able to select optimal protein α-helix lengths in gradual steps without losing the soliton mechanism. Next, we have found that soliton collisions could lead to either persistent pinned solitons, or to generating intermittent peaks of concentrated amide I energy through constructive quantum interference (Section 4.3). This provides a quantum physical mechanism for delivery of concentrated peaks of energy at protein active centers. The free energy delivered by the soliton to the active protein center may then be utilized to do physical work triggering chemical reactions and classical processes on nano-, micro- or millisecond timescales. The physical mechanisms for amplification of individual quantum processes to trigger macroscopic classical events at these slower timescales, which will effectively constitute the quantum-to-classical transition, were not covered in the current study and deserve further theoretical modeling along the lines of macromolecular electron clouds interacting with each other or with the electromagnetic field in biological systems.[68,105,106] Elaborate macro-quantum modeling of multiple proteins and their interaction with the quantized electromagnetic field would be particularly relevant for the function of protein voltage-gated ion channels incorporated in neuronal plasma membranes, where the electric field may reach values on the order of 10^7 V/m during the course of action potentials fired by the neurons.[21,107]

In addition to participation in quantum interference phenomena, the quantum nature of Davydov solitons was also illustrated by their capability to tunnel through massive barriers applied by external protein clamps (Section 4.4). This allows proteins to accomplish otherwise classically impossible tasks. Lastly, we have demonstrated the importance of the biological order for soliton stability through the regular helical geometry that ensures uniform dipole–dipole coupling energies between neighboring amide I oscillators (Section 4.5). Interestingly, the presence of disorder in the dipole–dipole coupling energies did not cause direct soliton disintegration, but instead resulted in compartmentalization of the solitons within protein segments flanked with low dipole–dipole couplings. This may have been valuable for the natural evolution of protein function because random coil domains could evolve into α-helices, or vice versa, thereby sculpturing-out either energy transmission lines or protein active centers for utilization of the delivered energy.

The present findings may also shed new light on the abiotic origin of life where randomly assembled polypeptides could have found a way to preserve existing seeds of order in the first replicators. Of course, the origin of life is the most challenging problem of all time[108] and we may never find out how exactly life started as a consequence of the very nature of the evolutionary process—we can only see the best survivors having lost forever all those less fit ancestors that were replaced in natural history. From living organisms that can fossilize, we have been lucky to have a glimpse at what life was in past eras. To look further back to the dawn of precellular life with the first biomolecular replicators, however, is an uncertain task since single biomolecules leave no traces, and the conditions on the early surface of the planet (Earth) are likely to have been quite different from those which we currently imagine.[109] There is evidence, however, that a network of cross-replicating molecules is more robust, and operates faster than a single self-replicator.[110,111] Regardless of whether polypeptides were present from the outset of life,[112] or at a later stage joined up with the biomolecular repertoire of living systems, our results on the quantum dynamics of Davydov solitons in protein α-helices provide evidence for the nontrivial role of quantum effects in maintaining life, and therefore highlight the prominent role of biological order to foster those quantum effects. These events also contend that the first replicators should have had lines for highly efficient quantum transmission of energy, capable of trapping energy quanta in active centers, and the prowess to utilize seeds of order to sustain nontrivial quantum dynamics at biochemically relevant timescales.

References

1. Oparin, A. I. *The Origin of Life on the Earth*; Academic Press: New York, 1957.
2. Koshland, D. E. The Seven Pillars of Life. *Science* **2002**, *295*(5563), 2215–2216. https://doi.org/10.1126/science.1068489.
3. Trifonov, E. N. Vocabulary of Definitions of Life Suggests a Definition. *J. Biomol. Struct. Dyn.* **2011**, *29*(2), 259–266. https://doi.org/10.1080/073911011010524992.
4. Kuhlman, B.; Bradley, P. Advances in Protein Structure Prediction and Design. *Nat. Rev. Mol. Cell Biol.* **2019**, *20*(11), 681–697. https://doi.org/10.1038/s41580-019-0163-x.
5. Atkins, J. F.; Gesteland, R. The 22nd Amino Acid. *Science* **2002**, *296*(5572), 1409–1410. https://doi.org/10.1126/science.1073339.
6. Pauling, L.; Corey, R. B.; Branson, H. R. The Structure of Proteins: Two Hydrogen-Bonded Helical Configurations of the Polypeptide Chain. *Proc. Natl. Acad. Sci.* **1951**, *37*(4), 205–211. https://doi.org/10.1073/pnas.37.4.205.
7. Pauling, L.; Corey, R. B. The Pleated Sheet, a New Layer Configuration of Polypeptide Chains. *Proc. Natl. Acad. Sci.* **1951**, *37*(5), 251–256. https://doi.org/10.1073/pnas.37.5.251.
8. Vieira-Pires, R. S.; Morais-Cabral, J. H. 3_{10} Helices in Channels and Other Membrane Proteins. *J. Gen. Physiol.* **2010**, *136*(6), 585–592. https://doi.org/10.1085/jgp.201010508.
9. Kumar, P.; Bansal, M. Dissecting π-Helices: Sequence, Structure and Function. *FEBS J.* **2015**, *282*(22), 4415–4432. https://doi.org/10.1111/febs.13507.
10. Guseva, E.; Zuckermann, R. N.; Dill, K. A. Foldamer Hypothesis for the Growth and Sequence Differentiation of Prebiotic Polymers. *Proc. Natl. Acad. Sci.* **2017**, *114*(36), E7460–E7468. https://doi.org/10.1073/pnas.1620179114.
11. Schopf, J. W.; Kitajima, K.; Spicuzza, M. J.; Kudryavtsev, A. B.; Valley, J. W. SIMS Analyses of the Oldest Known Assemblage of Microfossils Document Their Taxon-Correlated Carbon Isotope Compositions. *Proc. Natl. Acad. Sci.* **2018**, *115*(1), 53–58. https://doi.org/10.1073/pnas.1718063115.
12. Schnitzer, M. J.; Block, S. M. Kinesin Hydrolyses One ATP Per 8-nm Step. *Nature* **1997**, *388*(6640), 386–390. https://doi.org/10.1038/41111.
13. Coy, D. L.; Wagenbach, M.; Howard, J. Kinesin Takes One 8-nm Step for Each ATP That it Hydrolyzes. *J. Biol. Chem.* **1999**, *274*(6), 3667–3671. https://doi.org/10.1074/jbc.274.6.3667.
14. Ahn, M.; Beacham, D.; Westenbroek, R. E.; Scheuer, T.; Catterall, W. A. Regulation of Na_V1.2 Channels by Brain-Derived Neurotrophic Factor, TrkB, and Associated Fyn Kinase. *J. Neurosci.* **2007**, *27*(43), 11533–11542. https://doi.org/10.1523/jneurosci.5005-06.2007.
15. Tucker, K.; Fadool, D. A. Neurotrophin Modulation of Voltage-Gated Potassium Channels in Rat Through TrkB Receptors is Time and Sensory Experience Dependent. *J. Physiol.* **2002**, *542*(2), 413–429. https://doi.org/10.1113/jphysiol.2002.017376.
16. Szöke, A.; Hajdu, J. Energy Utilization in Fluctuating Biological Energy Converters. *Struct. Dyn.* **2016**, *3*(3), 034701. https://doi.org/10.1063/1.4945792.
17. Matsuno, K. Forming and Maintaining a Heat Engine for Quantum Biology. *Biosystems* **2006**, *85*(1), 23–29. https://doi.org/10.1016/j.biosystems.2006.02.002.
18. Matsuno, K. From Quantum Measurement to Biology Via Retrocausality. *Prog. Biophys. Mol. Biol.* **2017**, *131*, 131–140. https://doi.org/10.1016/j.pbiomolbio.2017.06.012.
19. Melkikh, A. V.; Khrennikov, A. Nontrivial Quantum and Quantum-Like Effects in Biosystems: Unsolved Questions and Paradoxes. *Prog. Biophys. Mol. Biol.* **2015**, *119*(2), 137–161. https://doi.org/10.1016/j.pbiomolbio.2015.07.001.

20. Marais, A.; Adams, B.; Ringsmuth, A. K.; Ferretti, M.; Gruber, J. M.; Hendrikx, R.; Schuld, M.; Smith, S. L.; Sinayskiy, I.; Krüger, T. P. J.; Petruccione, F.; van Grondelle, R. The Future of Quantum Biology. *J. R. Soc. Interface* **2018**, *15*(148). https://doi.org/10.1098/rsif.2018.0640.
21. Georgiev, D. D. *Quantum Information and Consciousness: A Gentle Introduction*; CRC Press: Boca Raton, 2017. https://doi.org/10.1201/9780203732519.
22. Georgiev, D. D. Inner Privacy of Conscious Experiences and Quantum Information. *Biosystems* **2020**, *187*, 104051. https://doi.org/10.1016/j.biosystems.2019.104051.
23. Georgiev, D. D.; Glazebrook, J. F. Subneuronal Processing of Information by Solitary Waves and Stochastic Processes. In: *Nano and Molecular Electronics Handbook*; Lyshevski, S. E. Ed.; CRC Press: Boca Raton, 2007; pp 1–41 (chapter 17).
24. Georgiev, D. D.; Glazebrook, J. F. Quasiparticle Tunneling in Neurotransmitter Release. In: *Handbook of Nanoscience, Engineering, and Technology*, Goddard III, W. A., Brenner, D., Lyshevski, S. E., Iafrate, G. J., Eds.; 3rd ed. CRC Press: Boca Raton, 2012; pp 983–1016 (chapter 30).
25. Schrödinger, E. *What is Life? The Physical Aspect of the Living Cell & Mind and Matter*; Cambridge University Press: Cambridge, 1977.
26. Lambert, N.; Chen, Y. N.; Cheng, Y. C.; Li, C. M.; Chen, G. Y.; Nori, F. Quantum Biology. *Nat. Phys.* **2013**, *9*(1), 10–18. https://doi.org/10.1038/nphys2474.
27. Brookes, J. C. Quantum Effects in Biology: Golden Rule in Enzymes, Olfaction, Photosynthesis and Magnetodetection. *Proc. R. Soc. A* **2017**, *473*, 20160822. https://doi.org/10.1098/rspa.2016.0822.
28. De Vault, D. *Quantum-Mechanical Tunneling in Biological Systems*; Vol. 2. Cambridge University Press: Cambridge, UK, 1984.
29. Sutcliffe, M. J.; Scrutton, N. S. Enzyme Catalysis: Over-The-Barrier or Through-The-Barrier? *Trends Biochem. Sci.* **2000**, *25*(9), 405–408. https://doi.org/10.1016/S0968-0004(00)01642-X.
30. Basran, J.; Patel, S.; Sutcliffe, M. J.; Scrutton, N. S. Importance of Barrier Shape in Enzyme-Catalyzed Reactions: Vibrationally Assisted Hydrogen Tunneling in Tryptophan Tryptophylquinone-Dependent Amine Dehydrogenases. *J. Biol. Chem.* **2001**, *276*(9), 6234–6242. https://doi.org/10.1074/jbc.M008141200.
31. Fleming, G. R.; Scholes, G. D.; Cheng, Y. C. Quantum Effects in Biology. *Procedia Chem.* **2011**, *3*, 38–57.
32. Ichinose, S. Soliton Excitations in Alpha-Helical Protein Structures. *Chaos Solitons Fractals* **1991**, *1*(6), 501–509. https://doi.org/10.1016/0960-0779(91)90040-G.
33. Edison, A. S. Linus Pauling and the Planar Peptide Bond. *Nat. Struct. Biol.* **2001**, *8*(3), 201–202. https://doi.org/10.1038/84921.
34. Kovac, J. A Weird Insult From Norway: Linus Pauling as Public Intellectual. *Soundings Interdiscip. J.* **1999**, *82*(1–2), 91–106.
35. Barth, A. Infrared Spectroscopy of Proteins. *Biochim. Biophys. Acta Bioenerg.* **2007**, *1767*(9), 1073–1101. https://doi.org/10.1016/j.bbabio.2007.06.004.
36. Sholl, D. S.; Steckel, J. A. *Density Functional Theory: A Practical Introduction*; John Wiley & Sons: Hoboken, NJ, 2009.
37. Kolev, S. K.; Petkov, P. S.; Rangelov, M.; Vayssilov, G. N. Ab Initio Molecular Dynamics of Na^+ and Mg^{2+} Countercations at the Backbone of RNA in Water Solution. *ACS Chem. Biol.* **2013**, *8*(7), 1576–1589. https://doi.org/10.1021/cb300463h.
38. Kolev, S. K.; Petkov, P. S.; Rangelov, M. A.; Trifonov, D. V.; Milenov, T. I.; Vayssilov, G. N. Interaction of Na^+, K^+, Mg^{2+} and Ca^{2+} Counter Cations With RNA. *Metallomics* **2018**, *10*(5), 659–678. https://doi.org/10.1039/c8mt00043c.
39. Pace, C. N.; Fu, H.; Lee Fryar, K.; Landua, J.; Trevino, S. R.; Schell, D.; Thurlkill, R. L.; Imura, S.; Scholtz, J. M.; Gajiwala, K.; Sevcik, J.; Urbanikova, L.;

Myers, J. K.; Takano, K.; Hebert, E. J.; Shirley, B. A.; Grimsley, G. R. Contribution of Hydrogen Bonds to Protein Stability. *Protein Sci.* **2014**, *23*(5), 652–661. https://doi.org/10.1002/pro.2449.

40. Myshakina, N. S.; Ahmed, Z.; Asher, S. A. Dependence of Amide Vibrations on Hydrogen Bonding. *J. Phys. Chem. B* **2008**, *112*(38), 11873–11877. https://doi.org/10.1021/jp8057355.
41. Li, X. Z.; Walker, B.; Michaelides, A. Quantum Nature of the Hydrogen Bond. *Proc. Natl. Acad. Sci.* **2011**, *108*(16), 6369–6373. https://doi.org/10.1073/pnas.1016653108.
42. Pusuluk, O.; Farrow, T.; Deliduman, C.; Burnett, K.; Vedral, V. Proton Tunnelling in Hydrogen Bonds and Its Implications in an Induced-Fit Model of Enzyme Catalysis. *Proc. R. Soc. A* **2018**, *474*(2218), 20180037. https://doi.org/10.1098/rspa.2018.0037.
43. Pusuluk, O.; Torun, G.; Deliduman, C. Quantum Entanglement Shared in Hydrogen Bonds and Its Usage as a Resource in Molecular Recognition. *Mod. Phys. Lett. B* **2018**, *32*(26), 1850308. https://doi.org/10.1142/S0217984918503086.
44. Krimm, S.; Bandekar, J. Vibrational Spectroscopy and Conformation of Peptides, Polypeptides, and Proteins. *Adv. Protein Chem.* **1986**, *38*, 181–364. https://doi.org/10.1016/S0065-3233(08)60528-8.
45. Hamm, P.; Lim, M.; Hochstrasser, R. M. Structure of the Amide I Band of Peptides Measured by Femtosecond Nonlinear-Infrared Spectroscopy. *J. Phys. Chem. B* **1998**, *102*(31), 6123–6138. https://doi.org/10.1021/jp9813286.
46. Manas, E. S.; Getahun, Z.; Wright, W. W.; DeGrado, W. F.; Vanderkooi, J. M. Infrared Spectra of Amide Groups in α-Helical Proteins: Evidence for Hydrogen Bonding Between Helices and Water. *J. Am. Chem. Soc.* **2000**, *122*(41), 9883–9890. https://doi.org/10.1021/ja001782z.
47. Wiggins, P. M. Role of Water in Some Biological Processes. *Microbiol. Rev.* **1990**, *54*(4), 432–449.
48. Wiggins, P. M. Life Depends on Two Kinds of Water. *PLoS ONE* **2008**, *3*(1), e1406. https://doi.org/10.1371/journal.pone.0001406.
49. Yang, H.; Yang, S.; Kong, J.; Dong, A.; Yu, S. Obtaining Information About Protein Secondary Structures in Aqueous Solution Using Fourier Transform IR Spectroscopy. *Nat. Protoc.* **2015**, *10*(3), 382–396. https://doi.org/10.1038/nprot.2015.024.
50. Edler, J.; Hamm, P.; Scott, A. C. Femtosecond Study of Self-Trapped Vibrational Excitons in Crystalline Acetanilide. *Phys. Rev. Lett.* **2002**, *88*(6), 067403. https://doi.org/10.1103/PhysRevLett.88.067403.
51. Edler, J.; Hamm, P. Self-Trapping of the Amide I Band in a Peptide Model Crystal. *J. Chem. Phys.* **2002**, *117*(5), 2415–2424. https://doi.org/10.1063/1.1487376.
52. Edler, J.; Pfister, R.; Pouthier, V.; Falvo, C.; Hamm, P. Direct Observation of Self-Trapped Vibrational States in α-Helices. *Phys. Rev. Lett.* **2004**, *93*(10), 106405. https://doi.org/10.1103/PhysRevLett.93.106405.
53. Edler, J.; Hamm, P. Spectral Response of Crystalline Acetanilide and *N*-Methylacetamide: Vibrational Self-Trapping in Hydrogen-Bonded Crystals. *Phys. Rev. B* **2004**, *69*(21), 214301. https://doi.org/10.1103/PhysRevB.69.214301.
54. Georgiev, D. D.; Glazebrook, J. F. On the Quantum Dynamics of Davydov Solitons in Protein α-Helices. *Physica A* **2019**, *517*, 257–269. https://doi.org/10.1016/j.physa.2018.11.026.
55. Davydov, A. S. The Theory of Contraction of Proteins Under Their Excitation. *J. Theor. Biol.* **1973**, *38*(3), 559–569. https://doi.org/10.1016/0022-5193(73)90256-7.
56. Davydov, A. S.; Kislukha, N. I. Solitons in One-Dimensional Molecular Chains. *Phys. Status Solidi B* **1976**, *75*(2), 735–742. https://doi.org/10.1002/pssb.2220750238.
57. Davydov, A. S. Solitons in Molecular Systems. *Phys. Scr.* **1979**, *20*(3–4), 387–394. https://doi.org/10.1088/0031-8949/20/3-4/013.

58. Davydov, A. S. Solitons in Quasi-One-Dimensional Molecular Structures. *Sov. Phys. Uspekhi* **1982**, *25*(12), 898–918. https://doi.org/10.1070/pu1982v025n12abeh005012.
59. Davydov, A. S. Quantum Theory of the Motion of a Quasi-Particle in a Molecular Chain With Thermal Vibrations Taken Into Account. *Phys. Status Solidi B* **1986**, *138*(2), 559–576. https://doi.org/10.1002/pssb.2221380221.
60. Davydov, A. S.; Ermakov, V. N. Linear and Nonlinear Resonance Electron Tunneling Through a System of Potential Barriers. *Physica D* **1987**, *28*(1), 168–180. https://doi.org/10.1016/0167-2789(87)90127-8.
61. Davydov, A. S.; Ermakov, V. N. Soliton Generation at the Boundary of a Molecular Chain. *Physica D* **1988**, *32*(2), 318–323. https://doi.org/10.1016/0167-2789(88)90059-0.
62. Brizhik, L. S.; Davydov, A. S. Soliton Excitations in One-Dimensional Molecular Systems. *Phys. Status Solidi B* **1983**, *115*(2), 615–630. https://doi.org/10.1002/pssb.2221150233.
63. Brizhik, L. S.; Gaididei, Y. B.; Vakhnenko, A. A.; Vakhnenko, V. A. Soliton Generation in Semi-Infinite Molecular Chains. *Phys. Status Solidi B* **1988**, *146*(2), 605–612. https://doi.org/10.1002/pssb.2221460221.
64. Brizhik, L. S. Soliton Generation in Molecular Chains. *Phys. Rev. B* **1993**, *48*(5), 3142–3144. https://doi.org/10.1103/PhysRevB.48.3142.
65. Brizhik, L. S.; Eremko, A. A. Electron Autolocalized States in Molecular Chains. *Physica D* **1995**, *81*(3), 295–304. https://doi.org/10.1016/0167-2789(94)00206-6.
66. Brizhik, L. S.; Eremko, A.; Piette, B.; Zakrzewski, W. Solitons in α-Helical Proteins. *Phys. Rev. E* **2004**, *70*(3), 031914. https://doi.org/10.1103/PhysRevE.70.031914.
67. Brizhik, L.; Eremko, A.; Piette, B.; Zakrzewski, W. Charge and Energy Transfer by Solitons in Low-Dimensional Nanosystems With Helical Structure. *Chem. Phys.* **2006**, *324*(1), 259–266. https://doi.org/10.1016/j.chemphys.2006.01.033.
68. Brizhik, L.; Eremko, A.; Piette, B.; Zakrzewski, W. Ratchet Effect of Davydov's Solitons in Nonlinear Low-Dimensional Nanosystems. *Int. J. Quantum Chem.* **2010**, *110*(1), 25–37. https://doi.org/10.1002/qua.22083.
69. Cruzeiro, L.; Halding, J.; Christiansen, P. L.; Skovgaard, O.; Scott, A. C. Temperature Effects on the Davydov Soliton. *Phys. Rev. A* **1988**, *37*(3), 880–887. https://doi.org/10.1103/PhysRevA.37.880.
70. Cruzeiro-Hansson, L.; Kenkre, V. M. Localized Versus Delocalized Ground States of the Semiclassical Holstein Hamiltonian. *Phys. Lett. A* **1994**, *190*(1), 59–64. https://doi.org/10.1016/0375-9601(94)90366-2.
71. Cruzeiro-Hansson, L.; Takeno, S. Davydov Model: The Quantum, Mixed Quantum-Classical, and Full Classical Systems. *Phys. Rev. E* **1997**, *56*(1), 894–906. https://doi.org/10.1103/PhysRevE.56.894.
72. Cruzeiro, L. The Davydov/Scott Model for Energy Storage and Transport in Proteins. *J. Biol. Phys.* **2009**, *35*(1), 43–55. https://doi.org/10.1007/s10867-009-9129-0.
73. Kerr, W. C.; Lomdahl, P. S. Quantum-Mechanical Derivation of the Equations of Motion for Davydov Solitons. *Phys. Rev. B* **1987**, *35*(7), 3629–3632. https://doi.org/10.1103/PhysRevB.35.3629.
74. Kerr, W. C.; Lomdahl, P. S. Quantum-Mechanical Derivation of the Davydov Equations for Multi-Quanta States. In: *Davydov's Soliton Revisited: Self-Trapping of Vibrational Energy in Protein*; Christiansen, P. L., Scott, A. C., Eds.; Springer: New York, 1990; pp 23–30. https://doi.org/10.1007/978-1-4757-9948-4_2.
75. MacNeil, L.; Scott, A. C. Launching a Davydov Soliton: II. Numerical Studies. *Phys. Scr.* **1984**, *29*(3), 284–287. https://doi.org/10.1088/0031-8949/29/3/017.
76. Scott, A. C. Launching a Davydov Soliton: I. Soliton Analysis. *Phys. Scr.* **1984**, *29*(3), 279–283. https://doi.org/10.1088/0031-8949/29/3/016.

77. Scott, A. C. Davydov Solitons in Polypeptides. *Philos. Trans. R. Soc. London, Ser. A Math. Phys. Sci.* **1985**, *315*(1533), 423–436. https://doi.org/10.1098/rsta.1985.0049.
78. Scott, A. C. Davydov's Soliton. *Phys. Rep.* **1992**, *217*(1), 1–67. https://doi.org/10.1016/0370-1573(92)90093-F.
79. Luo, B.; Ye, J.; Zhao, Y. Variational Study of Polaron Dynamics With the Davydov Ansätze. *Phys. Status Solidi C* **2011**, *8*(1), 70–73. https://doi.org/10.1002/pssc.201000721.
80. Luo, J.; Piette, B. A Generalised Davydov-Scott Model for Polarons in Linear Peptide Chains. *Eur. Phys. J. B* **2017**, *90*(8), 155. https://doi.org/10.1140/epjb/e2017-80209-2.
81. Sun, J.; Luo, B.; Zhao, Y. Dynamics of a One-Dimensional Holstein Polaron With the Davydov Ansätze. *Phys. Rev. B* **2010**, *82*(1), 014305. https://doi.org/10.1103/PhysRevB.82.014305.
82. Hennig, D. Energy Transport in α-Helical Protein Models: One-Strand Versus Three-Strand Systems. *Phys. Rev. B* **2002**, *65*(17), 174302. https://doi.org/10.1103/PhysRevB.65.174302.
83. Brizhik, L. S.; Luo, J.; Piette, B. M. A. G.; Zakrzewski, W. J. Long-Range Donor-Acceptor Electron Transport Mediated by α-Helices. *Phys. Rev. E* **2019**, *100*(6), 062205. https://doi.org/10.1103/PhysRevE.100.062205.
84. Holstein, T. Studies of polaron motion: Part I. The Molecular-Crystal Model. *Ann. Phys.* **1959**, *8*(3), 325–342. https://doi.org/10.1016/0003-4916(59)90002-8.
85. Holstein, T. Studies of Polaron Motion: Part II. The "Small" Polaron. *Ann. Phys.* **1959**, *8*(3), 343–389. https://doi.org/10.1016/0003-4916(59)90003-X.
86. Georgiev, D. D.; Glazebrook, J. F. Quantum Tunneling of Davydov Solitons Through Massive Barriers. *Chaos Solitons Fractals* **2019**, *123*, 275–293. https://doi.org/10.1016/j.chaos.2019.04.013.
87. Tauber, S. On Multinomial Coefficients. *Am. Math. Mon.* **1963**, *70*(10), 1058–1063. https://doi.org/10.1080/00029890.1963.11992172.
88. Dirac, P. A. M. *The Principles of Quantum Mechanics*, 4th ed.; Oxford University Press: Oxford, 1967.
89. Georgiev, D. D.; Glazebrook, J. F. Neurotransmitter Release and Conformational Changes Within the SNARE Protein Complex. In: *Nanoengineering, Quantum Science, and, Nanotechnology Handbook*; Lyshevski, S. E. Ed.; CRC Press: Boca Raton, 2019; pp 375–404.
90. Kivshar, Y. S.; Malomed, B. A. Dynamics of Solitons in Nearly Integrable Systems. *Rev. Mod. Phys.* **1989**, *61*(4), 763–915. https://doi.org/10.1103/RevModPhys.61.763.
91. Förner, W.; Ladik, J. Influence of Heat Bath and Disorder on Davydov Solitons. In: *Davydov's Soliton Revisited: Self-Trapping of Vibrational Energy in Protein*; Christiansen, P. L., Scott, A. C., Eds.; Springer: New York, 1990; pp 267–283. https://doi.org/10.1007/978-1-4757-9948-4_20.
92. Förner, W. Quantum and Disorder Effects in Davydov Soliton Theory. *Phys. Rev. A* **1991**, *44*(4), 2694–2708. https://doi.org/10.1103/PhysRevA.44.2694.
93. Itoh, K.; Shimanouchi, T. Vibrational Spectra of Crystalline Formamide. *J. Mol. Spectrosc.* **1972**, *42*(1), 86–99. https://doi.org/10.1016/0022-2852(72)90146-4.
94. Nevskaya, N. A.; Chirgadze, Y. N. Infrared Spectra and Resonance Interactions of Amide-I and II Vibrations of α-Helix. *Biopolymers* **1976**, *15*(4), 637–648. https://doi.org/10.1002/bip.1976.360150404.
95. Landau, L. D.; Lifshitz, E. M. *Quantum Mechanics. Course of Theoretical Physics*; Vol. 3; equations (33) and (34) for multiple. Pergamon Press: Oxford, 1965.
96. Georgiev, D. D.; Glazebrook, J. F. The Quantum Physics of Synaptic Communication via the SNARE Protein Complex. *Prog. Biophys. Mol. Biol.* **2018**, *135*, 16–29. https://doi.org/10.1016/j.pbiomolbio.2018.01.006.

97. Jordan, P. *Die Physik und das Geheimnis des Organischen Lebens*. Friedrich Vieweg & Sohn: Braunschweig, Germany, 1941. https://repository.aip.org/islandora/object/nbla:8133.
98. Beyler, R. H. Targeting the Organism: The Scientific and Cultural Context of Pascual Jordan's Quantum Biology, 1932-1947. *Isis* **1996**, *87*(2), 248–273.
99. Born, M.; Oppenheimer, R. Zur Quantentheorie der Molekeln. *Ann. Phys.* **1927**, *389*(20), 457–484. https://doi.org/10.1002/andp.19273892002.
100. Lowe, J. P.; Peterson, K. *Quantum Chemistry*, 3rd ed.; Academic Press: Amsterdam, 2005.
101. Ganesan, S. J.; Matysiak, S. Role of Backbone Dipole Interactions in the Formation of Secondary and Supersecondary Structures of Proteins. *J. Chem. Theory Comput.* **2014**, *10*(6), 2569–2576. https://doi.org/10.1021/ct401087a.
102. Cruzeiro, L. The VES Hypothesis and Protein Misfolding. *Discrete Contin. Dynam. Systems S* **2011**, *4*(5), 1033–1046. https://doi.org/10.3934/dcdss.2011.4.1033.
103. Alberts, B.; Bray, D.; Hopkin, K.; Johnson, A. D.; Lewis, J.; Raff, M.; Roberts, K.; Walter, P. *Essential Cell Biology*, 4th ed.; Garland Science: New York, 2013.
104. Milner-White, E. J. Protein Three-Dimensional Structures at the Origin of Life. *Interface Focus* **2019**, *9*(6), 20190057. https://doi.org/10.1098/rsfs.2019.0057.
105. Poznanski, R. R.; Cacha, L. A.; Latif, A. Z. A.; Salleh, S. H.; Ali, J.; Yupapin, P.; Tuszynski, J. A.; Ariff, T. M. Molecular Orbitals of Delocalized Electron Clouds in Neuronal Domains. *Biosystems* **2019**, *183*, 103982. https://doi.org/10.1016/j.biosystems.2019.103982.
106. Brizhik, L. S. Influence of Electromagnetic Field on Soliton-Mediated Charge Transport in Biological Systems. *Electromagn. Biol. Med.* **2015**, *34*(2), 123–132. https://doi.org/10.3109/15368378.2015.1036071.
107. Georgiev, D. D. Monte Carlo Simulation of Quantum Zeno Effect in the Brain. *Int. J. Mod. Phys. B* **2015**, *29*(7), 1550039. https://doi.org/10.1142/S0217979215500393.
108. Pross, A.; Pascal, R. The Origin of Life: What We Know, What We Can Know and What We Will Never Know. *Open Biol.* **2013**, *3*(3), 120190. https://doi.org/10.1098/rsob.120190.
109. Kitadai, N.; Maruyama, S. Origins of Building Blocks of Life: A Review. *Geosci. Front.* **2018**, *9*(4), 1117–1153. https://doi.org/10.1016/j.gsf.2017.07.007.
110. Lee, D. H.; Severin, K.; Yokobayashi, Y.; Ghadiri, M. R. Emergence of Symbiosis in Peptide Self-Replication Through a Hypercyclic Network. *Nature* **1997**, *390*(6660), 591–594. https://doi.org/10.1038/37569.
111. Duim, H.; Otto, S. Towards Open-Ended Evolution in Self-Replicating Molecular Systems. *Beilstein J. Org. Chem.* **2017**, *13*, 1189–1203. https://doi.org/10.3762/bjoc.13.118.
112. Andras, P.; Andras, C. The Origins of Life-the 'Protein Interaction World' Hypothesis: Protein Interactions Were the First Form of Self-Reproducing Life and Nucleic Acids Evolved Later as Memory Molecules. *Med. Hypotheses* **2005**, *64*(4), 678–688. https://doi.org/10.1016/j.mehy.2004.11.029.

CHAPTER NINE

Panexperiential materialism: A physical exploration of qualitativeness in the brain

Roman R. Poznański[a,*] and Erkki J. Brändas[b]
[a]Faculty of Informatics and Computing, Universiti Sultan Zainal Abidin, Terengganu, Malaysia
[b]Department of Chemistry, Uppsala University, Uppsala, Sweden
*Corresponding author: e-mail address: romanrichard@unisza.edu.my

Contents

Abstract

It is a century-old view that experiential philosophies are not compatible with materialism. In the contextual inconsistency with the reality, that *matter* is inertly acquiring only a single physical state, philosophers have gained ground in metaphysical beliefs, including dualism, monism, and idealism. We show that a new foundational self-referential identity theory of the mind is needed to bridge the explanatory gap. Panexperiential materialism is a new materialistic framework originating in the spectral domain of matter-wave energy quanta transcending the barrier of thermoquantal information, isomorphically aligning with consciousness. The holistic nature of its instantiation is panexperiential due to the *composite states of non-inert matter*, depending crucially on their interrelations without embracing essentialist ontology, further entwined with epistemic teleofunctionalism and informational relationalism, and based on the research agenda, concepts, and shared values of quantum chemistry. Panexperiential materialism is characterized by a spectral

Advances in Quantum Chemistry, Volume 82
ISSN 0065-3276
https://doi.org/10.1016/bs.aiq.2020.08.004

matter-wave structure, which is conjugate to the prescriptive structural properties of the spacetime domain. Yet panexperiential materialism is not contrary to ordinary materialism, although the latter may be fundamentally grounded in molecular networks. The phenomenology of consciousness is not merely a mental reification in the first-person perspective. The proper guideline should be the reduction of conscious processes to non-reductive physical correlates in the brain. The wet and hot environment of the brain affords quantum-thermal correlations in a transcending energy processing zone where quantum and classical fluctuations are fused to thermoquantal information. The quantum chemical basis incorporates non-self-adjoint analytic extensions in Liouville space and associated Fourier-Laplace transforms that conjoin energy, time, entropy, and temperature. The transformation across hierarchical thermodynamical domains is caused by the negentropic gain wholly implicated by the entropy production arising in the energy exchange resulting in the transformation of information forming informational holarchies, driven by nonlocal teleological mechanisms. The information transformation from the objective to the subjective is a process that is quantum in nature. The process of non-integrated information, actualizing the information-based action as a teleological process of cognition in the entailment of preconscious experientialities, should not be conflated with the experience itself, but rather as an isomorphic connection between mind and brain via the Fourier-Laplace transformation. Our holistic viewpoint denies the existence of integrated information as an emergentist ontology, instead advocating the canonical transformations $\boldsymbol{B}$ and $\boldsymbol{B}\dagger$ as the syntax or universal grammar for intrinsic information (proto-communication). The irreducible character of an informational holarchy where the whole is affected non-synergistically by the non-integrated information is how intrinsic information encapsulates the energy transformation from fusing thermal and quantum fluctuations that result in long-range correlations (phase wave) that constitutes the fundamental dynamics of physical feelings. In panexperiential materialism, there is no issue dividing holists and reductionists, concerning the issue whether the whole or the discrete parts are primary, but rather their interrelations. This relationalism is pivotal in understanding how non-integrated information holistically concresce. Although we consider matter waves to be fundamental, one might say, avoiding the trap of eliminative materialism, that the brain is conjugate to the mind and vice versa.

1. Introduction

Quantum phenomena present a compelling alternative to the predominant conceptions in the neuro- and cognitive sciences, incepting neurons organized into networks as the primary constituents of the animate brain [1]. We do realize that quantum effects, such as linear wave superpositions, quantum entanglement, and the associated decoherence phenomenon, are not sufficient to tackle biological problems, such as the workings of the material brain. This is so simply because the relevant energy-time scales and the attendant temperature dependence, related to a teleological interpretation of consciousness, have as yet no counterpart in the pioneering formulation of quantum mechanics. Nevertheless, challenging paradoxes, such as those related to the mind-body problem, collapse interpretations of traditional quantum mechanics, and extensions to open quantum systems in Liouville space, the latter a space of superoperators with the ordinary Hilbert space as carrier space, often confronted by critics, are nowadays progressed to be legitimately resolved within a quantum chemistry paradigma.

A naturalistic explanation of consciousness requires an understanding of nontrivial quantum effects arising in the brain. Schrödinger's equation and its wave function solutions do not provide information about particle trajectories. According to Bohr, the particle and wave aspects of physical objects are complementary pairs. Still, de Broglie [2] had proposed a pilot wave construct to explain the observed wave-particle duality. In his view, each particle has a well-defined position and momentum but is guided by a wave function derived from the Schrödinger equation. In the pilot wave theory, the particle's motion is subject to a guiding equation or quantum potential. Dual roles of the wave function not only determine the likely location of a particle, but it also influences the location by exerting a force on the orbit-acting as a pilot wave that guides the particle. Therefore, it is not enough to deal only with the quantum dynamics of energy quanta of matter or solely with "matter waves." Our extension, e.g., to include temperature and entropy, puts teleology on an authentic scientific footing that standard quantum mechanics *à la* Bohr cannot accomplish.

Our approach is thermodynamic, leaning towards the brain correlated irreversible processes of organized complexity [3]. In hierarchical thermodynamics, each level is a complex system that can be decomposed into an ascending family of successive, more encompassing subsystems subordinated to the next level, i.e., into levels of organization. Each structure in the hierarchy constitutes a steady state with the entropy change $dS = 0$. The result is a negentropic gain in the system as a consequence of entropy production at each level. Although cognitive information processing displaying self-organization, it is crucial to understand that a holistic account of the brain is provided by the second law of thermodynamics, which supports the view of the evolution and development of the human brain as an indigenous thermodynamic process by nature [4].

It is important to note that living systems that include a central nervous system require, as said, a continuous dissipation of energy in organisms to maintain order through entropy export (*negentropic action*) [5]. In the context of *non-equilibrium thermodynamics*, there is a production of negentropy across hierarchical levels via a steady-state entropy production rate. More detailed, the active cycles operate within each ordered level under steady-state conditions, with no entropic exchange in between, harnessing the entropy produced from the irreversible processes inside the dissipative system. In other words, the entropic exchange within and between holarchical domains works as an "open" system, which is operating out of and often far from equilibrium [6]. Such open systems are not only rooted deep within the molecular structure of all neurons, but the essential constituents are processes "*in circular independence with a larger fabric*," which cannot be reconstructed piecewise through traditional reductive physics based on hierarchical emergence [7]. It is the subsystems at higher levels in the holarchy that is "conscious" of the degrees of freedom of those at the lower levels, where "conscious" stems not from emergence, which is a subset of synergism, but from the intrinsic nature of the holarchy that is the bedrock in the formulation of conjugate interrelationship. According to Arthur Koestler [8,9], such a holarchy is a generalized version of a hierarchy in a complex system with holons in a contextually nested relationship through a self-reinforcing "resonance" between different hierarchical levels.

The brain is self-referential; the self-referential observation settles potentialities into a transition of states related to various fluctuation-induced time scales. How do we use it in the context of quantum physics? The "act of self-observation" means that the observation is expressed through the negation of the classical potential energy given the quantum potential, Q, but this was not part of Bohm's ontological explanation of quantum mechanics [10] nor his theory of the mind encompassing an infinite tower of pilot waves, each controlling the level below in the context of an informational holarchy [11,12]. How could a quantum system in the brain refer to itself? The answer is that it is not possible to do so within standard quantum theory. One needs extensions that, however, preserve the most important quantum mechanical relations, yet with constitutive constraints and associated symmetry violations. We will discuss this both in a Liouvillian density matrix framework as well as in *molecular Bohmian mechanics*. The reader is cautioned that the application of quantum physics will address the characteristic bearing of self-reference without presenting a clear boundary between mind and matter [12], yet derive their conjugate interrelationship from first principles. Note that the conjugate partners energy-time and, similarly, momentum-space appear already in the classical formulation, as, e.g., in the Fourier transformation and its inverse, between time and frequency of the respective correlation functions [13].

There has appeared a lot of work in the past, portraying the importance of the classical Fourier transformation taking place in biological systems [14]. These ideas were mainly designed and prepared for the quantum regime [15]. They defined the holomovement or holoflux, of which we are part, as movement through space and time bounded by the frequency domain of energy-momentum processing. The latter, according to Pribram [14], generate vast quantities of information using "resonant photons" in a transformational process of information-coded-energies flowing back and forth between spectral and spacetime. This internalized "negentropic" flow is information in contrast to matter and momentum as an externalized form of flux determined by the distribution of energy and momentum and measured as spectral density [16].

Karl Pribram [14,17] called his approach "the holonomic brain theory," adapting the richness of holography theory [18] to the dendrites of cortical neurons. Our theory is similar to the holonomic brain theory, where the Fourier transform of the interfering waves provides the spectral receipt for the interference patterns created by phase differences. However, by incorporating "matter waves" [2] as the fundamental starting point, the conjugate relationship is yet imperative in limiting the domain accuracy. Already incorporated are the ripples of the matter waves as phase-waves in the momentum and frequency domain linked to spacetime. This link is analogous to that of a prism, where a light beam is broken up into a spectrum of colors—the Fourier transform breaks up the function into a spectrum of frequencies. However, our approach extends to situations where time scales and temperature dependences are crucial. As a result, time signals are not acausal but teleological, as they contain a stochastic bias, and this time directedness necessitates a combined Fourier-Laplace transform that points to spatiotemporal waveform patterns as possible generations of physical feelings.

In the quantum regime, the conjugate variables are replaced by operators and their associated non-commutation relations. Extension of quantum theories to dissipative system dynamics of relevance for the present theme must maintain this operator structure. Consequently, we have developed a theoretical framework that addresses the mind-brain interface through a constitution that we call "*panexperiential materialism.*" It is based on Freudian metapsychology of preconscious experience interacting with consciousness. Thus, it signifies the meaning of "pan" in panexperiential as representing all processes that enter awareness: unconscious, subconscious, and preconscious. Avoiding the medical definition of consciousness as conscious awareness, we focus on the fundamental level of preconscious experientialities, dispensing with Chalmers's [19] notion of consciousness as phenomenal or exclusively conscious experience. By preconscious experience, we mean the fundamental mechanism by which conscious experiences take form within the brain at fundamental levels. The latter comprises a framework for consciousness that aligns with the idea that the teleological formulation related to hierarchical thermodynamics of living organisms, through molecular evolution, has acquired a means of experiencing the world self-referentially [20–22].

The process philosophy of Alfred Whitehead [23] regarded consciousness as a sequence of discrete "occasions of experience" in natural evolution. This promoted possible interpretations, such as "*innatism*" of pre-existing ideas present in the mind, and "*nativism*" grounded in the genome as the source of the mind, the latter being duly criticized [24]. Panexperiential materialism is a repositioning of panexperientialism and related frameworks as a habitual materialistic stance. Panexperiential materialism is essentially a new ontology that forms the basis of a *self-referential identity theory* that differs from the identity theory or reductive materialism that brain states are mental states. The strength of the self-referential identity theory is its constitutive explanations for both subjectivity and informational content, i.e., semantics and possibly pragmatics, without speculating on the emergence of consciousness [25].

In *panexperiential materialism*, a holarchy, i.e., an organization ruled by self-containment, reflects upon informational descriptors as the fundamental agent of experience associated with feelings in humans or other higher life forms. The crux of the problem is how to physically define *informational descriptors of "experience"* in terms of isomorphic links between quantum holons [8] in integrative biology [26]. Many models of the mind use both philosophical and psychophysical definitions of the mathematical concept of isomorphism. What we are interested in is the description of "*meaning*" through isomorphism, not in the mathematical sense, but in terms of informational relationalism where isomorphisms between quantum holons are teleofunctions carrying "meaning" as an "*information function*" in which the purpose is to inform. These descriptors of protophenomenal properties form informational holarchies driven by *teleological processes*, governed by quantum holons as both hierarchical and self-referential evolving under steady-state conditions, as part of contextually nested relationships in the informational holarchy. In particular, thermoquantal *informational descriptors* define quantum-thermal correlations fused into one entity, which in the steady-state form holarchies from e.g. delocalized π-electron clouds to simple amphipathic molecules and micelles [27], thus extending quantum holism [28], while not facing combination problems [29].

Teleological processes often take place, not through the atomistic *emergence* of synthesis, but by the *self-referential amplification* [30] of the quantum potential energy as a *nonlocal teleological mechanism* governing simultaneity when viewed from a first-person perspective. Such energetic structures in the brain [31] being pivotal in understanding how non-integrated information holistically concresce. They reflect internal energy that suggests a direct relation with the *quantum potential energy* [32], which gives form to the kinetic energy in the frequency domain of the internal motion [33]. This is fundamentally orchestrated through energy exchanges resulting in isomorphic connections between brain and mind defined as information preserving transformations from the atomic scale to the molecular scale, where the phase differences of the phase-wave evolve through phase-space as a unified whole controlled by the information-based action of the "quantum force."

The essence of *the hard problem* of consciousness [19] is understanding how organisms have qualia that are created by the organization of energy and entropy in the brain [31]. The *hard problem* is hard because of a foreseeable unbridgeable explanatory gap [34]. Panexperiential materialism aims at understanding the *hard problem of consciousness* [19] as a problem of experience or "*qualitativeness*" [35]. In other words, every conscious experience is ingrained with a "*scintilla of subjectivity*" that exists when the experience happens. Hence "*qualitativeness*" is commensurate with "*subjectivity*" simply because, for there to be a qualitative feel, there must be self-awareness of that experience.

According to Musacchio [36], intrinsic information or internalized information at the submolecular scale contributes to the "qualitativeness" of experience. Thus, there is a hidden world where "*qualitativeness*" is referred to as "*preconscious experientialities*," and it is only through "*unity*" that the final articulation of conscious experience is possible [35]. The "unity" comes not from the perspective of *pantheism*, which is *constitutive holistic panpsychism*, but through *coherent phenomena* involving long-range phase-wave correlations and their quantum physical merging with thermal fluctuations.

All conscious experiences unfold as part of a single unified conscious field originating at a submolecular scale where intrinsic information, wholly entailed by potentiality arising from energy exchanges in the quantum domain, unfolds. Such "*unity*" illuminates those phases with the greater clarity and distinctness through *conscious cognition* that includes both "*qualitativeness*" and "*subjectivity*" originating from *preconscious experientialities*. However, regarding the latter, qualia being entropy-decreasing, are not considered to be physically observable classically in agreement with Maccone's argument [37]. The capacity of feeling is rooted in the *informational descriptors of preconscious experientialities*. The idea of preconsciousness implies that there is a conscious experience in terms of *unity*, so this refers to the notion of qualia as the content of *conscious* experience combined to form *unity*. The transparency of conscious experiences and the imperceptibility of the neurobiological processes that realize them are not intrinsic but relational. Conscious experiences provide phenomenal concepts associated with their "aboutness" as the experiences that refer to the internal and external environment consisting of the establishment of a relationship between conscious experiences [36].

Subjective experience—"*what it is like*," does not capture the essence of *experientiality* (or feeling) experienced in a first-person perspective [38,39]. For example, to perceive a scene, to endure pain, to entertain a thought, or to reflect on the experience itself [40] have been mostly linked to subcortical affective mechanisms that arise in the understanding of the *feeling aspects* of emotional processes [41,42]. The subjective experience is hidden from the external world in the subconscious and is brought into conscious experience as qualia. We know from the work of Gerald Edelman and colleagues [43] that qualia are the ingredients of consciousness, which means that qualia are contents of subjective, *conscious experience* as a phenomenological concept of potentiality. Phenomenal experience implies a specific character of subjective experience. Particular qualities of such experiences are typically called qualia, and functional explanations are, in this sense, "formally" non-phenomenal.

The underlying function of the cognitive availability of informational content, i.e., the "easy problem" of consciousness, is not the same as the "*hard problem*" of consciousness, the latter based on the false assumption that matter is essentially non-informational (or non-inert) [19]. The vital issue has been whether consciousness exists without content (or qualia). The argument, according to Merker [44], suggests that *consciousness* should not be a brain state that has content that is separated from the content of subjective experience. That is, if consciousness is a biological phenomenon, it must nevertheless be associated with brain states, which are always objective [35]. In the context of an informational holarchy [9] implicated in the quantum fabric of matter [45], i.e., instantiated in matter from the perspective of quantum chemistry/biology, the action of instantiating necessitates the transformation from the objective to the subjective, and this is only possible through non-integrated information. Consequently, consciousness cannot have quantity, but only quality, as the former is determined by the amount of integrated information generated by a system [46].

We do not subscribe to the existence of two interacting brain processes: the first one in charge of cognitive processes and the second mediating physiological feelings about cognitive contents where the two are coupled by feedback supporting conscious processes [47,48] in which the reasoning behind affective processes focuses on the neural bases of emotions [49,50]. Consciousness anticipated by cognition, is similar to a two-factor approach based on proto-consciousness and its activating agents (or conscious perception) [51] and is a trap for the existence of a fictitious concept of "*pure consciousness*" that does not necessarily mean a "thing" [52], but only something not derived from experience/feeling. Instead, to avoid the trap of "*pure consciousness*," consciousness becomes a composite of hierarchically organized and functional processes [36] rooted at the fundamental level that forms complex enough molecular building blocks, all based on the law of self-references [21].

If one were to make a clear distinction between *sentience* and experience, then the animate matter has not all the potential for experience. In hierarchal organizations that have evolved brains, such *sentience* is "experienced" by higher-order phenomena in what has become the essence of conscious processes. Accordingly, a consequence of hierarchical organizations is that the novel emergent function cannot be explained by the parts alone, but rather must be explained by the properties of the parts and their interactions [7]. However, *sentience* is not enough, although always being part of consciousness. The ultimate defining irreducible characteristic of consciousness, however, is the "qualitative feel" of experience. At the most fundamental level, the *experience* can be unconscious as a potentiality and only takes place at the "boundaries." This is an example of the philosophical idea that experience is always relational as it needs to be felt, higher-up. If experience is not relational, then *sentience* is but only a reflex arc and not even the minimal capacity to have subjective experience. The stimulus-response reaction that is endemic to most natural processes begins with rudimentary prehension (i.e., sensing through cellular walls). This is not *experience* but *sentience.* The notion of "*experience*" is intrinsic to the brain [53,54], and such intrinsicness stems within a holarchy.

Philosophers assert *nonphysical intrinsic properties to experience*. For example, when according to David Longinotti [55], the energy transduces to phenomenal energy, it becomes responsible for phenomenal experiences and other *mystical* phenomena, such as *phenomenal information* and phenomenal consciousness [56,57]. However, the phenomenology of consciousness is a mistake [58]. The experiential nature of consciousness is not the same as conscious experience, since the latter uses phenomenology, while the former embrace preconscious experientialities. The phenomenological fallacy comes from the assumption that our conscious experiences are primarily descriptions. We do not perceive the world outside of us but only perceive internal objects that act as schemas of the objects in the external world.

Panexperiential materialism is opposed to the phenomenological reification of dual-world representationalism and further claims that the mental-immaterial realm, just like the physical world, can be derived from matter waves. In other words, realizing that we do not refer to our experience in terms of objects in the inner world of phenomenology, but perceive our experience by direct reference to external objects, one need not invoke the contents of conscious experiences through phenomenal consciousness, i.e., only from internal (first-person) nonphysical phenomenal information. A more elaborate explanation of the conscious experience in terms of brain processes rests on the conjugate relationship between fundamental observables constituting actual properties of our experience sustained as thermoquantal correlates of consciousness.

Recent research on the foundation of consciousness promotes the simultaneity of the brain's potentiality through negentropic entanglement [59], which exists in the preconscious as a function of affect [60]. It is assumed that a naturalistic explanation of consciousness stems from the thermoquantal information occurring with simultaneity at the subcellular scale as a precursor to the activity of vast numbers of neurons forming neuronal hierarchies of spiking networks [61]. Still, it is contingent on quantum effects in the neuronal environment, e.g., fatty acids [62]. If nontrivial quantum effects, which we believe, do play an active role, then the interplay between quantum mechanics and neuroscience promotes "*neuroquantology*" [63]. Importantly, however, the choice of the interpretation of quantum mechanics [64], the notion of wave/particle duality [65], and the query whether the quantum realm is distinct from classical reality [66] are all questions that arise at the nexus of philosophy and the mathematical treatment of the primary constituents of matter, i.e., the energy quanta of matter, as matter waves. It suggests the notion of "*quantum philosophy*," not be confused with the philosophical meaning of the quantum revolution [67]. Panexperiential materialism, as the epicenter of *neuroquantology*, *quantum philosophy*, and *neurophilosophy*, unifies the whole experience through quantum holism.

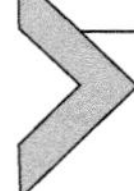

2. Is panexperiential materialism non-panpsychist materialism?

In the contemporary philosophy of mind, there is significant confusion between panpsychist materialism and panexperientialism [68]. The former claims that some fundamental physical entities, i.e., electrons, are "*protoconscious*" [69]. The latter considers fundamental physical entities, such as electrons, to be parts of the holistic property that leads to preconscious experientiality. The idea is that protoconsciousness, i.e., precursors to consciousness, and its fundamental physical entities, are protoconscious, while collectively constitute consciousness in larger systems.

Panpsychist materialism claims that microphysical entities, i.e., particles of matter, are composed of conscious episodes actualized in the physical realm. It considers consciousness to be a fundamental unit in itself, and not a holistic property nor an emergent property, yet physically reductive [70]. Furthermore, panpsychist materialism considers matter and consciousness to be both fundamental, yielding a dualistic touch to the concept. Panpsychist materialism is metaphysical because consciousness is undiscoverable. Some further problems with panpsychist materialism include a lack of causal efficacies of consciousness in terms of a reductive physicalist paradigm. Materialistic panpsychism advocates precursors of consciousness to be in particles of matter.

Panexperiential materialism considers matter waves to be fundamental in the sense that material substances and their movement are what we experience. Protoconsciousness in the noumena are nonspatial and nontemporal, and not dependent on the structure but on the transformation from the spectral domain to the spacetime domain of consciousness that is possible through the non-inert nature of matter, cf. the metaphysical contrast between phenomena and noumena [13]. In panexperiential materialism, a holistic view is an essential component of the panexperiential materialistic ontology.

Panexperiential materialism, however, considers matter only in the context of quantum holism. It is a holistic tenet, which, most importantly, goes deeper than the widely used integrative theory of consciousness at the system level [71]. The holistic theory of consciousness is, however, not proto-panpsychist, since the latter does not consider the correlates of consciousness to be protoconscious. Yet, it is conceived to be a fundamental aspect of nature. Proto-panpsychism is inseparable from the physical and the informational, nevertheless depending on the activity of a specific mechanism present in some kinds of systems, typically living systems, but not in others, to become actualized.

Like panexperientialism, *panprotopsychism* is of the view that fundamental microphysical entities are protoconscious. That is, they have specific unique properties that are precursors to consciousness. So unlike panpsychism, where structural properties are protoconscious, *panprotopsychism* is then the view that collectively constitutes protoconscious when arranged in the right structure. This suggests that in *panprotopsychism*, not all fundamental entities are protoconscious as compared to *panpsychism*, where all micro-physical entities are protoconscious. Still, these two frameworks are not directly associated with *panexperiential materialism* that goes beyond that stage and claims protoconsciousness as informational descriptors of *preconscious experientialities*, which are constituted from the holarchic relational organization of the non-inert matter.

There is, however, the third type of framework known as *non-panpsychist materialism* [72]. Is panexperiential materialism identical to non-panpsychist materialism? This is the crux of developing panexperiential materialism as a separate framework. As argued by David Chalmers [72], there is no longer an ontological "explanatory" gap between the physical and the experiential, but a possible epistemic argument arising from a difference between informational and the experiential. Yet, such differences are no longer tenable in the context of *panexperiential materialism* for the following reasons: (i) the special attributes that noumenal properties are explained to be non-inert matter, i.e., noumenal features are distinct from structural properties, and (ii) there is *a priori* entailment from noumena to phenomena. The problem with Chalmers [72] is the exclusion of ordinary type-A materialism (which grounds phenomenal properties in structural properties) from point (i) above. This is a significant reason why *panexperiential materialism* is proposed to be a new framework in the philosophy of mind stemming from the difficulty between these views that have a "*panprotopsychist*" flavor yet, according to Papineau [73], lack the flavor.

In summary, the consciousness process is mostly noumenal (in the subconscious part of the mind) representing protoconsciousness. Thus, the noumenal is not irreducible to the phenomenal, which refers to anything that can be apprehended by the senses. The noumenal *is non-trivially integrated* through Koestler's holarchy, and the nature of the relationship was proposed by Pribram [74].

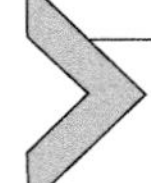

3. Defeating various monistic doctrines and the "*ghost in the machine*"

In the wide definition of monism, there is a phenomenal domain of conscious experience. It confers unity (mind and brain are not a distinct substance), but not the identity that mind and brain are identical ontologically. It implies that all aspects of reality (including the dual existence of mind and matter) are complementary, forming a differentiated unity. The fallacy of the various versions of monism lies in its denial of a dual relationship between the mental and physical phenomena. In passing, one should note the confusing use of the word "dual" above and the concept of duality practiced in mathematics and theoretical physics. Dual aspect theories require the mental and the physical to be inseparable and mutually irreducible. Self-referential identity theory, on the other hand, derives from the conjugate relationship between mind-body and claims that there are no (Cartesian) dualistic aspects and that mentalism is a phenomenological fallacy of the dual world representationalism [58].

Monist physicalism is the metaphysics of materialism in certain limited cases [75]. Neutral monism claims there is a single neutral entity, which is neither experiential nor physical from which both the experiential and physical arise. The major problem is to differentiate between experiential and physical entities arising from this aspect-less neutral entity. Chalmers [19], through the double-aspect theory of information, suggests that experiential entities are informational yet non-physical. Therefore, both the double-aspect theory of information and neutral monism are unable to accommodate experience in its framework without implying a "*ghost in the machine*" [8].

Natural monism, of which the Russellian version is the most prominent [76,77], claims that mental phenomena arise through physical interactions. Although they disregard non-inert matter as the source of these physical interactions, there remains the underlying question of the origin of experientiality from matter, thus leaving a "*ghost in the machine.*" Also, some monists claim matter and energy are two aspects because they treat matter as condensed matter (or inert matter). Frameworks based on aspect monism (which refer to all frameworks that include latent experiential entities in addition to the physical, e.g., dual-aspect, triple-aspect, and multiple-aspect) postulate the "mental" aspect of being "potentiality" as primary, yet a fundamental aspect belonging to the basic reality of Cosmic Energy that carries "mental states" or "mental properties" [78]. However, there is here no mental/physical incongruity, since the mental and the physical are epistemological rather than ontological aspects and therefore give impetus for their elimination (i.e., eliminativism), including the dismissal of aspect monism frameworks based on nonphysical and nonmaterial aspects.

With anomalous monism, a type of *property dualism* posits that there is no "natural laws" governing the interaction between "mental" and physical processes but only a causal connection through the notion of "supervenience" [79]. Anomalous monism assumes "supervenience" of "mental" over the physical. In the light of Chalmers' [80] work, conscious processes supervene upon the physical processes in the brain. However, in "biological naturalism"

John Searle [81] tries to explain that there is nothing in the brain that supervenes upon the physical processes but fails to give physical support to the claim that "mental" concepts are reducible to biology. In *anomalous monism*, there is a "*ghost in the machine*" referred to as "*supervenience*," while in *panexperiential materialism*, it is unpacked in self-referential identity theory.

As a final comment of this section, we also mention a more direct coupling between the material and the immaterial world, i.e., the famous commutation relations in quantum mechanics between material entities, such as momentum-energy and their immaterial conjugate partners space-time. A straightforward matrix analysis displays a direct link with Gödel's self-referential paradox, indicating a possible panexperiential scenario, where material interactions correlate with proto-communicative processes [20–22]. A more precise characterization of the mind-matter problem as an isomorphic connection between mind and brain is provided via the Fourier-Laplace transformation. Although our approach derives from a fundamental material stance, it is necessary to emphasize modern quantum mechanics and extensions as practiced in quantum chemistry and chemical physics, build on the Dirac kets:

$$|\vec{x}, ict\rangle, |\vec{p}, iE/c\rangle$$

with an obvious space-time and momentum energy notation, which includes their inherent irreducible future contingency.

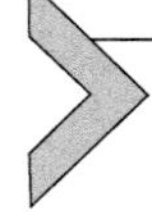

4. Dialectical materialism versus panexperiential materialism

The partisan conjecture that consciousness is an illusion [82,83] has been considered to be the result of the way we see, sense, and conceive our world as a consequence of the lengthy process of evolution, which, even if there was a more basic material scientific level beset with fundamental physical laws, does not design intention and understanding. The subsequent emphasis on evolution, through natural selection, nevertheless, leads to many interesting and useful conclusions, such as the concept of memes, Richard Dawkins [84], the origin of language, Noam Chomsky [85], and finally the recapturing of the word 'artificial design' to the world of science, Daniel Dennett [83].

However, the approach above does not connect consciousness with a fundamental dimension, since consciousness favors a physical, nondialectical, and affective process of energy exchange at the most fundamental level. Our view is that consciousness, in contrast to being emergent, is nondialectical, and hence, we cannot utilize *dialectical materialism as a philosophy of change* [86]. *Dialectical materialism* encompasses actions and their resultant effects on causality, from the specific motions (emergent interactionism) of mind-brain interactions through the dynamics of energy transfer [87]. Dialectical materialism comprises (i) nonreductive (biology), (ii) dynamicism, (iii) emergent interactionism, and (iv) phase transition from quantity to quality. Along with Max Born [88], we surmise that the development of physics cannot be used as a paradigma for the truth of dialectical materialism.

Dialectical materialism, in its opposition to reductive physicalism, is by and large also the view of *biological naturalism* and *panexperiential materialism*. *Panexperiential materialism* is the natural extension of *biological naturalism* [81] that takes into consideration experience as the main ingredient of the conscious process. However, while *dialectical materialism* is where action, interaction, and change between material entities govern brain states through *emergent interactionism*, the latter is not part of *panexperiential materialism*. Consciousness is not an emergent property, but a teleological brain process in the entailment of experienceability. Even "radical" emergence, i.e., the detection of some property not dissociated into underlying terms, would render consciousness to be a "thing" [52], conjuring a trap for the existence of "*pure consciousness*." This, and see more below, is the basic difference between dialectical materialism and panexperiential materialism.

The mechanisms of emergence are not entailed in the conceptual understanding of holistic epigenesis [89], nor is a panpsychist materialistic framework assumed to take fold [70]. In this context, a complex combination of proto-consciousness that contains the "*combination problem*" for panpsychism [90,91] is no longer a problem when it comes to holistic epigenesis. Emergence is the rising of a higher-order phenomenon from a lower level [92,93]. This upward causation can feedback to lower levels [94]. Moreover, consciousness as a state of matter [95] is an *emergentist philosophy* that is not "closed to efficient causation." Dialectical processes, including phase transitions, i.e., from quantity to quality [96], are all emergentist.

On the other hand, nondialectical processes are closed to efficient causation according to hierarchical thermodynamics at the atomic level; there are no feedback loops, and the mode of influence is the *affective drive*, which is the new mode of influence between information-based action controlling force-based action due to the thermal mixing of quantum states. This conflicts with Juan Roederer [97], who distinguishes between an abiotic world, where information plays no role, only entering through the physical interaction with a living thing.

We state here that *intrinsic information* is an attributed function at a lower, fundamental scale, but that it does not lead to causal functional effects higher-up, because the intrinsic content is spectral and functionally nonlocal. So the holistic brain function manifests through nonlocal functional interactions. Nevertheless, its holistic nature does not claim that new properties can "emerge" from the components of a system advocating an ontological claim for continuity. Essentially this is what continuity brings to natural *holistic evolution*, which by necessity circumvents any need for emergence since we can identify the boundaries of various systems (e.g., physical, biological, social) further back in evolution.

The relational co-existence of raw feelings achieved in the entailment organization of quantum holons is what makes the holarchical relationship that exists between preconscious experientialities as physical feelings, to be "protoconscious." Through holism as a property of self-reference, i.e., every quantum holon has a dual tendency to preserve its individuality as a quasi-autonomous whole. The relational co-existence of quantum holons within the holarchy is a crucial distinction in comparison with materialistic dialectics based on emergent causal effects [91] treating the system as an indissoluble whole [89], not merely as a collection of parts.

5. Quantum holism and classicality

Georg Northoff [54] suggests an alternative to classical localization in which a holistic view of brain function is presented. There are, moreover, several events that do take place in brains at the submolecular level, pointing to quantum holism, which of course, does not reflect on quantum entanglement or any quantum interpretations and considerations of wave function realism. Quantum holism [29] is committed to an ontological interpretation of quantum theory, which brings it to tell us something about the microphysical level of the nature of the brain, independently of experimental arrangements and measurements. It is a balance between the classical potential energy (U) and the quantum potential (Q) without an etiological discussion of implicate or explicate order [11,12]. It is the same, whether in soft or hard condensed matter, where U consists of all potential interactions. The overall thermo-quantum chemical potential reads:

$$\mu = U + Q - TS \tag{1}$$

where T is the temperature, S is the entropy in units of [energy]/[temperature], Q is the quantum potential energy. The quantum effects become negligible if $T \gg Q/S$, while in living systems, S is small with Q/S large. As a consequence, *quantum effects* are present and conceivably decisive at high temperatures.

Decoherence is implicit in brain thermodynamics, where thermal noise and environmental influences cause the onset of the apparent wave function collapse [98]. Although there have been many suggestions and counter arguments regarding decoherence, it is nevertheless obvious that thermal noise at about 0.025 eV at the human body temperature of 310 K washes away more subtle quantum effects that might be of fundamental importance in the study of conscious processes. Yet, with the concept of decoherence as defined by Wojciech Zurek [66], one usually associates "*decoherence*" with loss of (quantum) coherence, i.e., the system loses phase information as it finds itself in classical thermodynamic equilibrium. However, there is, in our theory, a gradual loss of coherence within a steady-state configuration, where partial phase information is kept and importantly provides negentropic information.

It is well known that the crossover between quantum and classical regimes requires a quantum temperature to be equivalent to body temperature under quasi-equilibrium conditions. One can define such a quantum temperature, T^*, in Kelvin [99]:

$$T^* = \frac{hf}{k_B} = 4.8 \times 10^{-11}/t \qquad (2)$$

where $h = 6.626 \times 10^{-34}$ Js (Planck constant), and $k_B = 1.381 \times 10^{-23}$ J/K (Boltzmann constant). At the human body temperature at 310 K, we experience time scales on the millisecond scale for apt processing in the brain, which are orders of magnitude larger than the thermal scale of about 1.5×10^{-13} s. This would imply that for conscious life forms, any quantum effects such as intrinsic magnetism, far-infrared absorption and rotational relaxation of polar molecules in liquids, proton transfer in aqueous solutions, protein folding dynamics, etc., should be totally "washed out" by the thermal noise unless one finds a rigorous way to deal with quantum thermal interactions. Although standard quantum chemistry, based on π-electron molecular structures on an energy scale of 0.5 eV, will survive, their correlated dynamics will be significantly affected by the thermal noise, as well as all other quantum effects occurring at, or smaller than, about 0.025 eV at normothermia.

Efforts to overcome the discrepancy above have promoted speculations of introducing a reduced Planck's constant, γ, as a "fundamental" ingredient to be utilized in a so-called *macroscopic Schrödinger equation*, i.e., here a neural wave equation governing the associated conjugated system of the specific membrane [100]. For instance, one might here attempt to define a *quantum wave function* at the submolecular scale, the *wave function* ψ is complex, having a state of reality used to measure potentialities [101].

The wave function is obtained from the neural wave equation by substituting the Madelung transformation: $\psi = \sqrt{\sigma(x,t)}\, e^{i\Phi/\gamma}$, upon separating imaginary and real parts, for the definitions of the parameters, see below. The real part describes the shape of the enveloping molecular field density σ(x, t) (dimensionless) as represented by the phase differences of the oscillating π-electrons expressed in terms of the Hamilton-Jacobi equation [100]:

$$\frac{\partial \Phi}{\partial t} + \frac{1}{2m}(\nabla \Phi)^2 + U + Q = 0 \tag{3}$$

where m is the mass of the model system in units of [mass], γ is the reduced Planck's constant [99], $\Phi = \frac{i}{2}\gamma \ln \frac{\psi^*}{\psi}$ is the action function, related to the phase differences of the oscillating molecular dipole-bound electrons, in units of [energy][time], ψ^* is the complex conjugate of the wave function and ∇ the gradient.

The *wave function* ψ is considered to be an approximation of the wave function of a molecule containing a large number of electrons within the familiar Born-Oppenheimer approximation [102]. As well known, delocalized electrons are tethered to follow the motion of the molecular dipole that acts as a pseudo-nucleus in the hydrophobic region of the lipid membrane [103].

While such models may not be unusable, they will not carry any fundamental meaning (isomorphisms) unless the wave function is decomposed so that the interference regions of atomic orbital wave functions in molecular orbitals can be computed. Unfortunately, we observe that the usual adiabatic model, i.e., electrons adapting to slow nuclear adjustments, is not justified when entering the domain, where quantum fluctuations and thermal noise coalesce. Fortunately, there is a simple mirroring relation between electronic and nuclear motion [20–22,104], which have been useful for the theoretical understanding at quantum-thermal regions.

A more fundamental approach can be elucidated. For instance, realistic time-temperature scales and correlated dissipative order have been derived from first principles in a Liouvillian subdynamics framework, which applied to dissipative systems yields the relation:

$$T = T^* = \frac{nh}{k_B t_{\rm rel}} \tag{4}$$

where t_{rel} is the proper relaxation time, commensurate with the physical temperature, and n being the system quality number corresponding to the size of the dissipative system, such as a cell, a DNA fragment, a selected protein, or other molecular aggregates. In the case above one finds $n \approx 10^{11}$. For more details, we refer to Brändas [20–22] and references therein.

In order to expand on the notion of quantum holism, we turn to de Broglie's wave theory, as the latter is professed to represent a more fundamental theory, where an electron is a particle at its core, with an active field surrounding it, and where the latter guides the motion of the electron as a guiding wave [15]. In its traditional form, this active field represents internal energy. It is due to "hidden" thermodynamics, according to de Broglie's original wave theory, with Bohm [10] introducing the energy through his "quantum potential."

The de Broglie-Bohm theory is considered on par with pioneering quantum mechanics. It is thus applicable from the sub-atomic scale (10^{-15} m) to the atomic scale of 0.1 nm (10^{-10} m) and above. However, position and momentum are not treated symmetrically, i.e., an experiment yielding a particle's position contextualize the momentum as a "measurement" property. The induced asymmetry in the treatment of canonically conjugate quantities makes the position more fundamental than momentum. Nevertheless, extensions of the pilot wave theory might represent an intriguing framework for modeling consciousness at the submolecular scale where microscopic interactions (interference regions) involving extra quality of energy in terms of the quantum potential is self-referentially amplified, becoming separated from the delocalized electrons as concealed motion [33].

The subtle changes inherent in oscillations of delocalized π-electron clouds depends on whether quantum fluctuations in the electron density dominate over thermal fluctuations (Brownian motion). In passing, we note that electronic motion and nuclear vibrations are coupled, i.e., the electronic- and the nuclear systems develop mirroring dynamics [20–22]. This dualism becomes of crucial importance in the low energy description of light energy carriers in a nuclear environment at the human body temperate zone. The critical scale for the entry of quantum effects, usually referring to the heavy particles in the system, is typically controlled by the thermal de Broglie wavelength, Λ, see definition Eq. (5), which is a quantity that provides the conventional condition that separates the classical from the quantum domain.

The classical regime holds when the mean separation, ℓ, between the particles, usually a quasifree ion, satisfies $\ell \gg \Lambda$, and to ensure that quantum conditions prevail, the simple memory rule is that ℓ must be smaller than Λ. However, since we are talking about thermal processes with typical length scales of the order of several nanometers or more, quantum correlations may violate the simple thumb rule above. In classical systems, at the molecular level, the de Broglie wavelength is pivotal in deciphering the difference between quantum fluctuations and thermal fluctuations. The standard classification of fluctuations as quantum or thermal depends on the thermal de Broglie wavelength, expressed as follows:

$$\Lambda = \sqrt{\frac{2\pi\hbar^2}{mk_{\mathrm{B}}T}} \tag{5}$$

where $\hbar = \frac{h}{2\pi}$ is the reduced Planck constant, m is the mass of the particles in question, and T the temperature.

For example, if m is the mass of a quasifree proton, or H^+ ion, in a liquid at room temperature, the thermal de Broglie length is about 0.1 nm. However, proton transfer processes in aqueous solutions display anomalous behavior. The high excess proton conductivity will develop via Grotthuss-like mechanisms, supporting quantum delocalization effects over geometric sizes d_{min} that scales as [105]:

$$d_{\mathrm{min}} \propto n\Lambda \tag{6}$$

Hence if the average molecular distance satisfies, $\ell \leq \Lambda$, i.e., limiting the conventional quantum domain accordingly, quantum-thermal correlations and quantum transport may reach over considerable regions, Eq. (6), which might traditionally have been considered classical.

According to Wojciech Zurek [66], quantum systems interact with their environment, leaving an ambient record. By extrapolating Zurek's ideas to dissipative systems, one might try to explain the diffusion of quantum states [106], where the thermo-quantum diffusion is governed by a neural wave equation containing the quantum potential, see the discussion above, as a function of the thermal entropy [107,108]. However, one needs to extend quantum mechanics to open systems [109], or by extending the Hamiltonian function to preserve the conjugate relationship between generalized momenta and coordinates [99,110]. Depending on the system's memory attributes, heat proportional to the temperature may, or may not, be dissipated. This assessment includes the case when only a density matrix is available, which needs the latter to be built into the fundamental postulates, see, e.g., W-Bohmian mechanics [109]. This realm borders quantum physics and quantum chemistry, and it can be quantified to be the scale where quantum processes prevail over classical processes.

6. The self-amplification mechanism of simultaneity

The hydrodynamical formulation of quantum mechanics [111] can be used in the context of propagating ensembles of quantum trajectories, as a probability fluid or quantum fluid. For open quantum systems, trajectories in phase-space are addressed by the Liouville equation [112], where the probability fluid evolves through phase-space as a unified whole. The physical process entails diffusion of quantum states expressing the reality of "matter waves" toward a complex enough, soft condensed matter, requiring the self-referential amplifications of interrelational information structures as nonlocal teleological mechanisms, through the quantum potential energy as a concealed motion [33]. This leads to *energy exchanges* that result in a "quantum force" based on the quantum hydrodynamic description of quantum diffusion [107,113,114], provided the conditions, as warranted by hierarchical thermodynamics, are satisfied on account of its high instability to small perturbations [115].

This self-amplification mechanism invokes a "quantum force":

$$f = -\nabla Q$$

where Q is the quantum potential energy that guides the actualization of the appearing phase differences [108]. Unlike the microscopic quantum potential of Bohmian mechanics were "pilot waves" guide the particle; the properties of Q involve non-Bohmian nonlocality. The nonlocal teleological mechanism entails intrinsic information, picking up the part of Q that shapes the phase-wave. This extends interconnectedness between *nonlocal functional interactions* that exist in the classical realm [26] to the quantum realm without involving superluminal signaling. The nonlocal teleological mechanism is partly an electronic spin-mediated process, i.e., subjected to fermionic statistics, which involves couplings to dissipative systems that possess long-range order that, despite thermalization, partially will keep phase differences of the π-electrons moving in the self-produced molecular field [31,116,117]. Possibly there are also quantum effects that can guide these delocalized electrons during interferences between delocalized electrons bringing on extra quality of energy exchange, providing the transformation of information that creates information-based action as a teleological formulation related to consciousness.

This is a hypothesis claiming that the components of the wave function are the results of the system being coupled to a coherent, partially coherent, or correlated, environment acquiring phases from its immediate surroundings. The wave function is established, and it is assumed to stay coherent even at the global level. This is conceivable under rather exceptional circumstances, for example, when "chaos" becomes the seed for a new wave function [118]. This might sound speculative, as mentioned earlier, that chaos, i.e., irreversible thermodynamic processes, could become the seed for an originally reversible process. However, Bohmian, or W-Bohmian, mechanics allows "quantum chaos" to follow thermodynamically [119]. The obvious absence of a self-organizing force, acting on the quantum system of matter waves, should by the second law of thermodynamics lead to dissipation rather than to an assemblage into a new structure. However, thermodynamic constraints, such as temperature dependences and associated time scales, influence the hierarchical nature of brain thermodynamics, where non-dissipative processes are coherent and coupled to irreversible dissipative processes—all within the second law.

An alternative way for understanding "force-free" information-based action has been addressed through the spin-dependent quantum potential energy in quantized EM theories of consciousness based on the physical properties of the vector potentials that were considered by analogy with the well-known Aharonov-Bohm and Josephson effects. The possibility of using the vector potential for "force-free" information-based action [120] suggests coherent systems at work. The simultaneity of interrelational information structures can increase the coherent phase dimensions of the magnetic dipoles (or the spin polarization orientation) through its concealed motion, which couples to the mesoscopic level via a "*gauge transformation*" that embodies the nature of the quantum to classical transition where quantum mechanics "end" and thermal fluctuations take over [121], or rather mix within an overlapping regime [104]. The quantum-classical transition, therefore, is a continuous onset of classicality via the build-up of thermal correlations that gradually swamp the quantum correlations. It should be emphasized that the quantum-classical dichotomy is not the same as the micro-macro distinction, however, it is the quantum boundary of life at the mesoscopic level. An interesting example is a transition from the microscopic to the mesoscopic level via the so-called magnetic phase transition that might contribute in a subtle way to *intrinsic magnetism*. Still, such phase transitions do not occur in macromolecules [122].

It has been shown that anesthetic drugs affect the electronic spin polarization of the protein molecules in fruit flies, causing changes to the electron current [123]. The relationship between consciousness and spin, despite regulating that external magnetic fields will play no role if the total molecular spin is zero, is yet fundamental if only indirect. A recent hypothesis is that electron spin networks bridge classical neural activity, serving as input via magnetic influences on ionic flow in the electro-ionic brain [118]. The claim is to have a realization of a "spiral-like" structure represented as a matter-wave. Such dissipative systems are driven by continuously transforming quantum systems interacting with its environment, where a fractal network of charges merging to form a "*charged-induced*" *quantum potential* [124].

Recent studies indicate that the *fractality of space* could be effected by the addition of quantum-type potentials to the classical equations, i.e., speculated that these extra energy contributions do emerge in biological-like processes [125]. However, in a semiclassical regime, where matter is treated quantally, and spacetime is represented classically, there appears a problem when referring to the notion of *fractal spacetime*. The dilemma arises since the underlying biological medium for the emergence of quantum potential energy is not matching up with the presence of a classical potential energy, since the latter is not commensurate with the necessary capacity to establish phase coherence [108]. The *phase coherency* underlies the localization of the *quantum potential energy* in the dissipative system, which is a subtle property that exists already in the absence of the classical potential energy. Therefore, the quantum potential energy cannot emerge classically at the submolecular scale. In other words, the quantum potential energy at the submolecular scale is the result of the self-amplification of the spin-dependent microscopic quantum potential energy due to concealed motion [33] resulting in energy exchange at the quantum-classical transition that embodies interrelational information structures. This makes sense if brains consume energy exchange to include holistic brain functions like raw and physical feelings, and qualia.

Intrinsic magnetism should not be confused with intermittent magnetization, as discussed by Arnold Mandell [126] or magnetization as a change of the spin orientation, in which there is a permanent spin-feature, which due to a temperature drop, may increase the coherent spin domain, corresponding to an emergent *magnetic field*. For example, Mandell [126] has proposed criticality in brain magnetic fields of neocortical pyramidal cells. This is not the EM field generated by neuronal activity in the electro-ionic brain, but the magnetic flux responsible for electrostatic magnetization and the incorporation of intrinsic magnetic effects of delocalized electrons (π-electrons) in the EM brain. The strategy of examining criticality in magnetic chaotic systems depends on the idea to relate scale transformations to changes in the temperature (renormalization). Two main attractors are corresponding to $T=0$ and $T=\infty$, and the boundary between these two domains occurs at the celebrated Curie temperature. It is doubtful whether the general concept of criticality will be specific enough to generate proper brain dynamics without additional ingredients. EM fields can also occur in nonpolar regions that might play a different role, such as self-referential amplification of *intrinsic magnetism* leading to conscious experiences through mechanisms of correlative interaction in the endogenous EM field of the brain. What are these mechanisms, and how do they mix with thermal correlations?

The existence of these subtle, intermittent magnetic fields might be "hidden" or are intangible since there appear to be no *macroscopic effects* subsumed to spin projections in molecules (but see [118]). Spin pairing in molecules is subordinate to Pauli's renowned exclusion principle. However, to comprehend the EM brain as composed of a hierarchy of EM fields [127] necessitates the amplification of *intrinsic magnetism* [103] in the EM model, which is quite different from the classical EM model developed by Johnjoe McFadden [128]. The EM brain in the two-brains hypothesis [129] consists of a subtle magnetic field that is caused by the electron's orbital motion interacting with the spin magnetic moment. The magnetic field is generated by the spin-orbit interaction of an electron within the molecule itself, known as an "*internal Zeeman effect*" in the absence of an external magnetic field. The intrinsic magnetism in the EM brain must reflect upon the limited molecular pairing of spin carried by delocalized π-electrons in their internal motion as well as their orbital motion as measured by phase differences [108].

The "*internal Zeeman effect*" is a magnetic interaction with a basic *magnetic field* generated by the orbital motions of the electrons. This corresponds to an intrinsic *magnetic field* of about 0.4 T, which is equivalent to an energy difference of about 0.000045 eV, far below the normal thermal noise of 0.025 eV. The latter would cause random flips in the direction of the magnetic moments (dipoles) subject to the electronic spin angular momentum in the molecular orbitals, "*washing away*" the effects of the *internal Zeeman effect* in the formation of a *magnetic field*, the latter as a consequence of the mixing of macromolecular quantum states [130]. Yet the quantum potential energy is produced by "internal motions" associated with spin, i.e., linked to the spin orientation [131,132]. To avoid decoherence, one would need to apply a realistic density matrix scenario.

7. Hierarchical thermodynamics via density matrices

The microscopic internal energy controls mesoscopic systems by irreversible subdynamical evolution and self-referential amplification. At this point, some thermodynamic constraints, related to the entropy lowering balancing entropy production, anticipating the environment at normothermia, are exacting due to the hierarchical nature of the thermodynamics of biological systems. The conundrum is that the lower level microscopic system, subjected to negligible dissipation, such as in superconductors, etc. [133], transduce to time-directed evolution at the mesoscopic level. However, the dissipation of the quantum state at the higher-level scale is not extremely low, as in the materials above. Still, according to *hierarchical thermodynamics*, such phenomena are possible when self-referential amplification of microscopic internal energy is coherently resulting in macroscopicity [3].

Herbert Fröhlich [134] did hypothesize that the supply of energy from the metabolic processes in the environment would result in a non-thermalized organization into a single collective mode. Fröhlich's "brainwave" was assumed to be a realization of the activated states of the enzymes distributed throughout the neuronal membranes with the connection between microphysics, through the equations of quantum mechanics, and delicately coupled to macrophysics, via the resulting nonlinear kinetic equations, the latter being the average of the former over the non-equilibrium ensemble [135]. However, Reimers and colleagues [136] showed that Fröhlich's condensates might be conceivable only at high energies, but not at those corresponding to body temperatures under equilibrium heath-bath conditions.

A solution to this is the replacement of the "heat bath" with a microscopic quantum potential, or a density matrix, as concealed or correlated motions contribute to long-range phase coherences associated with a Bose-Einstein-like "condensate" of *matter wave* energy quanta. It is important to realize that these long-range phase correlations should not be compared with those of *superconductivity* [137,138], which manifests themselves under conditions of low-temperature thermodynamic equilibrium. At the same time, the holistic basis for a biomolecular understanding of consciousness is born out of the non-equilibrium aspects of *hierarchical thermodynamics*. The latter concerns correlated dissipative order where each level evolves under steady-state conditions and where each level is successfully subordinated into levels of organization.

The unique aspect of living system non-equilibrium thermodynamics is the sparsity of non-dissipative processes, which are coherent and irreversible. The energetic closure enables the creation of a quantum system, which depends on quantum observables and the formulation of quantum statistics. Such a system can, if the temperature is low enough, be represented in terms of a wave function [139]. The quantum coherence is very delicate, reminding of correlated responses of phase-differences extending throughout the whole system. For this to happen nonlocally, it must be envisaged that each part should have a collective phase-order showing in all of its activities. That is, all components of the wave function should be strongly coupled coherently, acquiring a build-up of phases from their immediate surroundings. A coherent wave function, represented by a large eigenvalue in the second order reduced fermionic density matrix, shows the onset of Yang's Off-Diagonal Long-Range Order, ODLRO [140], and the foundation of extreme state super phenomena, such as superconductivity, etc. [141,142].

Negentropy is always, by definition, nonnegative, i.e., it signifies the difference between the maximum entropy at equilibrium and any given distribution [143]. The change of entropy arises from the internal change within an open quantum system, originally formulated as the somewhat controversial thermodynamics of the isolated particle, or de Broglie's hidden thermodynamics interacting with a hidden thermostat [65]. The integration of the open quantum system with its environment involves a mixture of quantum-thermal mechanisms requiring a density matrix framework. We further allow the open quantum system to be represented by a density matrix, which, although displaying properties of delocalization, cannot be represented by a pure wave function. This leads to the vital problem of how to formulate and incorporate quantum and thermal correlations consistently in dissipative (open system) dynamics [21].

The theoretical challenge is to find the implication of the quantum formulation, when a system, described by a wave function, interacts with an environment characterized only partially. For instance, if the surroundings are only given at a specific temperature, the onus is to find the appropriate density matrix. This mandates the role of density matrices also in Bohmian mechanics [109]. Various molecular identities will have subsystems with their wave function representations, nontrivially evolving according to their precise subdynamics. For example, the *Ehrenfest force* referred to as the EM force "felt" by a delocalized electron, due to the presence of other electrons as well as the motion of the nuclei, involving first and second-order reduced density matrices of the electrostatic molecular field, could in principle be pursued in a Bohmian setting [109,144]. However, this topology and the subsequent outcomes for defining the atom in the molecule is not the approach we will take here [145].

Alternatively, one might utilize the well-researched machinery of density matrices [141,142,146], where it has been mainly exploited in the domain of quantum chemistry as defined and developed by Per-Olov Löwdin [147], which provides a more general way to represent a mixed quantum state [148]. The second-order reduced density matrix, referring to correlated electronic motion "ODLRO," describes the coherent electronic motion in the system, and the electric current develops through its "large eigenvalue," which if the temperature is low enough, produces superconductivity and the Meissner effect.

Localization dynamics formulated classically applies to situations where the confines take place over macroscopic distances in classical phase-space. In contrast, i.e., in quantum situations, the absence of diffusion of waves in a disordered material, the so-called Anderson localization [149], is considered to be a general wave phenomenon. In strongly correlated situations, a wave function can be identified by means of Coleman's extremum condition satisfied by the homogeneous wave function-representable density matrix [141]. However, at, for example, room temperatures, thermal correlations destroy the coherent situation, and a pure wave function formulation is not possible. The mechanics of the classical system in phase-space follows a Liouville equation [150]. It has, in addition, also a quantum analogue, and a common formulation has been successfully achieved in the Prigogine sub-dynamics formulation [26,151]. For instance, the (quantum) Liouville operator $\mathcal{L}$ is a superoperator, working on an operator space of density matrices, and is defined as the commutator with the density matrix ρ, i.e., $\mathcal{L}\rho = [H,\rho] = H\rho - \rho H$ with the Liouville equation given by:

$$\mathcal{L}\rho = i\hbar\left(\frac{\partial\rho}{\partial t}\right) \tag{7}$$

For a pure density matrix, $\rho = |\psi\rangle\langle\psi|$, Eq. (7) is analogous to the Schrödinger equation $H\psi = i\hbar(\partial\psi/\partial t)$, where ψ is the wave function. However, since the density operator belongs to a convex set of system operators, there are other solutions to the Liouville equation that corresponds to inhomogeneous ensembles, e.g., the canonical ensemble. In the thermal case, one can show that Bloch thermalization of the Coleman-Sasaki representation of the density matrix, equipped with appropriate temperature-time boundary conditions, yields the so-called Correlated Dissipative Ensemble, CDE, which through its representation:

$$\delta\rho = \rho_{\text{tr}} = \sum_{k=1}^{n-1} |\psi_k\rangle\langle\psi_{k+1}| = |\boldsymbol{\psi}\rangle\,\boldsymbol{J}\langle\boldsymbol{\psi}| \tag{8}$$

where $\boldsymbol{J}$ is the classical Jordan form of the nilpotent operator J, with zeroes everywhere except with ones immediately above the main diagonal, the dimension n and the states $\{\psi_k\}$ are uniquely given by the CDE, exhibits a transition density matrix, commensurate with a thermodynamically steady state $dS = 0$.

It is important to realize that the gain of information associated with Eq. (8) is thermoquantal information generated by the entropy production balancing the steady state condition $dS = 0$, for more details, see further below. For instance, the transition density Eq. (8), generates non-classical long-range phase information as well as negentropic gain during steady-state evolution. Starting from a zero-entropy coherent ground state, the second order reduced density matrix is generated by averaging over all degrees of freedom except the delocalized fermion pair or the correlated average nuclear oscillator, being in a mirror relationship violating the adiabatic assumption. Further reduction of information comes from Bloch thermalization. The latter, following open system dynamics, must adhere to complex symmetry, due to the obvious biorthogonal construction, i.e., $\langle\psi_k|\psi_k\rangle \rightarrow \langle\psi_k^*|\psi_k\rangle$, with the "*" denoting a member of a dual Hilbert space, it is usually denoting complex conjugation.

Finally, the thermalized "state," masked as a complex symmetric representation of a Jordan block, Eq. (8), is uncovered by a unitary transformation to classical canonical form. This unitarity, leading to traditional diagonal matrix representations of pioneering quantum chemistry, contains the eigenvectors of the matrix problem, the latter baring the physical properties and information regarding the system under consideration. In the degenerate case, leading from the Bloch thermalized complex symmetric form to Eq. (8), it becomes $\boldsymbol{B}^{\dagger}$, i.e., with its unitary partner defined as

$$\boldsymbol{B}=\frac{\mathbf{1}}{\sqrt{n}}\begin{pmatrix} 1 & \omega & \omega^2 & \bullet & \omega^{n-1} \\ 1 & \omega^3 & \omega^6 & \bullet & \omega^{3(n-1)} \\ \bullet & \bullet & \bullet & \bullet & \bullet \\ \bullet & \bullet & \bullet & \bullet & \bullet \\ 1 & \omega^{2n-1} & \omega^{2(2n-1)} & \bullet & \omega^{(n-1)(2n-1)} \end{pmatrix}$$

with $\omega=e^{i(\pi/n)}$. The $\boldsymbol{B}$, $\boldsymbol{B}^{\dagger}$ pair has interesting symmetries. For instance, $\boldsymbol{B}$, as can be noticed by direct inspection, has a very interesting factoring property. An archetype example will be discussed, see Eq. (11) below, and also Brändas [20] for more details.

It is easy to see that $T_{\mathrm{op}}=\varepsilon\ I$, where the degenerate resonance energy, $\varepsilon=E-i/\tau$, $E=0$ for simplicity, the operator I being the n-dimensional unit representation, and with the complex part proportional to the inverse life time τ, generates the usual time decay $e^{-t/\tau}$. Furthermore, with the addition of the irreducible part J one gets the Poisson time evolution, which introduces self-organizational attributes, delaying the basic decay rule:

$$\mathrm{e}^{-iJ\frac{t}{\tau}}\mathrm{e}^{-\frac{t}{\tau}}=\mathrm{e}^{-\frac{t}{\tau}}\sum_{l=0}^{n-1}\left(\frac{t}{\tau}\right)^{l}\frac{1}{l!}J^{l}$$

Hence, ρ_{tr} exhibits a Poissonian time evolution reminding of the statistics of the number of calls per unit time arriving at a call center. The entropy change, from the zero entropy ground state to the Poisson stochastic development, including the formation of the degenerate population and the zero trace nilpotent addition Eq. (8), is given by

$$\Delta\rho = d\rho + \delta\rho;\quad d\rho \propto I$$

where $d\rho$ is a degenerate matrix proportional to the identity operator I with a finite trace equal to n.

It is interesting to note that the free energy relation $F=E\text{-}TS$, discussed above, leads to some remarkable but straightforward consequences. Starting with a (quasi)-equilibrium situation with a degenerate population, one finds for the steady-state, where the incoming and outgoing energy-entropy flow (for T constant) are the same, i.e.,

$$dF = dE - TdS = 0$$

Also, one gets that the changes in the free energy F, internal energy E, and the total entropy S, also satisfy $dF = dE = dS = 0$. This inference is compatible with

$$dS = -k_{\mathrm{B}}\mathrm{Tr}\{\Delta\rho \log \Delta\rho\} \propto \mathrm{Tr}\{\delta\rho(I + \text{higher order terms})\} = 0$$

where the number of higher-order terms is finite. Also, the total entropy change dS can be decomposed into the sum of two contributions [152]:

$$dS = d_eS + d_iS$$

where d_eS is the entropy flux with environment and $d_iS > 0$ is the entropy production due to irreversible processes inside the system. This implies that $dS = 0$ yields, see, Eq. (4) and the discussion of entropy production further below,

$$d_eS = -d_iS = -\left(\frac{dT}{T}\right) < 0 \tag{9}$$

For instance, entropy decrease in a cell, under the steady-state condition, leads to a reorganization of energy contributions of the free energy, i.e.,

$$dF = dF_e + dF_i = -Td_eS - Td_iS = 0$$

Hence nondialectical processes, such as the *affective drive*, will not change the free energy, i.e., $dF = 0$, but according to Eq. (9) one obtains a negentropic contribution to thermodynamic order given by

$$dF_e = -dF_i = dT > 0 \tag{10}$$

In conclusion, one might say that nondialectical/teleological processes are physically scrutinized by Eq. (10), commensurate with an experiential entropy decrease. Yet the experience is not physically observable through total changes in the *free energy F*, only through the reorganization, above.

The Liouvillian formulation, including Bloch thermalization, the transformation to the classical canonical Jordan form, described above, provides the key to the formulation of the self-referential natural law. In Section 12, we will be taking ρ_{tr} in Eq. (8) as the input for a higher-order hierarchical formulation, adapting the degenerate problem to a Liouvillian with a modified irreducible J, repeating the Poisson evolution to a higher-order level, commensurate with self-organization and the dependences on temperatures and time scales. Note that the standard relation between the resolvent and propagator in quantum mechanics is here generalized in essentially two ways: (i) the necessity, due to the generalized spectral properties of analytic operator families, to find the appropriate paths in the complex energy plane, (ii) to account for the multiplicities of the singularities in the Green's function. The appearance of these high order singularities is crucial for the self-referential traits of evolving life forms.

One can show that the formulation respects Gödel's incompleteness theorem as further recognized in that the unitary transformation $\boldsymbol{B}$, see also Eq. (11) below, generates nested factorizations. As a consequence, Gödel-numbering of cognitive information appears as a result of self-references, which yields a more complex thermo-qubit for communication, reminding of the cognitive units of information, *cuinfo*, advocated by Gerard Marx and Chaim Gilon [153]. For instance, a transformation representing communication between $n=12$ locations in the brain correlates according to the factorized paragon:

$$\begin{matrix}1\\1\\1\\1\\1\\1\\1\\1\\1\\1\\1\\1\end{matrix}\;12\;\begin{matrix}6\\6\end{matrix}\;\begin{matrix}4\\4\\4\end{matrix}\;\begin{matrix}3\\3\\3\\3\end{matrix}\;12\;\begin{matrix}2\\2\\2\\2\\2\\2\end{matrix}\;12\;\begin{matrix}3\\3\\3\\3\end{matrix}\;\begin{matrix}4\\4\\4\end{matrix}\;\begin{matrix}6\\6\end{matrix}\;12 \qquad (11)$$

where the blocks indicate spatiotemporal symmetries that reflect the perception-cognition details of neuron and mirror neuron images of the cognitive system. A remarkable by-product of the transformation process above is that it can be used as channels for communication [21].

8. The whole of nature and abiogenesis

Natural evolution is a continuous process based on physical laws where non-inert matter itself evolves. The non-inert matter is what de Broglie [2] defined as "*matter waves.*" Louis de Broglie showed in 1925 [65] that his electron phase wave has a number of unique qualities. Although the phase wave is relatively stationary (eigenstate), it is non-inert. Quantum holism [29] defines a minimal requirement that a system at a time is in an eigenstate of the Schrödinger equation. The mass of the nucleus has no effective role here, and it is inert. The mass of the electron, however, is quintessential for the eigenstates of the phase wave, by the de Broglie relation:

$$p = h/\lambda$$

where λ is the de Broglie wavelength—not the same as the average thermal wavelength Λ. Thus, the electron wave forms the equilibrium between mass (in p, momentum) and energy (also implied in p). Note the numerator h (Planck constant = unit of action) in the equation. The actual *phase wave* itself is the *Energy-in-Action*; that is, the momentum p, which in quantum physics, becomes a matter wave. As displayed in Fig. 1, matter waves evolve from the

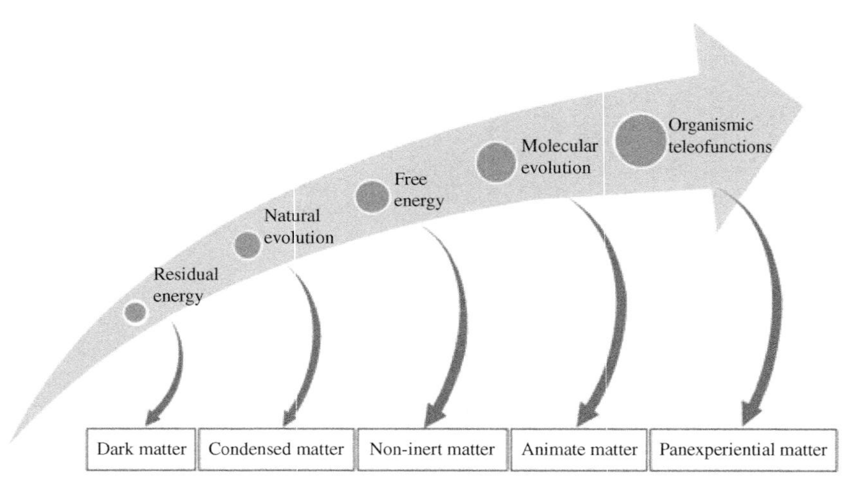

Fig. 1 An illustration of the essentialness of abiogenesis. Residual energy is not considered to be fundamental since it also encompasses the natural evolution of inanimate matter separated into the condensed matter and non-inert matter. The latter includes free-energy as stored energy via hierarchical thermodynamics. The non-inert matter is fundamental since residual energy in natural evolution can evolve both condensed or inanimate matter and dark matter in the cosmos as complex enough matter. It is the potentiality in the animate matter that is holistic due to a more substantial organized complexity arising from nonreductive physical processes leading to panexperiential matter carrying organismic teleofunctions.

microscopic level, forming the world of chemistry as a biological complex enough system subjected to the self-referential physical law of nature [20].

In panexperiential materialism, there is no issue dividing holists and reductionists, concerning the problem whether the whole of nature or the discrete parts is primary. Still, the relations between the parts—this relationalism depends crucially on their interrelations without embracing essentialist ontology, which involves *sui generis* causal powers irreducible to their components. This agrees with the dictum that the *whole of nature* considers the natural evolution of matter as part of the *process of reality* [23].

Molecular catalysis is central in molecular evolution that imparts atomic and molecular interactions, through long-range order and thermalization in open systems, entails a gradual teleological structure that is wholly within natural evolution. Yet, it has a new agenda that matters significantly in life and also in the expression of catalytic processes as intrinsic to the organism. For instance, thermally activated electron transfer is the mechanism underlying electron transfer carrying enzyme-catalyzed redox reactions by way of wave-based facilitation depending on specifics of the excited electronic states involved, e.g., vibrational or rovibrational modes of the enzyme-substrate complex. However, thermally-activated electron transfer does not take precedence in hydrophobic regions where bridging between chains of π-electrons commensurate with bound state solutions of the Schrödinger equation entails the movement of *potentiality* across scales that embody information transformation. The latter, being conjugate to the brain, evolves together with the material complex organization. Hence life depends on molecular evolution simply because molecular abundance grows from complex quantum chemical processes. Molecular evolution is, therefore, essential, suggesting that life and mind are ontologically inseparable, sharing a conjugate relationship [22].

Brains evolve through molecular evolution leading to a holistic organization based on an authentic foundation of reality [54,154]. Molecular evolution depends on hierarchical thermodynamics for its revitalization as "*panexperiential matter*," i.e., a composite state of animate matter out of equilibrium originating from "order above inhomogeneity" [155], or negentropic order that has biologically evolved from a degenerate population of biological complexes in general and *neuroenergetics* in particular [156,157]. An interesting coincidence is that the present self-referential identity theory arises from a fundamental analysis of degenerate spectral representations yielding cellular quality degeneracy indices for natural selection and assortment in the hierarchy of life forms [20–22].

The idea has been to analyze abiogenesis in terms of Gibbs free energy $G = E + PV - TS$, where E is the internal energy, P the pressure, V the volume, T the temperature, and S the entropy. John von Neumann established a rigorous quantum-theoretical framework for S appropriate for the microscopic level. The interesting equation concerns the entropy change dS, valid also for non-equilibrium situations, where the change in G (for PV constant G equals the Helmholtz free energy) is essentially caused by the self-organizing degrees of freedom n, see Eq. (4), i.e.

$$G = \mu n = nk_{\mathrm{B}}T; \quad S = \left(\frac{\partial G}{\partial T}\right)_n = nk_{\mathrm{B}}$$

where μ is the chemical potential. For a steady-state, i.e., with $dS = 0$, one finds, via Eq. (4) relating the characteristic time scales t_{rel} of a biological open system during the process of evolution at temperature T, that

$$dS = \frac{Sdn}{n} = \frac{dt_{\mathrm{rel}}}{t_{\mathrm{rel}}} + \frac{dT}{T} = 0$$

and a negentropy of dT/T in terms of the attendant entropy production, see more below and Brändas [22]. Hence a cell, or neuron, represented by the dimension n or set of numbers n, reflecting its responsibilities in the system, respects a delicate balance between anticipating time scales and the consumption of heat.

This is not in conflict with traditional physical chemistry, as free energy stored via traditional thermodynamics:

$$\text{Free energy} = \text{Internal energy } (Q) - \text{Thermal entropy } \left(S_{\mathrm{Q}}\right) \times \text{Temperature } (T)$$

Note that the thermal entropy S_{Q} is not the quantum version—the latter known as the von Neumann entropy with the limits $T \to 0$, $S_{\mathrm{Q}} \to 0$. The internal energy (Q) is separated in scale from the macroscopically ordered energy. As already said, one can transcend to the atomic scale introducing quantum statistical mechanics of dynamical processes under quasi-equilibrium conditions. The definition of entropy, as used in information theory, is directly analogous to the definition used in statistical thermodynamics. When the free-energy is positive, the process will proceed spontaneously in the reverse direction to increase potentiality [108].

In summary, we ask the question: is "*panexperiential matter*" a composite physical state of inert matter? According to panexperiential materialism, preconscious experientialities are self-localized in the inner structure of atoms and molecules yet are not "protoconscious". They are properties of matter-waves represented by density matrices driven by diffusion, resembling Brownian motion of quantum states in the organization of energy. Detecting the overall qualities of different neural activity patterns in the brain would justify the presence of a *homunculus* unless the energy patterns in the composite states of inert-matter through asymmetrical actions or "*affective drive*" of a holarchical system down to the atomic scale [9] is transformed (not transferred), as for instance, *intrinsic information*.

The difference between structure forming condensed soft matter, such as water, molten salts, polymers, etc., and non-inert matter, including biological entities, such as DNA, RNA, genes, cells, etc., is fundamental. The latter is characterized as a unique organismic system, comprising, on the one hand, thermodynamically open systems organized at far from equilibrium situations, on the other, information-carrying correlated dissipative constituents, feeding on negentropy while disposing of the obligatory entropy production. Thus, at the most fundamental level, the physical processes entail diffusion of quantum states for the reality of de Broglie's "matter waves" to condensed soft matter where quantum mechanics end and thermal fluctuations take over. In the "grey zone," interesting phenomena occur, such as the effects of superposed long-range correlations of quantum states, based on matter waves of characteristic wavelengths and size [2]. Thus, one needs to describe the sub-dynamical qualities of general systems, where classical and quantum attributes are allowed to intermix according to the proper spatio-temporal conditions and temperature. This compels an extension to open state non-Hermitian quantum mechanics, which can be rigorously carried out via the spectral theorem for dilatation analytic operators [158].

As a final remark, our framework differs from *panexperiential physicalism* [91,159,160] because the "*panexperiential flow*" is outside of, or extrinsic, to "dull" or "inert" condensed matter. The fact that the intrinsic "mental" flow is distinctive to the outside material reality signals a dualistic quality to the concept. Panexperiential materialism, contrary to dualistic qualities, is based on open quantum systems [22], where the conjugate relationship between the body-mind leaves the "*panexperiential flow*" to be within the non-inert matter that constitutes holarchical systems [9] with the *free energy* conjoined to the internal energy of self-referential amplifications [30] of interrelational information structure via a nonlocal teleological mechanism governing simultaneity. The term *free energy* in the brain is valid in the context of *panexperiential matter*. It is suggestive of stemming from molecular evolution, where there is a tendency to increase free energy in living cells [161].

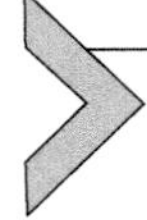

9. The teleological formulation of organismic teleofunctions

Teleological processes are goal-directed action-at-a-distance without the implication that the end is causally efficacious, i.e., a-causality often being perceived as teleological. It is an action without being effective causally or exerting any force or spending any energy, yet unitarily connected through an informational holarchy, extending quantum holism, i.e., holistic epigenesis, taking place in open quantum systems, and they actualize raw feelings and realize physical feelings. The latter is most likely energy patterns carrying thermoquantal information. Physical feelings are not causal, in the classical sense, and, therefore, not mechanical. Hence the teleological formulation of physical feelings unfolds in what can be described as a *discrete conscious field* through the "atomistic" conceptions of subjectivity as conscious experiences.

Conscious experiences are suggested to originate from the said *discrete conscious field*, made up of teleofunctions, unified through a functionally-linked teleosemantic hierarchy. This is also expected for the subjective "meaning" ascending causally through bottom-up brain processes while having no causal effect at the top-down line of the cognitive process [35]. According to the interpretation of the hard problem, they should not be characterized as *functional properties*, but rather as experiential ones. To circumvent the problem that experience is something *beyond the mere functional performance*, one should note that teleofunctions are distinct from the causal-role functions involved in functionalism. For example, with an ordinary function, the elements are irrelevant, while in teleofunction, they become important because the actual function concerns what something is *for* and the notion of what something was selected *for* counts. This is a "teleological" notion of function, which can be minimally defined as "teleofunctionality" and connected to the activity of the structure. Moreover, teleofunctions carry a value judgment or "meaning," unlike ordinary function, which brings only action or a resultant effect, which is observer-independent and experiencer-dependent semantics akin to subjective "meaning." The rise of subjective "meaning" stems from the perceptual binding of intercommunicative cognitive content to higher levels of awareness—the unification of subjective "meaning" across the whole relation through cognitive and perceptive processing of external information, due to random connectivity in the brain wiring, and thus are relative to receiving external information as a result of environmental influence, which defines subjectivity or conscious experience.

Organismic teleofunctions represent interconnected subjective entities within the *discrete conscious field* containing the attributes of elemental subjectivity [162]. They carry the "*subjective meaning*" as interrelated *informational* content that reconciles the epistemological qualities with higher-order causation mechanisms. The conceptual underpinning of organismic teleofunctions is teleology [163]. Mayr's notion of teleology characterizes processes that owe its goal-directedness to the influence of an evolved program. The idea invites the field of information theory [164] and the concept of semiotics, which comprises syntax, semantics, and pragmatics. An authentic scientific theory should strictly speak only concern syntax. For instance, the scientific explanation of the occurrence of the genetic code should produce its syntax, while the physical law behind it should generate its semantical possibilities [21–22].

William Lycan [165] developed the category of phenomena to consist of teleofunctions, and these are subjectively, but not objectively, observable. Specifically, capacities have been built into us by natural selection, which naturally leads to a teleological notion of function resting on a teleological functionalist epistemology where "teleofunction" becomes synonymous with teleological functions [165]. There is meaning attached to the interconnectedness of teleofunctions, i.e., the synchronicity of teleological processes.

The teleological process mimics an ongoing computation that never ends, yet communicating teleosemantically, while upholding dual relationships between conjugate entities, such as momentum-energy and spacetime. Although conjugate observables also appearing in a classical setting, as related through the Fourier transformation, there is no contradiction to assert that matter is fundamental in the sense that it is what we observe. The German philosopher Immanuel Kant, almost 250 years ago, said that *space and time are the two essential forms of human sensibility*, and a little more than 100 years ago, Einstein found that general relativity imparts that matter influence spacetime and vice versa. Hence everything we as evolved life forms experience concerns phenomena set in this perspective.

In a "*homuncular functionalist*" version of materialism, it becomes teleofunctionalism, the framework of choice [166]. This begs the question: how does epistemic teleofunctionalism give rise to conscious experience? The root cause embedded in the notion of teleofunctions becomes conscious experiences, reflecting a global workspace function, and requiring cognitive content. The latter formulates a *global workspace*, which has established a functional hub for binding and propagation in a population of loosely coupled signaling proponents that mirrors on conscious cognition [167].

How they are implicated in the humanities, psychology, and art, even religion [168], falls into the category of teleosemantics and the integration of teleofunctionality in cognition [169].

Panexperiential materialism is a materialistic ontology that grounds consciousness as a teleological process through an epistemic teleofunctionalism and informational relationalism. It has answers regarding the explanation of the fundamental human experiences in terms of long-range correlations of phase differences in the molecular orbitals, such as the perturbed π-electron structures or their mirroring nuclear disturbances, see Eq. (8). However, explaining teleosemantics through self-awareness, according to *panexperiential materialism*, does require consciousness *a priori*, ensuring teleosemantics and necessitating the presence of "qualia," which in turn makes clear why it is comparatively demanding to give functional accounts of teleosemantics without teleofunctionality.

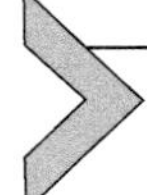

10. Disruption of free energy through non-covalent bonds

When consciousness fades from the intrinsic perspective of experiencing subject, as it does in dreamless sleep, the entire world vanishes. The origin of consciousness is believed to be in the midbrain reticular system brainstem leading to both cortical hemispheres. However, loss of consciousness may be the result of neural transmitters being released from the brainstem. The unleashing of an overpowering neural signal amongst cholinergic, noradrenergic, and serotonergic neurons are causing consciousness to fade. During anesthesia, localized effects on the brainstem can result in loss of consciousness due to gamma-aminobutyric acid, GABA, a receptor-active anesthetic [170]. During dreamless sleep, the brainstem prevents the free energy that is available to the cortical areas through a variety of neural transmitters [171]. Likewise, arousal and awareness disruptions of consciousness in the cortical regions, specifically the left, ventral, anterior insula, and anterior cingulate cortex regions of comatose patients, is regulated by the brainstem through the *rostral dorsolateral pontine tegmentum* [172]. Similar events take place during diabetic conditions and head trauma, where the loss of metabolic energy, due to the disturbance of the synthesis of adenosine triphosphoric acid (ATP), can result in free energy changes.

A potentially critical hypothesis is the loss of consciousness caused by the interference of *free energy* within the cortical areas due to homeostasis in the upper brainstem [173]. The above-localized events disrupt homeostasis at the level of neurons causing loss of consciousness. To elucidate the associated physical mechanism, one needs to determine the smallest scale at which consciousness loss occurs under natural circumstances. Globally spread anesthesia works by perturbing the van der Waals forces causing the inability of *free energy* to traverse regions, such as hydrophobic regions [174] that may silence consciousness. This is caused by disruption of the free energy through non-covalent bonds, such as the conjugated bonds harboring delocalized electrons, with the conjugation thermodynamically stable. Hence carbonyl groups attached to arachidonic acids provide bridging between chains of π-electrons (see Fig. 2).

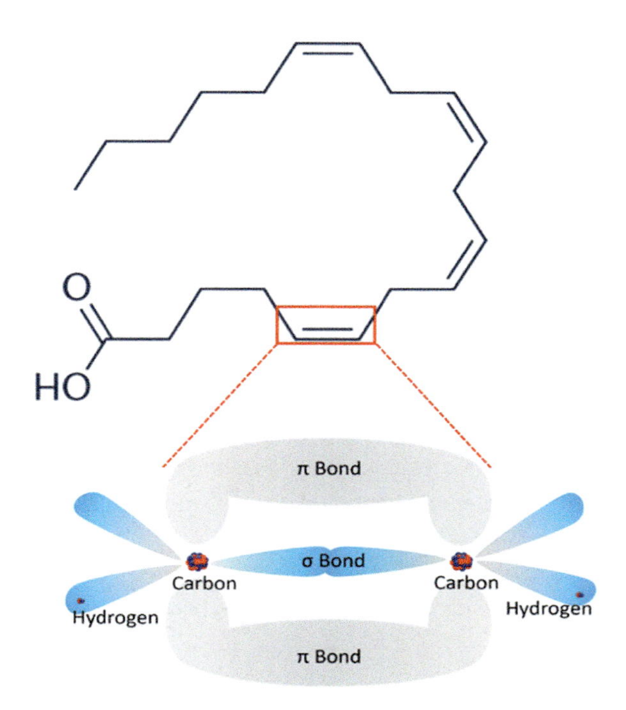

Fig. 2 Illustration of an amphiphilic biomolecular structure of membrane phospholipids, e.g., arachidonic acids. The inset shows the composite structure that represents the distribution of the delocalized π-electron density clouds of the membrane phospholipids. The electron clouds at the 132 pico-meter scale [185] are comparable to most chemistry transformations and measurements. Note that hydrophobic lipids self-assemble in an aqueous environment and have aromatic side chains with its double carbon bonds forming, in addition to the σ-binding structure, the delocalized π-electron bonds contributing to the familiar organization known as a conjugated system.

Due to the inherent uncertainty in electron localization, the weakest type of van der Waals force, the so-called London dispersion forces, are distinctive quantum effects that act on the borderline, where thermal interactions may become increasingly important. These forces, due to the charge separation, originating, e.g., from induced dipole-induced dipole couplings, result in attractive Coulomb interactions occurring between nonpolar, delocalized π-electron clouds of two or more neutral atoms, molecules, or macromolecules. Nonpolar hydrophobicity signifies an absence of electrostatic forces and hydrogen bonding.

Stuart Hameroff [175] has hypothesized that anesthetics are caused by binding in nonpolar hydrophobic regions of cytoskeletal proteins, dispersing endogenous London forces by inhibiting delocalized π-electron clouds. Hameroff [176] further postulated that London forces bound dipoles in nonpolar hydrophobic regions of cytoplasmic proteins couple and oscillate coherently and that this correlation suggested the bearing of *consciousness*, where *synchrony* conjures a sub-molecular basis for a quantum understanding of consciousness [176,177,178].

A handful of cytoplasmic proteins bind anesthetic molecules by fitting into π bonds, located in hydrophobic regions, blocking endogenous London forces, which are in some way responsible for the loss of consciousness [174]. Benzene rings are neutral, nonpolar molecules that can be packed to support short-range interactions involving London dispersion forces, believed to be the collective "glue" of unfolding living systems by forming "*instantaneous dipoles*" [174,179]. In nucleic acids and, but less commonly in proteins, "benzene-rings" are stacked precisely on top of each other hence supporting short-range interactions through London forces. It is obvious that ring-like structures like "benzene-rings" that supports oscillatory motion back and forth due to delocalized π-electron clouds do play a vital role in the human body in general and the brain and its hot and wet milieu in particular.

As is well known, anesthetic molecules also bind to the hydrophobic regions of fatty acids, such as docosahexaenoic acid [180,181], linoleic acid [182], and arachidonic acids, which are part of the neuronal membrane. According to the Meyer-Overton theory, the rate of anesthetic effects seemed to indicate a nearly linear correlation between lipid solubility and anesthetic potency. Even if other factors may interfere, the implication is clear, i.e., it is the hydrophobic regions of membrane phospholipids, in particular, that is responsible for the anesthetic drug to bond and interact with the constituent membrane causing loss of consciousness. Lipid-based theories of anesthesia fail when proposed targets of aesthetic action are proteins. However, the Meyer-Overton theory is still valid since lipophilic proteins are embedded/integrated to membrane lipids via London forces to the hydrophobic sections of the phospholipid [183].

Michael Crawford and colleagues [180,181] have indicated that docosahexaenoic acid, chosen predominately for its role in molecular evolution, has π bonds interrupted by methylene groups, preventing electron transfer as in a conjugated sequence. Thus, unlike the π bonds in benzene, that form resonating rings, or the conjugated bond sequence in the visual pigment, that creates bridging states between interacting π-electrons, the π bonds in membrane phospholipids, such as docosahexaenoic acid and arachidonic acids, the latter involving carbonyl groups, see Fig. 2, do not create resonance structures. The conjugation is inherently thermodynamically stable with conjugated π-bonds more stable than isolated π-bonds. Hence carbonyl groups attached to arachidonic acids provide bridging between chains of π-electrons without the need for a resonant state like in benzene rings in microtubules.

Nevertheless, the assumption [184] is that docosahexaenoic acid and arachidonic acids will display coherence by the effect of conjugated bonds that forms delocalized orbitals, enhancing the mobility of electrons. Electron transfer or tunneling in biological systems is related to coherent dissipative structures sensitive to pH and temperature. It involves a mixture of thermal and quantum fluctuations that create long-range correlations. Unlike the classical redox reaction thermally activated process dependent on classical Arrhenius behavior, this non-Arrhenius behavior occurs in the wet and warm brain. No classical mechanism can explain this behavior—and the eigenstates of the Schrödinger equation corresponding to these thermal energies are subject to decoherence, i.e., thermal decoherence.

Herbert Fröhlich [135] proposed that dipole oscillations propagate along the perimembranous region composed of actin filaments (proteins) without thermal loss maintained by electrons trapped and moving along the filamentous string. The dipole oscillations are sensitive to the macroscopic potential differences associated with the Josephson effect between upper and lower perimembranous regions leading to quantum effects. This "sensitivity" between the microscopic and macroscopic physical phenomena involves co-operative wave motions. Yet the problem with these effects, in cellular systems at room or body temperatures, rests with the understanding that biological processes at the actual energy ranges threaten to be destroyed by quantum decoherence. Although Fröhlich's proposition appeared ahead of its times, proper thermalization suggests that phase-wave correlations may offer alternatives by suggesting that de Broglie's "matter wave" in quantum theory remains comparable to a "*phase wave*" deep in the hydrophobic regions of lipid membranes amphiphilic biomolecular structure of membrane phospholipids, e.g., arachidonic acids, see Fig. 2. The majority of fatty acids will interact with other fatty acids via London dispersion forces. Still, there are also dipole-dipole interactions and hydrogen bonding at the head-end where oxygen is present.

Recent developments in wave function modeling are not thermodynamically closed. Instead, it reflects the steady-state condition, $dS=0$, without the need to incorporate macroscopic superconductivity of the Josephson effect. The self-referential nature of life originated in the cellular membrane of unicellular organisms[186,187]. It is not known whether traditional quantum mechanical treatments, involving long-range energy transfer along the entire neuronal membrane, will form a pathway of delocalized π-electrons along fatty acid chains in membranes. The conjugated bonds are the locations of delocalized electrons, and possibly a pilot wave applies as a facilitator of the bridging between chains of π-electrons that is based on a new interpretation of Bohmian mechanics, which includes temperature.

Furthermore, the hypothesis is that the dielectric bridge (across the phospholipids) can be quantum tunneled through the lipophilic proteins embedded/integrated to hydrophobic sections along the phospholipid bilayer through the covalent bonds that are joined to hydrophilic proteins (polar, enzymatic), resulting in thermally activated processes described by electron transfer redox reactions. Depending on the states involved, this could, for instance, be excited electronic states resulting in a Davydov soliton wave that governs energy transfer in the cytoskeletons of the neurons embodying information of the internal free energy, which might assist the electrostatic coupling of quantum and classical charges [188] or a changeover between quantum and classical soliton wave interactions.

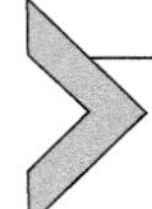

11. The partial gain of intrinsic information as negentropic entanglement

Teleological processes are suggested to be relational, providing informational thermoquantal descriptors of negentropic action. The resultant negentropic entanglement underlies the basis of what John Searle [81] calls a *unified conscious field* and which originates at the fundamental level reflecting upon the electron phase-wave in the membrane phospholipids. However, it is not the actual wave characteristics, such as the phase correlations, the amplitude, or the frequency that underlies the entropy production. The information is to be stored and encoded by the transformation properties exhibited by the extended quantum theoretic formulation.

Thus, teleology, in the broader sense, must be associated with negentropy. Quantum systems are fundamentally acausal—the opposite of causal. However, the continuous build-up of a constructive thermoquantal fabric displays a gradual changeover from teleological to causal phenomena. *Teleological* systems operate based on the final goal as attribution of an information-based communication, originally situated in quantum logic, becoming Bayesian during the semantic stage.

Negentropic action from disorder to order that reflects upon a "*teleological pull*" of negentropic entanglement, where intrinsic information picks up the part of the quantum potential that shapes the in-phase correlated molecular orbitals of delocalized π-electrons in hydrophobic regions of brain membrane phospholipids, e.g., arachidonic acids. It is suggested that the long-range order and thermalization in open quantum systems derived from fusing thermal and quantum fluctuations resulting in a transient density matrix, i.e., subject to thermal decoherence entails a gradual teleological structure that provides a stochastic background for higher-order correlations, i.e., proto-communication.

The long-range *information transformation* is a quantum process in nature that occurs in *correlated dissipative order*. Accordingly, the transformation process is not merely a quantitative description of causal relations between measured events (subjective) but, more importantly, has a physical consequence in thermal entropy production, which is an objective property of the material brain. This can be expressed in terms of the *negentropy principle of information* [189] that claims an increase of information at the expense of entropy increases in the thermal bath. Hence, *negentropic entanglement* becomes a teleological process that gives "*meaning*" to intrinsic inter-relational information structures at distant locations in the brain at the sub-molecular scale of the brain's potentiality.

Arieh Warshel and Michael Levitt [188] have used hybrid quantum mechanical and molecular mechanical methods to investigate electrostatic effects in enzymes. One can extend these ideas to show that the electrostatic coupling not only between London forces but also through the intervening amino acid side chains during enzyme catalysis [190] do result in classical ions and quantum charge distributions through the phospholipid layer membrane, which with the aid of embedded lipophilic proteins, will lead to charge transfer. The build-up of significant charge may be a conduit for such energy exchanges to take place, i.e., a transfer of electrostatic potential energy by carrying away the kinetic energy of the electrons, here giving rise to potentiality as a qualitative aspect of conscious experience, namely subjective "meaning." By *potentiality*, we mean the usual potential energy $U(x)$.

Alternative formulations are based on superfluidity as the zero-temperature characteristic property of the quantum fluid, without any loss in kinetic energy, due to a continuous supply of kinetic energy from the energy exchange at the quantum-classical transition. The "*teleological pull*," caused by the *negentropic entanglement* from the gain in intrinsic information, which is directly linked to the kinetic energy, perturbed by the negentropic term $\frac{\gamma^2}{2m}[\nabla S_Q]^2$. The connecting relation follows from the definition of the dimensionless *Boltzmann-Gibbs thermal entropy* [110]:

$$S_Q = -\frac{1}{2}\ln\ (\sigma)$$

where σ is the enveloping molecular field density distribution of the π-electrons moving under the influence of the nuclei. However, in the context of an alternative formulation, based on a quantum state diffusion [106,113,114], one finds that the quantum potential energy is proportional to the Fisher's information (theoretic) entropy and can be determined from the gradient of the local *Boltzmann-Gibbs* thermal entropy ∇S_Q [114]:

$$Q = -\frac{\gamma^2}{2m}[\nabla S_Q]^2 + \frac{\gamma^2}{2m}\nabla^2 S_Q \tag{12}$$

where m is the mass of the model system in units of [mass], and the action parameter, γ, carries dimensions of [energy][time].

In Eq. (12) the first-term on the RHS is interpreted as the quantum-type integrator or "corrector" of the kinetic energy term. In contrast, the second term influences the classical potential energy U. The kinetic energy, see Eqs. (3, 12), becomes $\frac{1}{2m}(\nabla\Phi\,)^2 - \frac{\gamma^2}{2m}[\nabla S_Q]^2$ while the classical potential energy becomes $U + \frac{\gamma^2}{2\mathrm{m}}\nabla^2 S_Q = 0$. This implies that the kinetic energy of the molecular orbital electrons contains a negentropic term $\frac{\gamma^2}{2m}[\nabla S_Q]^2$ which informs or communicates "meaning" connected with the quantum wave function. The reference to "meaning" alludes to the activity of intrinsic information/energy as preconscious experientialities, and the word "informs," on account of the potential $U = -\frac{\gamma^2}{2m}\nabla^2 S_Q$ canceling the classical potential energy.

In passing we note that $(\nabla\Phi)^2$ has the unit of ([energy][time]/[length])^2 =[mass][energy]. Moreover, given that $(\nabla E)/(\nabla S) = T$ (*Tsekov, personal communication*), where $E = -\frac{\partial\Phi}{\partial t}$ is the total energy, it is possible to find the quantum potential energy as a function of T, and since the *Boltzmann-Gibbs thermal entropy* is dimensionless, the connection is established by multiplication of the Boltzmann constant k_B.

In summary, the neural wave equation [100]:

$$i\gamma\frac{\partial\psi}{\partial t} = -\frac{\gamma^2}{2m}\nabla^2\psi + U\psi \tag{13}$$

where m is the mass of the model system in units of [mass], and the action parameter, γ, carries dimensions of [energy][time]. Eq. (13) describes an internal system in constant interaction with its environment, represented through the macroscopic quantum potential Q and the classical potential energy U. The quantized system is maintained as long as $Q \gg U$. Given that our Q is now dependent on T, this is possible. Finally, we stress that

(i) the T-dependence can be further reinforced by the inclusion of density matrices [109]; see also the next section;

(ii) the extended Hamiltonian function preserves the conjugate relationship between generalized momenta and coordinates [110];

(iii) teleological phenomena where cognitive appraisal as the end goal precedes intrinsic information using the quantum potential for "force-free" action at the quantum boundary of life in the "grey zone".

12. Thermoquantal information as the physicality of quality

Popular theories of consciousness are based on quantum-state reduction in spacetime geometry, the so-called Orch OR theory [191], integrated information theory (IIT) [46], and the free energy principle [192]. On closer inspection, the quintessential definition of brain-based consciousness remains a hyperbole. As we assume that consciousness has an irreducible subjective ontology, the result is that we do experience and articulate it at the cognitive level. Therefore, epistemologically objective claims can be made about teleological phenomena that are ontologically subjective.

The human brain is a holarchical system through successive transformations of the functional and hierarchical organization. Functional interactions depend on nonsymmetry and nonlocality. The former is contingent on the physical aspect because of its structure, and the latter rests on the informational aspect because of its function. Physical interactions that rely on force-based action that operates when the medium is homogenous (local), concern only with the structural organization of molecular neural networks without affecting its functional organization. Yet, each hierarchical level is dependent on the other through its functional organization. Integrative brain functions manifest by way of functional connectivity, i.e., the connections that span the functional organization. In contrast, holistic brain function manifests through nonlocal functional interactions that depend on structural discontinuity (non-symmetry), with the mode of interaction informationally relying on information-based action.

At the fundamental level, structural discontinuities can apply to fractal space-time geometry itself as in the Orch OR theory in physics or molecular systems composed of atoms with electrostatic interactions (London dispersion forces) [188]. A fractal too far beyond quantum biology is nonbiological and, as such, needs to be accommodated through "*macroscopic*" *quantum potentials* [129] and the role they play in electrostatic interactions within the molecular layer of neurons. However, fractality of space cannot support long-range order in the phase wave. Therefore, ripples in space-time geometry must be trivial for brain-based consciousness, as they occur in an entirely different energy range compared to those of relevance for brain-based consciousness.

Is consciousness an integrative process encapsulating integrated information theory? The anomaly placed upon the notion of integration espouses the existence of emergent interactionism that is based on emergent causal effects and the transference between systems. In contrast, an information-based action or "*affective drive*" where information is transformed (not transferred), for instance, *intrinsic information*, has spectral content that can be used in the transformation of thermoquantal information. The latter are *de facto* correlates of consciousness via information-based actions of panexperiential matter, controlled by conjugate force-based actions in our material environment, experienced through our senses as Shannon information entropy. As a result, there must be entities that interact, correlate, communicate, and evolve into life-forms that might develop ethical conduct through our mental representations derived from Kant's *essential forms of human sensibility in space and time*.

The Fourier-Laplace transform, as discussed above, provides an extension and combination of the classical integral transforms used to convert variables (operators) such as material-momentum observables onto space-time [193]. For instance, with a Liouvillian $\mathcal{L}$, putting $\hbar = 1$, the degenerate frequency, the unit and nilpotent operator respectively, given by ω_0, I, J one obtains in the retarded-advanced representation:

$$\mathcal{L} = (\omega_0 - i/\tau(I - J))$$
$$\mathcal{G}^{\pm}(t) = \mp i\vartheta(\pm t)\mathrm{e}^{-i\mathcal{L}t}; \mathcal{G}(z) = (zI - \mathcal{L})^{-1}$$

where ϑ is the Heaviside step-function, whose value is one for positive arguments and zero otherwise, the following generalized transformations:

$$\mathcal{G}^{\pm}(t) = \frac{1}{2\pi}\int_{C^{\pm}} \mathcal{G}(z)e^{-izt}dz$$
$$\mathcal{G}(z) = \int_{-\infty}^{+\infty} \mathcal{G}^{\pm}(t)e^{izt}dt; \begin{cases} \text{upper sign} \Longrightarrow \text{im}z > 0 \\ \text{lower sign} \Longrightarrow \text{im}z < 0 \end{cases}$$

with the contour $C_{\pm}$ going in the upper-lower complex energy half planes and closed in the lower-upper half where the integrand converges to zero, the latter due to the occurrences of complex resonance energies in the lower-upper Riemann sheet respectively. Note that the integration contours will be suitably deformed during analytic continuation. Combining time-temperature, conjugate to energy-entropy, yields further generalizations as the resonance poles are no longer simple with the actual degeneracy given by Eq. (8), resulting in the subsequent Poisson stochastic evolution.

These conjugate relationships take place at the quantum boundary of life admitting thermoquantal information as the given natural process. This is in contrast to inner world simulations that contain perspective projections [194,195], which is nothing but a dual-world representation without self-identity and hence subject to phenomenological fallacies. The given natural process stems from the premise that "subjectivity comes from "intrinsic information," such that *there is something intrinsically it is like* to be in this experience, is a self-referential motif that should be verifiable by reference to Gödel's incompleteness theorem [196]. It demonstrates that any formal system, broadly defined as any well-defined system of abstract thought, based on the model of mathematics, will be incomplete because there will always be statements that cannot be proven. Computers, e.g., are formal systems that judge the truth of statements. No formal system is powerful enough to prove its consistency and is therefore incomplete. Minds are not the same as computers because the former can perform tasks that are *non-computable*, and this is the key *to understanding the difference between computers and minds* [197].

Although, computational properties are physical properties—are computations "intrinsic to physics"? Mathematical modeling is only an approximation of reality when the reality is itself approximated. In reality, computation is not discovered in physics, but it is assigned to it. The laws of natural processes are merely contingently computational because the mathematical language we use to succinctly formulate physical laws is biased towards being computational, e.g., Schrödinger's equation. Thus, since physics is "knowledge of nature," computations appear to be a way to disseminate this knowledge. Still, due to the *intrinsicness problem*, the computations are not about semantics but a mere syntax for mathematical language. Note: we do not adhere to the Platonic notion since the materialistic evolution, together with its conjugate partner, including human cognition, expressed through language, mathematics, poetry, music, etc., evolves commensurate with abiogenetic life processes originating from a self-referential quantum chemistry life principle [22].

A solution to this *intrinsicness problem* is to ensure the open quantum system is not a formal system in the sense of Gödel. For example, transforming a truth-functional proposition calculus to linear algebra reveals a surprising connection between the provability of Gödel's famous sentence and the self-referential character of general gravity and the exceptional interpretation of the appearing singular point, see the Appendix in Brändas [198]. Alternatively, one can show from this that teleological processes entail a state of consciousness to be composed of intrinsic information that is non-integrated information manifesting information-based action as opposed to integrated information theory. This makes sense since the latter fosters the mind as a computational functionalist machine or a formal system, while the former renders the mind qualitative relying on nonlocal functional interactions. Hence the system is not formal, which tells us whether this *likeness* is intrinsic or contingent. In other words, we can resolve the *intrinsicness problem* by the irreducibility of the informational holarchy due to quantum holism.

The understanding of consciousness is to be based on quantum holism and, therefore, on global phenomena in the brain. Mathematical modeling enables us to move from local to global phenomena, and in this sense, global phenomena are explained by the functional integration of local phenomena. The vehicle from local to global is based on *nonlocal functional interactions*, which entails information-based action dependent on structural discontinuity (non-symmetry) [199]. Therefore, we surmise that consciousness is a global phenomenon or a holistic brain function that is dependent on structural discontinuities, and therefore outside of any structures suggestive of *panprotopsychism*.

Furthermore, a quantum universe can be made of *n* bosonic material particles in a multi-dimensional configuration space. Within this framework, one might expand on the notion of biological nonlocality in the non-Euclidean geometry of biophysical neural networks [200,201]. The three-dimensional space would be inappropriate, just as well as classical graph theory of discrete mathematics since it also would be insufficient to model the corresponding "connectome" of consciousness [202]. An abstract space of all possible configurations is needed to appropriately model the fundamental levels of network neuroscience based on its underlying nonsymmetry and nonlocal functional interactions.

Steven Bodovitz [203,204] claims that consciousness is discontinuous and disintegrates without temporal integration. If so, then how does consciousness become a unified conscious experience? The unification of consciousness as a perception of continuity is needed for cognitive performance, binding together a disparate collection of thoughts in molecular networks. We suggest that the cognitive binding problem [205] applies to conscious cognition. Consciousness is discrete and unified through cognition. Cognition does not arise from the linking of symbols but the linking of meanings. Cognitive neural networks should not only be associative networks of neurons, but they should also amount to unraveling the nature of the isomorphic connections, which give the meaning of the formal rules. As per the Searle's Chinese Room Argument, neural networks operate formal rules leading to the capture of symbols, but not the capture of meanings of raw feelings. Thus, reality isomorphically defined in terms of the *intrinsicness problem* can be modeled to actualize the preconscious experientialities as meanings. Still, preconscious experientialities concur synchrony as a precondition of binding, where simultaneity carries meaning by way of the physical process of negentropic entanglement. This is evidenced through non-Bohmian nonlocality, in which Q is a function of the position of all delocalized electrons, thus triggering vast numbers of neural processes within the brain without an *integration process*. In molecular networks, there must exist some non-Bohmian nonlocality, since the latter exhibits no *integration process*. These networks form a structural discontinuity, which is needed for delocalization, and are suggested to lower the network entropy [137]. However, they do not invoke the *integration process* that is required for unitary consciousness. Yet, it is difficult to imagine how quantum coherence can be maintained across neurons.

A recent hypothesis suggests that networks of networks involving biochemical reactions are subject to *quantum phase transitions* that recognize a macroscopic quantum coherent phase that might avoid temperature quantum decoherence by way of a *Feshbach resonance* [137]. However, this is not a feasible mechanism, since the *superconductivity* property, or ODLRO [140], develops at an entirely different temperature range, which is extremely sensitive to decoherence, than those that cover hierarchical structures constituting biological networks. It was shown that macroscopic "long-range order" exists only in the absence of classical potential energy [108]. In other words, the quantum potential energy is the result of self-referential amplification by way of a "quantum force" that is a force-free information-based action as concealed quantum-thermal correlated motion, resulting in entropy-energy exchanges at the quantum-classical transition. The gradual change from pure QM via thermal perturbation offers a gradual transition from quantum logic to Bayesian inferences, i.e., from telicity to causality.

To be fundamental, the free-energy principle [192] would entail going beyond Bayesian statistics. This does not suggest Quantum Bayesianism (QBism), although a quantum mechanical interpretation of the latter, in relation to the understanding of the mind-matter problem, would strengthen its plausibility. Nevertheless, we reject the idea of experience as fundamental that is the cornerstone of QBism [206]. In contrast, panexperiential materialism follows an epistemic teleofunctionalism and informational relationalism where *meanings* create flows of transitions equipped with time scales and precise temperature dependencies building up the steady-state and, importantly, not based on Bayesian statistics. Through the lowering of entropy, it suggests information or rather a proto-communication hypothesis deriving the syntax for *Communication Simpliciter*, i.e., communication pure and simple, straightforwardly stressing the development of physical interactions and correlations through coding and Shannon's theory of communication, to the macroscopic ranks including humans, societies, ecology, and cosmos, yet defining communication in a restricted sense, cf. the difference between semiotics and semiology [21].

What are the teleological processes of physical feelings in our understanding of the nature of consciousness? If consciousness is discrete at the most foundational level, i.e., the "quantum underground" where life and conscious processes originate [207], then its role as a *holistic brain function* can only be possible as an information-based action and not as a computational brain function based on integrated information theory. A summary of

Table 1 A summary of holistic brain functions.

Information-based action	Scale	Cause	Attribution
1. Microscopic level: Quantum realm			
(a) Active information	Atomic	Quantum potential	Potentialities
(b) Intrinsic information	Submolecular	Self-referential amplification "Quantum force" $(-\nabla Q)$[a]	Raw feelings (microfeels)
2. Mesoscopic level: Quantum boundary of life[b]			
(c) Thermoquantal information	Molecular	Energy-entropy dissipation	Physical feelings
3. Macroscopic level: Classical realm			
(d) Cognitive information[c]	Subcellular	Negentropic entanglement	Qualia

[a]"Quantum force" is a "force-free" information-based action.
[b]The quantum boundary of life arises in quantum biology when quantum uncertainty is replaced with (entropic) stochasticity.
[c]We consider cognitive information to be distinct from cognitive information processing where the latter is a type of self-organization, responsible for Free Will.

holistic brain functions associated with the attribution of information-based action is given in Table 1.

The *physicality of quality* is based on the relational co-existence of *raw feelings* in the unconscious mind [208] that act as analytical lenses through which the holarchical relationship that exists between preconscious experientialities as *physical feelings* are objective in the unconscious mind. Yet, the information transformation of *physical feelings* become qualia higher-up in the subconscious mind. The physical attribution relating to negentropy changes and hence influencing the internal energy across the information holarchy are subliminal feelings classified as subjective. Moreover, there is no category error if objective entities are encased by subjectivity in the subconscious from the unconscious, objective entities also encase subjectivity in the conscious mind, which depend on the informational holarchy. Subliminal feelings are informational and occur subconsciously through interaction with *cognitive content.*

However, emotions cannot be interchangeable with subliminal feelings (the core affect lacking cognitive appraisal). We claim that the underlying emotional system is affective but can also have cognitive aspects that are entirely expressed as feelings requiring a cortex. Affect, in our context, is the subliminal feeling in the subconscious of cortical functioning, i.e., appearing on a deeper teleological level. It is not the emotional varieties of affect, e.g., a sensory or "bodily" affect, but it is rather in the context of a psychological "qualitative" or "subjective" experience.

The confusion between feeling as *the capacity for consciousness* and *cognitive-emotional feeling experiences* began with the Dutch philosopher Baruch Spinoza. The physical aspect of feelings is uniquely placed in understanding how consciousness operates. However, there are cognitive manifestations of these physical feelings as cognitive-emotional (physiological) feelings, which may also be conceived as a *teleological process* that extends beyond the physical correlates of feeling. Emotion is the pragmatic result of a complex interplay between the amygdala and the cortex resulting from the physical activity of a vast number of neurons via cognitive firing of action potentials. Unlike emotions that are influenced by the environment and are fleeting, subliminal feelings in the unconscious may be recalled to the subconscious as preconscious experience. Moreover, physical feelings are shaped by *preconscious experientialities* that stems from informational descriptors isolated from perception, cognition, and self-awareness. We call the informational descriptors atomic "microfeels" [209]. Unlike "qualia," which is the feeling of experience (a subjective phenomenon), atomic "microfeels" are "raw" feelings' at the submolecular scale that act as analytical lenses through which subliminal feelings are fathomed.

The functional analysis of "qualia" in terms of physical feelings must, therefore, lead to subliminal feelings higher-up in the holarchy at the macroscopic level, which is composed of brain regions [210] and are the fundamental agents of preconscious experientialities. We defined preconscious experientialities as potentialities under "*consciousness guidance*" of a "quantum force" as a reflection of a "keycode" in the actualization process of the potentialities arising from the energy exchange to be raw feelings/microfeels [107]. That is why we have argued that consciousness is neither emergent nor a fundamental property of nature because it is intrinsic through the material organization of energetic and informational exchanges with the environment.

Preconscious experientiality is not conscious experience, but a crucial stage before the onset of self-awareness and conscious perception [211]. According to Antonio Damasio [210], feelings have an evolutionary origin, with their conscious experience arising from a vibrant spectrum of qualia in extremely variable configurations. The spectra of quale are not inherent in external objects. They have evolved in brains, transforming thermoquantal information to the next level. Due to the brain's holistic organization, it results in the holistic epigenesis of informational structures made up of negentropy. Thus, self-creation or self-realization of the organism wholly entailed by potentialities arising from energy exchange at the quantum domain, by granting the omnipresence of negentropic actions that sustains the image of evolution constituting the creative potential of an organism [212]. The self-creative processes of organisms contribute to the self-adaptation of the individual organism within its life span subsuming negentropic entanglement [59] and nonlocal functional interactions [25,202].

13. Discussion

Alfred Whitehead's panexperientialism makes it clear that "occasions of experience" are what move reality-processes along. For Whitehead [23], the experience is the *sine qua non* of reality, and each actual event (occurring every pico-second or so) experiences its moment of becoming then subjectively "perishes." It is only then that it turns into a part of authentic experiences. Thus, it will provide continuity to influence the new becoming of actual events in the process of reality. Note, that subjectivity is affected by the "*teleological pull*" just *before* it becomes part of objective reality and vice versa. This in no way puts the "material" before the experience, so Whitehead's process theory is most certainly, *not* materialist. The experience takes place throughout nature and not just in brains, where the cosmos is seen as a continuous basis of experience. Ideas based on David Bohm's quantum potential, Arthur Koestler's holarchy concept, and William Lycan's teleofunctionalism, puts the "*teleological pull*" into perspective with matter before the experience, reinforcing panexperiential materialism as a framework of choice in our understanding of consciousness as a teleological process of cognition created by the transformation of information within the functional and hierarchical organization of the brain.

Experience-in-itself through the physical material-energy world evolves at the end of the experiential cycle (Whitehead's "occasions of experience"). Still, in panexperiential materialism, the conceptualization of the enveloping fields forms the bases of non-inertness of panexperiential matter. The enveloping field of the π-electrons, and their thermal agglomeration, constitutes a native property of the membrane that emanated from molecular evolution. What evolves is the holarchical relationship of the enveloping fields. Hence the raw *experience-in-itself* does not require a centralized "*experiencer*" or a "*decoding homunculus*" or a "*homuncular functionalist*" version of teleofunctionalism [166] only because it contains the whole experience that will become the "subjective experience." That means that the whole is something else than the sum of the parts, but it is not an independent existence, i.e., is not "*gestalt information*" in the brain [213].

Panexperiential materialism has answers regarding the explanation of the fundamental human experiences in terms of "long-range" correlations of phase differences in the molecular orbitals, such as the perturbed π-electron structures or their mirroring nuclear disturbances, see Eq. (8). Such partial coherence allows for off-diagonal long-range correlative information, ODLCI (as an extension of Yang's ODLRO), and the explicit role of thermoquantal information based on the following points:

(i) Yang's concept of ODLRO, [140], relating superconductivity to the appearance of a large eigenvalue in the fermionic two-matrix, see [214], implies a coherent ground state of condensed matter.

(ii) Provided the temperature is lower than a critical temperature, the state becomes superconducting. A similar argument leads to superfluidity. This is a zero-entropy state. Above the critical temperature, thermal noise destroys the quantum state, which will decohere to equilibrium (maximum entropy) above the critical temperature.

(iii) Bloch thermalization of the quantum state entails an excited configuration that defines partial phase coherence, provided energy, and time scales essentially match, see, e.g., Eq. (4). This provides the fusion between quantum and thermal fluctuations.

(iv) The steady state, see Eq. (8) is a transition density matrix (Jordan block), which can be proven to fulfill the steady state condition $dS=0$.

(v) The relation between the temperature, T, and the relaxation time, t, yields straightforwardly Eqs. (9, 10) from Gibb's free energy. This gives an explicit relation between entropy production and negentropic gain.

(vi) The self-referential property obtains from the analogy with Gödel's incompleteness theorem, translated to linear algebra. The simple analogy between "Gödel" and gravity displays the importance of the conjugate relationship between energy-time and momentum-space.

(vii) The transformations $\boldsymbol{B}$ (and $\boldsymbol{B}^{\dagger}$), see Eq. (11), transforming the Bloch thermalized state to classical canonical form, provides the syntax for proto-communication.

(viii) The thermoquantal fluctuations generate the irreducible degenerate spectrum of the Liouvillian maintaining the conjugate relationship between energy-time through the Fourier-Laplace transform between the resolvent $(zI - \mathcal{L})^{-1}$ and the time propagator $e^{-i\mathcal{L}t}$. Since one has been careful to extend quantum mechanics via analytic continuation to dissipative systems, the space-momentum relationship has not been violated.

(ix) The irreducible degeneracy of the spectrum provides a stochastic Poisson time evolution that displays the self-organizational properties of the thermoquantal system.

(x) The dissipative dynamics extends Gibbs free energy to the steady state, $dS=0$, which provides the conjugate relationship between temperature and entropy. The latter is particularly significant since it relates directly to physical feelings through the reorganization of free energy.

We have defined *panexperiential materialism* as a new materialistic theory of mind that replaced "*emergent interactionism*" in *dialectical materialism* with quantum holism. We have pointed out that quantum holism is a way to bridge the "explanatory gap" [34] using the informational holarchies as a theoretical explanation of the physical self-referential mechanism that evolves preconsciously, originating from the spectral structure of phase waves. The process includes quantum energy exchange to take place, resulting in "long-range order" The loss of phase coherence occurs by quantum Brownian motion when the classical diffusive force exceeds the internal "quantum force." This indicates that "long-range order" exists only in the absence of the classical potential energy. When the quantum potential energy (Q) negates the classical potential energy, there is phase coherence. Therefore, the quantum potential energy cannot emerge classically [108].

Our holistic viewpoint denies the existence of integrated information theory as an emergentist ontology. Still, it advocates the irreducibility of the informational holarchy to fundamental levels due to quantum holism as a property of self-reference, i.e., every holon has a dual tendency to preserve its individuality as a quasi-autonomous whole. The compelling proposition in *panexperiential materialism* is that there needs to be no "global workspace" because of the irreducibility of the informational holarchy.

The thermodynamic cohesion between open quantum systems stems as a holarchical system of teleofunctional or informational structures that can be accounted for by information-based action that acts asymmetrically, resulting in the introspectability required for human-type conscious experience based on the following essential conditions: (i) Relational closure achieved in the entailment of the holarchy; (ii) Holistic epigenesis taking place through the electrostatic coupling of quantum and classical charges in proteinaceous structures, e.g., lipophilic proteins as a passage for cognition; (iii) Autopoietic processes supplying continuous metabolic energy instantiated by macroergic effects in the nuclei of neurons [215] needed for neurophysiological action in cognition and not information-based action. Therefore, preconscious experientialities can affect cognitive processes, but not vice versa.

Furthermore, we have directed our focus on more deep-seated issues, such as the interaction, collocation, and communication between complex enough atomic and molecular systems as governed by modern quantum chemical laws. This involves (i) the rejection of metaphysical approaches to consciousness which are grounded on a phenomenological fallacy, (ii) the key role of nonreductive physical processes underpinning the correlates of consciousness through the conjugate relationship between quantum observables, (iii) the continuation of quantum theory to open system dynamics, (iv) the extension to the Fourier-Laplace transformation, where the conjugate relationship augments entropy-temperature, (v) the fusing of quantum-thermal fluctuations to constructive thermoquantal correlations, (vi) the importance of simultaneity and coalescence through negentropic entanglement rather than unitary consciousness through binding, (vii) the holarchic descriptors at the fundamental microscopic level in principle responsible for the self-referential amplification of quantum holism beyond quantum eigenstates described by the pure wave functions of pioneering quantum mechanics, and (viii) the information transformation of teleological processes through negentropic entanglement that extracts 'meaning' into the context of cognitive information processing.

We claim that the higher-order holistic brain functions are not produced by chance. They evolve through fundamental natural laws and self-referential quantum chemical-physical principles originating at the fundamental level. Placing consciousness at the submolecular scale, not the subatomic scale, makes it an entirely different problem. The measurement problem fades, and wave function has collapsed as consciousness is brought to the subcellular scale via thermoquantal information "ripples" we define to be the subjective character of experience. In other words, it is "qualia" driven by energy and entropy exchange with the environment, the net result of the steady-state organization being the partial gain of intrinsic information providing negentropic entanglement for cognitive appraisal. The subcellular scale being a kind of interface between neural and quantum levels where classical dynamics of quantum localization prevails.

Panexperiential materialism is distancing itself from panpsychist materialism. It is the only framework that solves the mind-body problem without using eliminative arguments. Although it considers the matter-wave as fundamental [2], it is a modified Bohmian mechanics that includes a thermodynamic quantum potential as well as non-Bohmian nonlocality and intrinsic information as opposed to Bohmian "active" information [216,217]. The novelty of this framework is the avoidance of dialectism as a process of change and traps such as panpsychist materialism with protoconsciousness and unitary consciousness assumed in phenomenology. Panexperiential materialism fosters an understanding of brain-based consciousness originating within the quantum realm as potentialities with their actualization at the quantum boundary of life. Therefore, we stop at the boundary, but we must indeed include it; otherwise, no consciousness and no isomorphism are connecting the brain with the mind. So etiological issues arising from the origins of the quantum potential and holoinformationonal views of consciousness [218] are beyond the scope of this work.

Panexperiential materialism could be viewed as quantum vitalistic since it assumes the fundamental correlates of consciousness being derived from the states of coherent quantum entities at the subcellular scale. In quantum vitalism [219], there is a quantum force suggestive of force-based action without any explanation of its functional role, but in panexperiential materialism, it is the fabric of non-inert matter that organizes interferences between regions for the extra quality of energy-negentropy exchange providing the transformation of information that creates the information-based action as a teleological process of cognition. This is why panexperiential materialism goes deeper than quantum vitalism.

More realistic and down-to-earth, it should be interesting to develop and devise our self-referential identity theory of the crucial physical process-based phase differences that underlies "long-range order," and their role in the quantum-thermal regime. It is also evident that a closer understanding of the interrelations between the de Broglie-Bohm extensions and the density matrix approach could be worked out in some more detail to design practical methods in revealing the secrets of the brain in the energy regime of kT at $T=310$ K. In particular, one might study the pixel-like structure of the cognitive system and/or examine the stochastic nuclear background, serving as channels for the negentropic π-electron cloud, providing general teleological strategies for the chains of membrane phospholipids through new developments in electron phase microscopy at the picometer scale.

Finally, quantum effects in biology may have limited degrees of freedom, because lipophilic proteins are enzymes that can be considered to constrain the energy release, thus acting as a "*quantum boundary condition*" for constraining the release of free energy in a non-equilibrium process into a few degrees of freedom, causing a more delayed entropy production in the process and so more easily radiated from the organism to the environment.

Acknowledgment

The helpful hints from Dr. Roumen Tsekov of Sofia University (Bulgaria) are greatly acknowledged.

References

1. Crick, F.; Koch, C. Towards a neurobiological theory of consciousness. *Semin. Neurosci.* **1990**, *2*, 263–275.
2. de Broglie, L. Research on the quantum theory. *Ann. Phys.* **1925**, *2*, 22–128.
3. Ho, M.-W. Towards a theory of the organism. *Integ. Physiol. Behav. Sci.* **1997**, *32*, 343–363.
4. Varpula, S.; Annila, A.; Beck, C. Thoughts about thinking: cognition according to the second law of thermodynamics. *Adv. Stud. Biol.* **2013**, *5*, 135–149.
5. Prigogine, I. *From Being to Becoming*; W. H. Freeman & Company: San Francisco, 1980.
6. Del Castillo, L. F.; Vera-Cruz, P. Thermodynamic formulation of living systems and their evolution. *J. Mod. Phys.* **2011**, *2*, 379–391.
7. Feinberg, T. E.; Mallatt, J. Phenomenal consciousness and emergence: Eliminating the explanatory gap. *Front. Psychol.* **2020**, *11*, 1041.
8. Koestler, A. *The Ghost in the Machine*; Macmillan: New York, 1967.
9. Koestler, A.; Smythies, J. R. *Beyond Reductionism: New Perspectives in the Life Sciences*; Hutchinson: London, 1969.
10. Bohm, D. J. A suggested interpretation of the quantum theory in terms of hidden variables. *Phys. Rev.* **1952**, *85*, 166–193.
11. Bohm, D. J. A new theory of the relationship of mind and matter. *J. Am. Soc. Psych. Res.* **1986**, *80*, 113–135.
12. Bohm, D. J. A new theory of the relationship of mind and matter. *Philos. Psychol.* **1990**, *3*, 271–286.
13. Pribram, K. H. The deep and surface structure of memory and conscious learning: toward a 21st century model. In *Mind and Brain Sciences in the 21st Century*; Solso, R. C., Ed.; MIT Press: Cambridge, MA, 1997.
14. Pribram, K. H. *Languages of the Brain: Experimental Paradoxes and Principles in Neuropsychology*; Prentice-Hall: New Jersey, 1971.
15. Bohm, D. J.; Hiley, B. J. *The Undivided Universe: An Ontological Interpretation of Quantum Theory*; Routledge: London, 1993.
16. Pribram, K. H. Consciousness reassessed. *Mind Matter* **2004**, *2*, 7–35.
17. Pribram, K. H. Prolegomenon for a holonomic brain theory. In *Synergetics of Cognition*; Haken, H., Stadler, M., Eds.; Springer-Verlag: Berlin, 1990.
18. Gabor, D. Theory of communication. *J. Inst. Electric. Eng.* **1946**, *93*, 429–441.
19. Chalmers, D. J. Facing up to the problem of consciousness. *J. Conscious. Stud.* **1995**, *2*, 200–219.
20. Brändas, E. J. The origin and evolution of complex enough systems in biology. In *Advances in Quantum Systems in Chemistry, Physics, and Biology*; Tadjer, A., Pavlov, R., Maruani, J., Brändas, E. J., Delgado-Barrio. G., Eds.; Progress in Theoretical Chemistry and Physics, Vol. 30; Springer: Switzerland, 2017; pp. 409–437.
21. Brändas, E. J. Molecular foundation of evolution. In *Quantum Systems in Physics, Chemistry and Biology—Theory, Interpretation, and Results*; Jenkins, S., Kirk, S. R., Maruani, J., Brändas, E. J., Eds.; Advances in Quantum Chemistry, Vol. 78; Elsevier: UK, 2019; pp. 1–30.

22. Brändas, E. J. Abiogenesis and the second law of thermodynamics. In *Advances in Quantum Systems in Chemistry, Physics, and Biology*; Mammino, L., Ceresoli, D., Maruani, J., Brändas, E. J., Eds.; Progress in Theoretical Chemistry and Physics, Vol. 32; Springer: Switzerland, 2020; pp. 393–436.
23. Whitehead, A. N. *Process and Reality: An Essay in Cosmology*; Macmillan: New York, 1929.
24. Baluška, F.; Yokawa, K.; Mancuso, S.; Baverstock, K. Understanding of anesthesia-why consciousness is essential for life and not based on genes. *Commun. Integr. Biol.* **2016**, *9*, e1238118.
25. Polak, M.; Marvan, T. Neural correlates of consciousness meet the theory of identity. *Front. Psychol.* **2018**, *9*, 1269.
26. Chauvet, G. A. Nonlocality in biological systems results from hierarchy: Application to the nervous system. *J. Math. Biol.* **1993**, *31*, 475–486.
27. Hameroff, S. R. The quantum origin of life: how the brain evolved to feel good. In *On Human Nature: Biology, Psychology, Ethics, Politics, and Religion*; Tibayrenc, M., Ayala, F. J., Eds.; Academic Press: San Diego, 2017.
28. Esfeld, M. Quantum holism and the philosophy of mind. *J. Conscious. Stud.* **1999**, *6*, 23–38.
29. Chalmers, D. J. The combination problem for panpsychism. In *Panpsychism*; Brüntrup, G., Jaskolla, L., Eds.; Oxford University Press: Oxford, 2017.
30. Woo, C. H. Consciousness and quantum interference-an experimental approach. *Found. Phys.* **1981**, *11*, 933–944.
31. Pepperell, R. Consciousness as a physical process caused by the organization of energy in the brain. *Front. Psychol.* **2018**, *9*, 2091.
32. Dennis, G.; de Gosson, M. A.; Hiley, B. J. Bohm's quantum potential as an internal energy. *Phys. Lett. A* **2015**, *379*, 1224–1227.
33. Holland, P. Quantum potential energy as concealed motion. *Found. Phys.* **2015**, *45*, 134–141.
34. Levine, J. Materialism and qualia: The explanatory gap. *Pac. Philos. Quart.* **1983**, *64*, 354–361.
35. Searle, J. R. Consciousness. *Ann. Rev. Neurosci.* **2000**, *23*, 557–578.
36. Musacchio, J. Why do qualia and the mind seem nonphysical? *Synthese* **2005**, *147*, 425–460.
37. Beshkar, M. A thermodynamic approach to the problem of consciousness. *Med. Hypotheses* **2018**, *113*, 15–16.
38. Carruthers, P.; Veillet, B. The phenomenal concept strategy. *J. Conscious. Stud.* **2007**, *14*, 9–10.
39. Levin, J. What is a phenomenal concept? In *Phenomenal Concepts and Phenomenal Knowledge. New essays on Consciousness and Physicalism*; Alter, T., Walter, S., Eds.; Oxford University Press: Oxford, 2006; pp. 87–110.
40. Tononi, G.; Boly, M.; Massimini, M.; Koch, C. Integrated information theory: From consciousness to its physical substrate. *Nat. Rev. Neurosci.* **2016**, *17*, 450–461.
41. Panksepp, J. On the subcortical sources of basic human emotions and the primacy of emotional-affective (action-perception) processes in human consciousness'. *Evol. Cognit.* **2001**, 7, 134–163.
42. Panksepp, J. Affective consciousness: Core emotional feelings in animals and humans. *Conscious. Cognit.* **2005**, *14*, 30–80.
43. Edelman, G. M.; Gally, J. A.; Baars, B. J. Biology of consciousness. *Front. Psychol.* **2011**, *2*, 4.
44. Merker, B. Consciousness without a cerebral cortex: a challenge for neuroscience. *Behav. Brain Sci.* **2007**, *30*, 63–134.
45. Maruani, J. Can quantum theory concepts shed light on biological evolution processes? In *Advances in Quantum Systems in Chemistry. Physics, and Biology*; Mammino, L., Ceresoli, D., Maruani, J., Brändas, E. J., Eds.; Progress in Theoretical Chemistry and Physics, Vol. 32; Springer: Switzerland, 2020; pp. 393–436.

46. Tononi, G. Consciousness as integrated information: a provisional manifesto. *Biol. Bull.* **2008**, *215*, 216–242.
47. Almada, L. F.; Pereira, A., Jr.; Carrara-Augustenborg, C. What affective neuroscience means for science of consciousness. *Mens Sana Monogr.* **2013**, *11*, 253–273.
48. Pace-Schott, E. F.; Arnole, M. C.; Aue, T.; et al. Physiological feelings. *Neurosci. Biobehav. Rev.* **2019**, *103*, 267–304.
49. Panksepp, J. *Affective Neuroscience: The Foundations of Human and Animal Emotions*; Oxford University Press: New York, 1998.
50. Panksepp, J. The neuro-evolutionary cusp between emotions and cognitions: Implications for understanding consciousness and the emergence of a unified mind science. *Evol. Cognit.* **2001**, 7, 141–163.
51. MacGregor, R. J.; Vimal, R. L. P. Consciousness and the structure of matter. *J. Integr. Neurosci.* **2008**, 7, 75–116.
52. Pockett, S. Consciousness is a thing, not a process. *Appl. Sci.* **2017**, 7, 1248.
53. Pribram, K. H. *The Form Within: My Point of View*; Prospecta Press: Westport, CT, 2013.
54. Northoff, G. Localization versus holism and intrinsic versus extrinsic views of the brain: a neurophilosophical approach. *Minerva Psichiatr.* **2014**, *55*, 1–15.
55. Longinotti, D. Agency, qualia and life: connecting mind and body biologically. In *Philosophy and Theory of Artificial Intelligence*; Müller, V. C., Ed.; Springer Nature: Switzerland, 2018.
56. Carruthers, P. *Phenomenal Consciousness: A Naturalistic Theory*; Cambridge University Press: Cambridge, UK, 2010.
57. Nagasawa, Y. *God and Phenomenal Consciousness: A Novel Approach to Knowledge Arguments*; Cambridge University Press: Cambridge, UK, 2016.
58. Place, U. T. Is consciousness a brain process? *Br. J. Psychol.* **1956**, *47*, 44–50.
59. Poznanski, R. R.; Cacha, L. A.; Latif, A. Z. A.; Salleh, S. H.; Ali, J.; Yupapin, P.; Tuszynski, J. A.; Tengku, M. A. Theorizing how the brain encodes consciousness based on negentropic entanglement. *J. Integr. Neurosci.* **2019**, *18*, 1–10.
60. Solms, M. A neuropsychoanalytical approach to the hard problem of consciousness. *J. Integr. Neurosci.* **2014**, *13*, 173–185.
61. Zhao, T.; Zhu, Y.; Tang, H.; Xie, R.; Zhu, J.; Zhang, J. H. Consciousness: New concepts and neural networks. *Front. Cell. Neurosci.* **2019**, *13*, 302.
62. Accorsi, P. A.; Mondo, E.; Cocchi, M. Did you know that your animals have consciousness? *J. Integr. Neurosci.* **2017**, *16*, S3–S11.
63. Tarlaci, S. A historical view of the relation between quantum mechanics and the brain: A neuroquantologic perspective. *NeuroQuantology* **2010**, *8*, 120–136.
64. Bohm, D. J.; Hiley, B. J. Unknown quantum realism, from microscopic to macroscopic levels. *Phys. Rev. Lett.* **1985**, *55*, 2511–2514.
65. de Broglie, L. The reinterpretation of wave mechanics. *Found. Phys.* **1970**, *1*, 5–15.
66. Zurek, W. H. Decoherence and the transition from quantum to classical. *Phys. Today* **1991**, *44*, 36–44.
67. d'Espagnat, B. *On Physics and Philosophy*; Princeton University Press: Princeton and Oxford, 2006.
68. Nixon, G. M. From panexperientialism to conscious experience: The continuum of experience. *J. Conscious. Explor. Res.* **2010**, *1*, 216–233.
69. Goff, P. *Galileo's Error: Foundations for a New Science of Consciousness*; Pantheon Books: New York, 2019.
70. Strawson, G. Realistic monism: Why physicalism entails panpsychism. *J. Conscious. Stud.* **2006**, *13*, 110–116.
71. de Sousa, A. Towards an integrative theory of consciousness, part 2 (An anthology of various other models). *Mens Sana Monogr.* **2013**, *11*, 151–209.

72. Chalmers, D. J. Panpsychism and panprotopsychism. In *Consciousness in the Physical World: Perspectives on Russellian Monism*; Alter, T., Nagasawa, Y., Eds.; Oxford University Press: Oxford, 2015.
73. Papineau, D. *Thinking about Consciousness*; Oxford University Press: Oxford, 2002.
74. Pribram, K. H. Quantum holography: Is it relevant to brain function? *Inform. Sci.* **1999**, *115*, 97–102.
75. Holman, E. Panpsychism, physicalism, neutral monism, and the Russellian theory of mind. *J. Conscious. Stud.* **2008**, *15*, 48–67.
76. Seager, W. E. Classical levels, Russellian monism and the implicate order. *Found. Phys.* **2013**, *43*, 548–567.
77. Jylkka, J.; Railo, H. Consciousness as a concrete physical phenomenon. *Conscious. Cogn.* **2019**, *74*, 102779.
78. Pereira, A., Jr. Triple-aspect monism: A conceptual framework for the science of human consciousness. In *The Unity of Mind, Brain, and World: Current Perspectives on a Science of Consciousness*; Pereira, A., Jr., Lehmann, D., Eds.; Cambridge University Press: Cambridge, 2014.
79. Davidson, D. *Mental Events, Reprinted in Essays on Actions and Events*; Clarendon Press: Oxford, 1970.
80. Chalmers, D. J. *The Conscious Mind: In Search of a Fundamental Theory*; Oxford University Press: Oxford, 1996.
81. Searle, J. R. Biological naturalism. In *The Blackwell Companion to Consciousness*; Velmans, M., Schneider, S., Eds.; Blackwell: Oxford, 2007.
82. Dennett, D. C. *Consciousness Explained*; Back Bay Books, Little, Brown and Company: New York, 1991.
83. Dennett, D. C. *From Bacteria to Bach and Back. The Evolution of Minds*; Penguin Random House: UK, 2017.
84. Dawkins, R. *The Extended Phenotype the Gene as a Unit of Selection*; Oxford University Press: Oxford, 1982.
85. Chomsky, N. *Syntactic Structures*; Mouton Publishers: The Hague, 1957.
86. Wan, P. Y.-Z. Dialectics, complexity, and the systemic approach: toward a critical reconciliation. *Philos. Soc. Sci.* **2012**, *43*, 411–452.
87. Jamali, M.; Golshani, M.; Jamali, Y. A proposed mechanism for mind-brain interaction using extended Bohmian quantum mechanics in Avicenna's monotheistic perspective. *Heliyon* **2019**, *5*, e02130.
88. Freire, O.; Lehner, C. 'Dialectical materialism and modern physics' an unpublished text by Max Born. *Notes Rec. R. Soc. J. Hist. Sci.* **2010**, *64*, 155–182.
89. Bunge, M. Emergence and the mind. *Neuroscience* **1977**, *2*, 501–509.
90. Seager, W. E. Consciousness, information, and panpsychism. *J. Conscious. Stud.* **1995**, *2*, 272–288.
91. Hunt, T. Kicking the psychophysical laws into gear: A new approach to the combination problem. *J. Conscious. Stud.* **2011**, *18*, 96–134.
92. Mallet, J.; Feinberg, T. E. Consciousness is not inherent in but emergent from life. *Anim. Sent.* **2007**, *1*, 15.
93. Bello-Morales, R.; Delgada-Garcia, J. M. The social neuroscience and the theory of integrative levels. *Front. Integr. Neurosci.* **2015**, *9*, 54.
94. Bassett, D. S.; Gazzaniga, M. S. Understanding complexity in the human brain. *Trends Cogn. Sci.* **2011**, *15*, 200–209.
95. Tegmark, M. Consciousness as a state of matter. *Chaos Soliton. Fract.* **2015**, *76*, 238–270.
96. Carneiro, R. L. The transition from quantity to quality: a neglected causal mechanism in accounting for social evolution. *Proc. Natl. Acad. Sci. U. S. A.* **2000**, *97*, 12926–12931.
97. Roederer, J. G. On the concept of information and its role in nature. *Entropy* **2003**, *5*, 3–33.

98. Tegmark, M. The importance of quantum decoherence in brain processes. *Phys. Rev.* **1999**, *E61*, 4194–4206.
99. Sbitnev, V. I. Quantum consciousness in warm, wet, and noisy brain. *Mod. Phys. Lett.* **2016**, *B30*, 1650329.
100. Gould, L. I. Quantum dynamics and neural dynamics: Analogies between the formalisms of Bohm and Pribram. In *Scale in Conscious Experience: Is the Brain Too Important To Be Left to Specialists to Study?*; King, J., Pribram, K. H., Eds.; Lawrence Erlbaum: New Jersey, 1995.
101. Heisenberg, W. *Physics and Philosophy: The Revolution in Modern Science*; George Aleen & Unwin: London, 1959.
102. Esposito, S.; Naddeo, A. The genesis of the quantum theory of the chemical bond. *Adv. Hist. Stud.* **2014**, *3*, 229–257.
103. Poznanski, R. R.; Cacha, L. A.; Latif, A. Z. A.; Salleh, S. H.; Ali, J.; Yupapin, P.; Tuszynski, J. A.; Ariff, T. M. Molecular orbitals of dipole-bound delocalized electron clouds in neuronal domains. *Biosystems* **2019**, *183*, 103982.
104. Brändas, E. J. Relaxation processes and coherent dissipative structures. In *Dynamics During Spectroscopic Transition*; Lippert, E., Macomber, J. D., Eds.; Springer-Verlag: Berlin, 1995; pp. 148–193.
105. Chatzidimitriou-Dreismann, C. A.; Brändas, E. J. Proton delocalization and thermally activated quantum correlations in water: complex scaling and new experimental results. *Ber. Bunsenges. Phys. Chem.* **1991**, *95*, 263–272.
106. Percival, I. *Quantum State Diffusion*; Cambridge University Press: Cambridge, 1998.
107. Tsekov, R. Thermo-quantum diffusion. *Int. J. Theor. Phys.* **2009**, *48*, 630–636.
108. Poznanski, R. R.; Cacha, L. A.; Latif, A. Z. A.; Salleh, S. H.; Ali, J.; Yupapin, P.; Tuszynski, J. A.; Tengku, M. A. Spontaneous potentiality as formative cause of thermo-quantum consciousness. *J. Integr. Neurosci.* **2018**, *187*, 371–385.
109. Dürr, D.; Goldstein, S.; Tumulka, R.; Zanghi, N. On the role of density matrices in Bohmian mechanics. *Found. Phys.* **2005**, *35*, 449–467.
110. Sbitnev, V. I. Bohmian trajectories and the path integral paradigm. Complexified Lagrangian mechanics. *Int. J. Bifurcation Chaos* **2009**, *19*, 2335–2346.
111. Wyatt, R. E. *Quantum Dynamics With Trajectories: Introduction to Quantum Hydrodynamics*; Springer: New York, 2005.
112. Donoso, A.; Martens, C. C. Classical trajectory-based approach to solving the quantum Liouville equation. *Int. J. Quant. Chem.* **2002**, *90*, 1348–1360.
113. Tsekov, R. Quantum diffusion. *Phys. Script.* **2011**, *83*, 035004.
114. Tsekov, R. Bohmian mechanics versus Madelung quantum hydrodynamics. In *Ananuire de l'Universite de Sofia*; Popov, V., Ed.; Faculte de Physique, Sofia University Press: Sofia, 2012.
115. Bohm, D. J.; Hiley, B. J. Unbroken quantum realism, from microscopic to macroscopic levels. *Phys. Rev. Lett.* **1985**, *55*, 2511–2514.
116. Hu, H.; Wu, M. Spin as primordial self-referential process driving quantum mechanics, spacetime dynamics, and consciousness. *NeuroQuantology* **2004**, *2*, 41–49.
117. Hu, H.; Wu, M. Quantum spin formalism on consciousness. In *Biophysics of Consciousness: A Foundational Approach*; Poznanski, R. R., Tuszynski, J. A., Feinberg, T. E., Eds.; World Scientific: Singapore, 2017.
118. Turner, P.; Nottale, L.; Zhao, J.; Pesquet, E. New insights into the physical processes that underpin cell division and the emergence of different cellular and multicellular structures. *Progr. Biophys. Mol. Biol.* **2020**, *150*, 13–42.
119. Dürr, D.; Goldstein, S.; Zanghi, N. Quantum chaos, classical randomness, and Bohmian mechanics. *J. Stat. Phys.* **1992**, *68*, 259–270.
120. Trukhan, E. M.; Anosov, V. N. Vector potential as a channel of informational effect on living objects. *Biofizika* **2007**, *52*, 376–381 (in Russian).

121. Bogan, J. R. *Spin: the Classical to Quantum Connection*; Arxiv.org 0212110, 2002.
122. Binhi, V. N. Nonspecific magnetic biological effects: a model assuming the spin-orbit coupling. *J. Chem. Phys.* **2019**, *151*, 204101.
123. Turin, L.; Skoulakis, E. M. C.; Horsfield, A. P. Electron spin changes during general anesthesia in Drosophila. *Proc. Natl. Acad. Sci. U. S. A.* **2014**, *111*, E3524–E3533.
124. Nottale, L.; Auffray, C. Scale relativity theory and integrative systems biology: 2 Macroscopic quantum-type mechanics. *Prog. Biophys. Mol. Biol.* **2008**, *97*, 115–157.
125. Nottale, L. Macroscopic quantum-type potentials in theoretical systems biology. *Cells* **2014**, *1*, 1–35.
126. Mandell, A. J. Can a metaphor of physics contribute to MEG neuroscience research? Intermittent turbulent eddies in brain magnetic fields. *Chaos Soliton. Fract.* **2013**, *55*, 95–101.
127. Fingelkurts, A. A.; Fingelkurts, A. A.; Neves, C. F. H. Consciousness as a phenomenon in the operational architectonics of brain organization: Criticality and self-organization considerations. *Chaos Soliton. Fract.* **2013**, *55*, 13–31.
128. McFadden, J. The CEMI field theory closing the loop. *J. Conscious. Stud.* **2013**, *20*, 153–168.
129. Bercovich, D.; Goodman, G.; Cacha, L. A.; Poznanski, R. R. The two-brains hypothesis: Implications for consciousness. In *Biophysics of Consciousness: A Foundational Approach*; Poznanski, R. R., Tuszynski, J. A., Feinberg, T. E., Eds.; World Scientific: Singapore, 2017.
130. Binhi, V. N.; Prato, F. S. Rotations of macromolecules affect nonspecific biological responses to magnetic fields. *Sci. Rep.* **2018**, *8*, 13495.
131. Recami, E.; Salesi, G. Kinematics and hydrodynamics of spinning particles. *Phys. Rev. A* **1998**, *57*, 98–105.
132. Eposito, S. On the role of spin in quantum mechanics. *Found. Phys. Lett.* **1999**, *12*, 165–177.
133. Leggett, A. J. The superposition principle in macroscopic systems. In *Quantum Concepts in Space and Time*; Penrose, R., Isham, C. J., Eds.; Clarendon Press: Oxford, 1986.
134. Fröhlich, H. Long-range coherence and energy storage in biological systems. *Int. J. Quant. Chem.* **1968**, *2*, 641–649.
135. Fröhlich, H. The connection between macro- and microphysics. *La Rivista del Nuovo Cimento* **1973**, *3*, 490–534.
136. Reimers, J. R.; Mcemmish, L. K.; McKenzie, R. H.; Mark, A. E.; Hush, N. S. Weak, strong and coherent regimes of Fröhlich condensation and their applications to terahertz medicine and quantum consciousness. *Proc. Natl. Acad. Sci. U. S. A.* **2009**, *106*, 4219–4224.
137. Poccia, N.; Ricci, A.; Innocenti, D.; Bianconi, A. A possible mechanism for evading temperature quantum decoherence in living matter by Feshbach resonance. *Int. J. Mol. Sci.* **2009**, *10*, 2084–2106.
138. Mikheenko, P. Possible superconductivity in the brain. *J. Superconduct. Novel Magnet.* **2019**, *32*, 1121–1134.
139. Fröhlich, F.; Hyland, G. J. Fröhlich coherence at the mind-brain interface. In *Scale in Conscious Experience: Is the Brain Too Important To Be Left to Specialists to Study?*; King, J., Pribram, K. H., Eds.; Lawrence Erlbaum Associates: Mahwah, NJ, 1993.
140. Yang, C. N. Concept of off-diagonal long-range order and the quantum phases of liquid helium and of superconductor. *Rev. Modern Phys.* **1962**, *34*, 694–704.
141. Coleman, A. J. Structure of fermion density matrices. *Rev. Modern Phys.* **1963**, *35* (686), 668.
142. Sasaki, F. Eigenvalues of fermion density matrices. *Phys. Rev.* **1965**, *138*, B1338–B1342.
143. Schrödinger, E. *What is Life?: The Physical Aspect of the Living Cell*; Cambridge University Press, 1944.

144. Maroney, O. J. E. The density matrix in the de Broglie-Bohm approach. *Found. Phys.* **2005**, *35*, 493–510.
145. Pendas, A. M.; Hernandez-Trujillo, J. The Ehrenfest field: Topology and consequences for the definition of an atom in a molecule. *J. Chem. Phys.* **2012**, *137*, 134101.
146. Löwdin, P. O. Quantum theory of many-particle systems I. Physical interpretations by means of density matrices, natural spin orbitals, and convergence problems in the method of configuration interactions. *Phys. Rev.* **1955**, *97*, 1474–1489.
147. Brändas, E. J. Per-Olov Löwdin—Father of quantum chemistry. *Mol. Phys.* **2017**, *115* (17-18), 1995–2024.
148. Busemeyer, J. R.; Bruza, P. D. *Quantum Models of Cognition & Decision*; Cambridge University Press: New York, 2014.
149. Anderson, P. W. Absence of diffusion in certain random lattices. *Phys. Rev.* **1958**, *109*, 1492–1505.
150. Percival, I. C.; Strunz, W. T. Classical dynamics of quantum localization. *J. Phys. A* **1998**, *31*, 1815–1830.
151. Prigogine, I. *The End of Certainty: Time, Chaos, and the New Laws of Nature*; The Free Press: New York, 1996.
152. Nicolis, G.; Prigogine, I. *Self-Organization in Nonequilibrium Systems. From Dissipative Structures to Order Through Fluctuations*; Wiley: New York, 1977.
153. Marx, G.; Gilon, C. The molecular basis of neural memory. Part 6: Chemical coding of logical and emotive modes. *Int. J. Neurol. Res.* **2016**, *2* (2), 259–1268.
154. Josephson, B. D. Biological organization as the true foundation of reality. In *Unified Field Mechanics 2:10th International Symposium in Honor of Mathematical Physicist Jean-Pierre Vigier*; Amoroso, R. L., Kauffman, L. H., Rowlands, P., Eds.; World Scientific: Singapore, 2018.
155. Elsasser, W. M. Acausal phenomena in physics and biology: A case for reconstruction. *Sci. Am.* **1969**, *57*, 502–516.
156. Edelman, G. M.; Gally, J. A. Degeneracy and complexity in biological systems. *Proc. Natl. Acad. Sci. U. S. A.* **2001**, *98*, 13763–13768.
157. Mason, P. H.; Dominquez, J. F.; Winter, D. B.; Grignolio, A. Hidden in plain view: Degeneracy in complex systems. *BioSystems* **2015**, *128*, 1–8.
158. Balslev, E.; Combes, J. M. Spectral properties of many-body Schrödinger operators with dilatation-analytic interactions. *Commun. Math. Phys.* **1971**, *22*, 280–294.
159. Griffin, D. R. Panexperiential physicalism and the mind-body problem. *J. Consciousness Stud.* **1997**, *4*, 248–268.
160. Coleman, S. Being realistic: Why physicalism may entail panexperientialism. *J. Conscious. Stud.* **2006**, *13*, 40–52.
161. Kompanichenko, V. Thermodynamic jump from prebiotic microsystems to primary living cells. *Sci* **2020**, *2*, 14.
162. Poznanski, R. R.; Cacha, L. A.; Tengku, M. A.; Ahmad Zubaidi, A. L.; Hussain, S.; Ali, J.; Tuszynski, J. A. On intrinsic information content of the physical mind in quantized space: Against externalism. *Axiomathes* **2019**, *29*, 127–137.
163. Mayr, E. Teleological and teleonomic: A new analysis. *Boston Stud. Philos. Sci.* **1974**, *14*, 91–117.
164. Shannon, C. E. A mathematical theory of communication. *Bell Syst. Tech. J.* **1948**, *27*, 379–423.
165. Lycan, W. G. *Consciousness and Experience*; MIT Press: Cambridge, MA, 1996.
166. Polger, T.; Flanagan, O. A decade of teleofunctionalism: Lycan's consciousness and consciousness and experience. *Minds Machines* **2001**, *11*, 113–126.
167. Baars, B. J.; Franklin, S.; Ramsoy, T. Z. Global workspace dynamics: Cortical "binding and propagation" enables conscious contents. *Front. Psychol.* **2013**, *4*, 200.

168. Holmgren, J. Human cultural evolution is completely immersed in natural evolution. In *Media Models to Foster Collective Human Coherence in the PSYCHecology*; Shafer, S. B., Ed.; IGI Global: Hershey, PA, 2019.
169. Macdonald, G.; Papineau, D. *Teleosemantics: New Philosophical Essays*; Clarendon Press: Oxford, 2006.
170. Minert, A.; Devor, M. Brainstem node for loss of consciousness due to $GABA_A$ receptor-active anesthetics. *Exp. Neurol.* **2016**, *275*, 38–45.
171. Kayama, Y.; Koyama, Y. Brainstem neural mechanisms of sleep and wakefulness. *Eur. Urol.* **1998**, *33*, 12–15.
172. Fischer, D. B.; Boes, A. D.; Demertzi, A.; Evrard, H. C.; Laureys, S.; Edlow, B. L.; Liu, H.; Saper, C. B.; Pascual-Leone, A.; Fox, M. D.; Geerling, J. C. A human brain network derived from coma-causing brainstem lesions. *Neurology* **2016**, *87*, 2427–2434.
173. Solms, M. The hard problem of consciousness and the free energy principle. *Front. Psychol.* **2019**, *10*, 2714.
174. Hameroff, S. R. That's life! The geometry of π electron resonance clouds. In *Quantum Aspects of Life*; Abbott, D., Davis, P. C. W., Pati, A. K., Eds.; World Scientific Publishers: Singapore, 2008.
175. Hameroff, S. R.; Watt, R. C. Do anesthetics act by altering electron mobility? *Anesthesia Analg.* **1983**, *62*, 936–940.
176. Hamerroff, S. R.; Watt, R. C.; Borel, J. D.; Carlson, G. General anesthetics directly inhibit electron mobility: Dipole dispersion theory of anesthetic action. *Physiol. Chem. Phys.* **1982**, *14*, 183–187.
177. Hameroff, S. R. Quantum coherence in microtubules: a neural basis for emergent consciousness. *J. Consciousness Stud.* **1994**, *1*, 91–118.
178. Tory Toole, J.; Kurian, P.; Craddock, T. J. A. Coherent energy transfer and the potential implications for consciousness. *J. Cogn. Sci.* **2018**, *19*, 115–124.
179. Hameroff, S. R. *Ultimate Computing: Biomolecular Consciousness and Nanotechnology*; North-Holland: Amsterdam, 1987.
180. Crawford, M. A.; Broadhurst, C. L.; Guest, M.; Nagar, A.; Wang, Y.; Ghebremeskel, K.; Schmidt, W. F. A quantum theory for the irreplaceable role of docosahexaenoic acid in neural cell signalling throughout evolution. *Prostaglandins, Leukot. Essent. Fatty Acids* **2013**, *88*, 5–13.
181. Crawford, M. A.; Thabet, M.; Wang, Y. An introduction to a theory on the role of π-electrons of docosahexaenoic acid in brain function. *OCL* **2018**, *25*, A402.
182. Cocchi, M.; Minuto, C.; Tonello, L.; Gabrielli, F.; Bernroider, G.; Tuszynski, J. A.; Cappello, F.; Rasenick, M. Linoleic acid: Is this the key that unlocks the quantum brain? *BMC Neurosci.* **2017**, *18*, 38.
183. Sandberg, W. S.; Miller, K. W. The Meyer-Overton relationship and its exceptions. In *Neural Mechanisms of Anesthesia. Contemporary Clinical Neuroscience*; Antognini, J. F., Carstens, E., Raines, D. E., Eds.; Humana Press: Totowa, NJ, 2003.
184. Crawford, M. A.; Broadhurst, C. L.; Galli, C.; Ghebremeskel, K.; Holmsen, H.; Saugstad, L. F.; Schmidt, W. F.; Sinclair, A. J.; Cunnane, S. C. The role of docosahexaenoic and arachidonic acids as determinants of evolution and hominid brain development. In *Fisheries for Global Welfare and Environment*; Tsukamoto, K., Kawamura, T., Takeuchi, T., Beard, T. D., Kaiser, M. J., Eds.; 5th World Fisheries Congress: Yokohama, Japan, 2008.
185. Applegate, K. R.; Glomset, J. A. Computer-based modeliing of the conformation and packing properties of docosahexaenoic acid. *J. Lipid Res.* **1986**, *27*, 658–680.
186. Torday, J. S. A central theory of biology. *Med. Hypothesis* **2015**, *85*, 49–57.
187. Mouritsen, O. G.; Bagatolli, L. A. *Life- As a Matter of Fat: Lipids in a Membrane Biophysics Perspective*; Springer: New York, 2016.

188. Warshel, A.; Levitt, M. Theoretical studies of enzymic reactions: Dielectric, electrostatic and steric stabilization of the carbonium ion in the reaction of lysozyme. *J. Math. Biol.* **1976**, *103*, 227–249.
189. Brillouin, L. The negentropy principle of information. *J. Appl. Phys.* **1953**, *24*, 1152–1163.
190. Gray, H. B.; Winkler, J. R. Electron flow through proteins. *Chem. Phys. Lett.* **2009**, *483*, 1–9.
191. Hameroff, S. R.; Penrose, R. Conscious events as orchestrated spacetime selections. *NeuroQuantology* **2003**, *1*, 10–35.
192. Friston, K. J. The free-energy principle: a unified brain theory? *Nat. Rev. Neurosci.* **2010**, *11*, 127–138.
193. Brändas, E. J. Resonances and dilatation analyticity in Liouville space. In *Advances in Chemical Physics*; Prigogine, I., Rice, S. A., Eds.; Vol. 99; Wiley: New York, 1997; pp. 211–244.
194. Williford, K.; Bennequin, D.; Friston, K.; Rudrauf, D. The projective consciousness model and phenomenal selfhood. *Front. Psychol.* **2018**, *9*, 2571.
195. Pereira, A., Jr. The projective theory of consciousness: from neuroscience to philosophical psychology. *Trans/Form/Ação* **2018**, *41*, 199–232.
196. Gödel, K. Über formal unentscheidbare sätze der principia matematica und verwandter Systeme. *Monat. Math. Phys.* **1931**, *38*, 173–198 (in German).
197. Penrose, S. *Shadows of the Mind: A Search for the Missing Science of Consciousness*; Oxford University Press: Oxford, 1994.
198. Brändas, E. J. Arrows of time and fundamental symmetries in chemical physics. *Int. J. Quant. Chem.* **2012**, *111*, 1321–1332.
199. Chauvet, G. A. *The Mathematical Nature of the Living World: The Power of Integration*; World Scientific: Singapore, 2004.
200. Pellionisz, A. The geometry of massively parallel neural interconnectedness in wired and wireless volume transmission. Ch. 45. In *Volume Transmission in the Brain*; Fuxe, K. J., Agnati, L., Eds.; Raven Press: New York, 1991; pp. 557–568.
201. Poznanski, R. R. *Biophysical Neural Networks: Foundations of Integrative Neuroscience*; Mary Ann Liebert: New York, 2001.
202. Grindrod, P. On human consciousness: A mathematical perspective. *Netw. Neurosci.* **2017**, *2*, 23–40.
203. Bodovitz, S. Consciousness is discontinuous: the perception of continuity requires conscious vectors and needs to be balanced with creativity. *Med. Hypoth.* **2004**, *62*, 1003–1005.
204. Bodovitz, S. Consciousness disintegrates without conscious vectors. *Med. Hypoth.* **2008**, *70*, 8–11.
205. Mashour, G. A. The cognitive binding problem: from Kant to quantum neurodynamics. *NeuroQuantology* **2004**, *2*, 29–38.
206. Fuchs, C. A.; Mermin, N. D.; Schack, R. An introduction to QBism with an application to the locality of quantum mechanics. *Am. J. Phys.* **2014**, *82*, 749–754.
207. Craddock, T. J. A.; Hameroff, S. R.; Tuszynski, J. A. The 'quantum underground': Where life and consciousness originate. In *Biophysics of Consciousness: A Foundational Approach*; Poznanski, R. R., Tuszynski, J. A., Feinberg, T. E., Eds.; World Scientific: Singapore, 2017.
208. Bargh, J. A.; Morsella, E. The unconscious mind. *Perspect. Psychol. Sci.* **2008**, *3*, 73–79.
209. Holmgren, J. Natural evolution and human consciousness. *Mens Sana Monogr.* **2014**, *12*, 127–138.
210. Damasio, A.; Carvalho, G. B. The nature of feelings: Evolutionary and neurobiological origins. *Nat. Rev. Neurosci.* **2013**, *14*, 143–152.

211. Schiffer, F. The physical nature of subjective experience and its interaction with the brain. *Med. Hypotheses* **2019**, *125*, 57–69.
212. Beaulieu, A. A.N. Whitehead as precursor of the new theories of self-organization. In *Alfred North Whitehead's Science and the Modern World*; Beets, F., Dupuis, M., Weber, M., Eds.; Ontos-Verlag: Heusenstamm, Germany, 2006.
213. Pockett, S. Problems with theories that equate consciousness with information or information processing. *Front. Syst. Neurosci.* **2014**, *8*, 225.
214. Brändas, E.; Dunne, L. J. Bardeen-Cooper-Scrieffer (BCS) theory and Yang's concept of off-diagonal long-range order (ODLRO). *Mol. Phys.* **2014**, *112*, 694–699.
215. Maturana, H. R.; Varela, F. J. *Autopoiesis and Cognition: The Realization of the Living*; Reidel: Boston, MA, 1980.
216. Hiley, B. J. From the Heisenberg picture to Bohm: A new perspective on active information and its relation to Shannon information. In *Quantum Theory: Reconsideration of Foundations*; Khrennikov, A., Ed.; Växjö University Press: Sweden, 2002.
217. Hiley, B. J.; Pylkkanen, P. Can mind affect matter via active information? *Mind Matter* **2005**, *3*, 8–27.
218. Di Biase, F. The unified field of consciousness. In *Unified Field Mechanics- Natural Sciences Beyond the Veil of Spacetime*; Amoroso, R. L., Kauffman, L. H., Rowlands, P., Eds.; World Scientific: Singapore, 2016.
219. Wendt, A. A quantum vitalism. In Wendt, A., Ed.;, *Quantum Mind and Social Science: Unifying Physical and Social Ontology*; Cambridge University Press: Cambridge, 2015; pp. 131–148.

Index

Note: Page numbers followed by "*f*" indicate figures.

D

E

F

M

N

O

P

R

S

-inted in the United States
-ookmasters